Social
Insects Volume IV

CONTRIBUTORS

Roger D. Akre
William H. Gotwald, Jr.
John H. Sudd
Neal A. Weber

Social
Insects Volume IV

Edited by **HENRY R. HERMANN**

Department of Entomology
University of Georgia
Athens, Georgia

1982

ACADEMIC PRESS

A Subsidiary of Harcourt Brace Jovanovich, Publishers
New York London
Paris San Diego San Francisco São Paulo Sydney Tokyo Toronto

ACADEMIC PRESS, INC.
111 Fifth Avenue, New York, New York 10003

United Kingdom Edition published by
ACADEMIC PRESS, INC. (LONDON) LTD.
24/28 Oval Road, London NW1 7DX

Library of Congress Cataloging in Publication Data
Main entry under title:

Social insects.

Includes bibliographies and index.
1. Insect societies. I. Hermann, Henry R.
QL496.S6   595.7'0524   78-4871
ISBN 0-12-342204-3 (v. 4)   AACR2

PRINTED IN THE UNITED STATES OF AMERICA

82 83 84 85   9 8 7 6 5 4 3 2 1

To My Wife, Lisa,
and Son, Brad

Contents

1. Social Wasps

Roger D. Akre

2. Ants: Foraging, Nesting, Broad Behavior, and Polyethism

John H. Sudd

3. Army Ants

William H. Gotwald, Jr.

4. Fungus Ants

Neal A. Weber

List of Contributors

Numbers in parentheses indicate the pages on which the authors' contributions begin.

Roger D. Akre (1), Department of Entomology, Washington State University, Pullman, Washington 99164

William H. Gotwald, Jr. (157), Department of Biology, Utica College of Syracuse University, Utica, New York 13502

John H. Sudd (107), Department of Zoology, The University of Hull, Hull H46 7RX, England

Neal A. Weber (255), Department of Biological Science, The Florida State University, Tallahassee, Florida 32306

Preface

This volume is dedicated to a treatment of wasps and their allies, the ar.ts. Social wasps are under a degree of scrutiny now unlike any of its kind in the past. There was an apparent increased interest in aculeate wasps in the 1950s, as indicated in Volume I, Chapter 1. There has, however, been a breathtaking renewal of interest in wasp biology since Hamilton's reevaluation of Mendelian genetics and Darwinian theory of natural selection in terms of social theory. Wasp biology, then, is currently undergoing careful observation in terms of modern theory, and attempts are being made to tie together concepts that were only theory just a few years ago.

The chapters on ants portray a group of insects quite diverse in habit—from group predaceous foraging to garden growing. Probably the most active research toward understanding this group of eusocial insects relates to their general biology and territorial concepts. These fields of endeavor have provided some interesting and very unusual information on intra- and interspecific relationships.

It is with sincerity that I express my thanks to the contributors of all four volumes. This treatise on the social biology of insects represents many hours of research for each author and for the subjects that were reviewed. It was our goal to cover the field of insect sociality as completely as possible without overly repeating what has been reported on in the past, and to collate the work of modern researchers. While there has been some overlap between materials presented by various authors, there also has been abundant new material, and we hope that these books serve the reader of social biology well.

Henry R. Hermann

Contents of Other Volumes

1

Social Wasps

ROGER D. AKRE

I. INTRODUCTION

Humans have always been involved with social wasps [see Spradbery (1973) for an interesting and complete historical account of wasps and humans]. The mass attacks and formidable stings of some species make them frightening adversaries. Conversely, many species are also recognized as unparalleled predators of arthropod pests by people who grow crops for subsistence or sale. A number of wasp species are large and brightly colored, many have populous colonies numbering into the thousands, and most build nests that are architectural masterpieces. Whatever the nature of the involvement, people everywhere are attracted to the social wasps and their nests and are fascinated by their diverse behaviors.

SOCIAL INSECTS, VOL. IV

Most people fear wasps because of their painful stings that can cause allergic reactions and even death. However, social wasps are also highly beneficial, serving as food for humans (honey and wasp brood), as predators of many insects directly pestiferous to humans, and as subjects for scientific investigations of social behavior. Actually these considerations are probably minor compared to the value of these insects as unknown, or at best, totally unappreciated biological control agents. While the number of species of wasps are few, individual colonies of vespines contain from 30 individuals to 50,000 individuals, and the number of colonies undoubtedly makes them one of the dominant predatory insects of many areas. All are carnivorous and require vast numbers of arthropod prey to feed their young. Since there are few studies of wasps as predators, much less as biological control agents of immediate importance and concern to humans, their value is unheralded and largely unknown, but their influence is assumed to be very great. In temperate forests where some information is available regarding the prey of the vespine genera *Vespula* and *Dolichovespula,* these wasps are dominant insects that probably rival the ants as protectors of the forest from defoliators and other forest pests.

At present slightly more than 800 species of wasps are known to be social. This is a relatively small number when compared to about 12,000–14,000 species of ants, less than 2000 species of social bees, and about 2200 species of termites. Considered from another perspective, there are approximately 50,000 aculeate Hymenoptera (having a sting) in the world, but only 15,000 of these are wasps and only about 5% of these are social. All social wasps are in the family Vespidae except for a single sphecid wasp, *Microstigmus comes* Krombein, which has been shown to have social behavior (Matthews, 1968a,b).

There are only about 3000 published reports on aspects of social wasp biology, behavior, and taxonomy. When this is compared to the vast amount of literature available on ants and bees (see Spradbery, 1973), the wasps are unquestionably the least known of all the social insects. Nesting habits and general biology have been reported for a number of species, at least in a fragmentary way, but most information on behavior within the nest is based on studies of *Polistes* and species in several other genera (e.g., *Mischocyttarus, Belonogaster*) which have relatively small colonies and make nests without an outer, enclosing envelope. This situation is slowly changing as techniques are developed to study species that construct large nests enclosed in an envelope, and much new information is now being published on these species (particularly among the Vespinae).

Several books were published during the 1970's which have considerably added to the information available on wasps. In particular, the following are recommended readings: the excellent books by Spradbery (1973) and Edwards (1980), dealing primarily with the Vespinae; Evans and West Eberhard (1970), who describe both solitary and social wasps; the classic and incomparable text on social insects by Wilson (1971). Since the older publications are generally available, this chapter concentrates on behavioral studies published since 1973.

In addition to the texts mentioned previously, Richards and Richards (1951), Richards (1971), and Jeanne (1975a) give good descriptions of the nest classification scheme of de Saussure (phragmocyttarus, astelocyttarus, stelocyttarus, etc.), and its application to the nests of social wasps. However, nests of many species do not fit the basic scheme without extensive modification and vast architectural differences are often encountered in the nests of species within a genus and even within a species. Therefore, this classification system was not used in this chapter.

For many Vespidae, no behavioral and little biological information is available, but the family includes species of extremely diverse habits, and behavior varies from presocial to eusocial. Colony cycles vary from perennial species with nests that last for years to the annual colonies of temperate vespines. Nearly all species construct nests containing at least some vegetable fiber. The nests are initially constructed by a single queen (haplometrosis) or by a number of queens (females) which cooperate (pleometrosis). All species are carnivorous, usually preying on live arthropods, although several species scavenge flesh from dead animals. Some also forage for sweets used by man (sugar, soft drinks, etc.) or other insects (honey).

Nearly all literature references to the Vespidae have used the classification of Bequaert (1918, 1928), but Richards (1962) divided the Vespidae into three families: Vespidae, Eumenidae, and Masaridae; the system used currently by most people. However, additional revisions may be imminent since van der Vecht (1977a), after studying all available information and specimens of Stenogastrinae (subfamily of Vespidae), suggested they should perhaps be made a separate family or a subfamily of the Eumenidae with which they are closely related. This is fairly indicative of the taxonomy of the Vespidae, where individual subfamilies, tribes, species groups, and even species will undoubtedly be revised and changed in the future. In this chapter brief taxonomic considerations are given with the discussion of the biology for each group.

II. GENERAL BIOLOGY AND BEHAVIOR

A. Sphecidae: Pemphredoninae

1. Taxonomy and Distribution

Among the 226 genera and 7634 species of described sphecid wasps in the world (Bohart and Menke, 1976), only one, *Microstigmus comes,* is known to be social (Matthews, 1968a,b), with a few other species displaying presocial behavior (Evans, 1966). This species is a member of the subfamily Pemphredoninae which is not closely related to the rest of the sphecids and contains some of the most beelike forms. It has been suggested that the bees and pemphredonines probably had common ancestors (see discussion in Bohart and Menke, 1976). *Microstigmus* Ducke contains nearly 50 species, all Neotropical, occurring from Paraguay north to Costa Rica. All species nest in relatively undisturbed tropical rain forest (West Eberhard, 1977).

2. Nests

Microstigmus comes was studied by Matthews (1968a,b) in Costa Rica where it constructs its nests on the underside of leaves of the palm *Crysophila guagara* Allen. In this position the fragile nests are protected from heavy rains. Most of the observed nests were located 2–3 m high, and many were located close together with 1–3 nests on a single leaflet. The small (about 17 × 12 mm), conical, baglike nest is suspended from the leaf by a spiraled nest pedicel. The pedicel usually has two coils, and when straightened measures 12–18 mm long. The nest is constructed of a waxy bloom coating the under surface of the leaf, and an area about 10 cm diameter about the pedicel attachment is scraped by the wasps to furnish this material. The nest entrance is to one side of the pedicel. One presumably unusual nest had two entrances, one on either side of the pedicel attachment. The upper half of the nest is hollow and houses adult wasps. The lower half is used to construct brood cells. These cells are constructed around the periphery first, and later cells are added in the middle. A translucent material covers the pedicel, the nest entrance lip, the hollow upper chamber, and the internal surface of the cells. It is believed that this material is secreted from glands connected to a setal brush on the terminal gastral tergum since females were observed stroking these areas of the nest with their abdomens. Presumably, the fragments comprising the bottom of the nest are also loosely held together by this material. This translucent material has been referred to as silklike in other species (Matthews, 1970). Nests contain few cells ($\overline{X} = 3.6$, $N = 39$ nests) and all but 7 nests contained fewer than 4 cells. Ten nests had only 1 cell, and the greatest number in a single nest was 18. Cells were mass provisioned with Collembola (mostly Entomobryidae), and evidence suggested that only 1 cell is constructed at a time. The egg is laid only when provisioning is complete, and then the cell is "closed" with a few loose, secreted strands. However, the cells are probably reopened periodically by adults to remove fecal pellets and perhaps other waste material.

3. Behavior

Twenty-seven nests were collected at night when all adults should have been present. Half of these nests contained two or more females; one contained ten adults. No external morphological differences to indicate caste could be detected between females of the same nest, but an examination of the ovaries indicated that one female lays most of the eggs; an indication of a reproductive division of labor. It was discovered that the females of a nest probably cooperate in provisioning the cell under construction at that time and also defend the nest against intruders such as parasites. Also important was the fact that in nests containing more than one female, one was always present to defend the nest. Brood care, at least to the extent of opening provisioned cells for cleaning, exists, as does overlap of generations. Nest distribution in any one area was distinctly clumped, with the nests quite close together, and since no new nests were present, it was assumed that offspring stay with the parent nest or join nearby nests probably constructed by close relatives. Thus, another criterion of social behavior, overlap of generations, is present. Therefore, this is undoubtedly the first known eusocial sphecid.

Other species of *Microstigmus* are not as well known, but some construct nests which have mud pellets incorporated into the walls. Others construct nests of pieces of rotten wood, while still others make nests of moss and lichen fragments (see discussion in Matthews, 1968b). All have a nest pedicel of varying length, and all hang baglike below leaves of various plants. Immature thrips, Collembola, and Homoptera (Cicadellidae) serve as prey for these small wasps, and prey type is species specific (West Eberhard, 1977). Although the social status of these wasps is unknown, evidence of three females in one nest of one species, *M. thripoctenus* Richards, suggests that at least some social behavior exists in other species (Matthews, 1970).

The studies of the species mentioned previously occupied only a very short period of time in the field, and thus there is still a paucity of biological and behavioral data for these and other members of the genus. More information is needed concerning predatory behavior, adult interactions inside and outside the nest, the possible use of pheromones in communication, and even information on colony cycles and parasitism. This is a very important and exciting area of research because studies of this relatively obscure group of sphecids will undoubtedly provide a better understanding of the underlying causes, origin, and advantages of sociality in insects. Even though one species appears to be fully social, its behavioral organization is primitive enough that studies may reveal previously uncovered and unanswered questions regarding social origins.

B. Vespidae

1. Stenogastrinae

a. Taxonomy and Distribution. The Stenogastrinae are restricted to the Papuan and Oriental Regions of southeast Asia (van der Vecht, 1965, 1975, 1977a) from India to south China and eastward to the Philippines, Celebes, and New Guinea (Spradbery, 1975) (Fig. 1). This relatively homogeneous

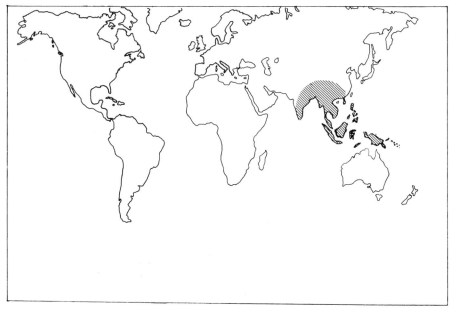

Fig. 1.   Distribution of the Stenogastrinae. (From van der Vecht, 1967.)

group evolved in this area and has not radiated outward into adjacent geographical areas probably because of their specialized adaptations to rain forests. In the Papuan Region the stenogastrines are represented by two genera, *Anischnogaster* van der Vecht and *Stenogaster* Guérin, with 17 species. The subfamily has diversified slightly more in the Oriental Region and contains about 50 species among the genera *Eustenogaster* van der Vecht, *Holischnogaster* van der Vecht, *Liostenogaster* van der Vecht, *Metischnogaster* van der Vecht, and *Parischnogaster* von Schulthess.

Although the stenogastrines are currently placed in the Vespidae (Richards, 1962), they are intermediate in many morphological and behavioral features between the Eumenidae and Vespidae (Yoshikawa *et al.,* 1969) and are so different from the Polistinae and Vespinae that van der Vecht (1977a) does not consider any taxon containing the three groups to be monophyletic. Indeed, the ancestor of the stenogastrines was probably a solitary cell builder closely related to Zethinae (Eumenidae). Therefore, van der Vecht (1977a) proposed that Stenogastrinae be made a separate family or a subfamily of the Eumenidae, or, preferably, that all diplopterous wasps (wings folded longitudinally at rest) again be placed in a single family with numerous subfamilies (see Richards, 1962). In this scheme, the stenogastrines would remain as a subfamily.

Stenogastrines are different from other vespids in that most are solitary or gregarious (presocial). In addition, although they are placed among the diplopterous wasps by basic morphological considerations, no species fold their wings lengthwise, either when living or dead (Yoshikawa *et al.,* 1969). Another distinctive difference is that stenogastrines never build a petiole on the first cell of the nest (van der Vecht, 1967; Yoshikawa *et al.,* 1969; Spradbery, 1975), whereas most higher vespids construct a petiole. This is extremely important as this structure is a prerequisite to building larger combs (Richards, 1971; van der Vecht, 1977b). This limits the size of the nest and may be one of the reasons for the small size of colonies of stenogastrines. This in turn probably limits the development of sociality in this group.

The behavior and biology of the stenogastrines are obviously very poorly known, placing the stenogastrines in a category somewhat similar to the species in the pemphredonine sphecids in the genus *Microstigmus.* Undescribed species undoubtedly exist (Iwata, 1976) and only fragmentary and incomplete biological studies have been reported for the described species. For example, the report by Spradbery (1975) on *Stenogaster concinna* van der Vecht contains the only information available on the behavior and biology of any papuan stenogastrine. The data presented in the following discussion are therefore necessarily based on brief reports of a few (in many cases) selected species. It is hoped that this still conveys a true rendition of the biology of these wasps as presently known.

b. Nests. These primitive wasps usually make their nests in extreme shade in dense forests, or near or over streams or waterfalls. They are also frequently found on overhanging banks and cliffs alongside established roadways. All these locations suggest that these are the most readily accessible areas to investigators enabling them to find nests in these areas. However, nests have not been found in open, sunlit areas and the wasps tend to forage in shade, presumably an adaptation to avoid predators, especially hornets (Spradbery, 1975).

Nests are constructed of mud, masticated rotten wood, bark, or leaves and a few use a combination of these materials. Nests may be fastened to a broad substrate (under a rock), attached to twigs or small roots of trees, or suspended from fungus threads. Nest types are extremely diverse (Yoshikawa *et al.,* 1969; Spradbery, 1975; Iwata, 1976) and vary from strings or clusters of individual cells to quite elaborate nests composed of a comb of cells surrounded by an envelope (Sakagami and Yoshikawa, 1968). Nests of a single species [e.g. *Parischnogaster mellyi* (Saussure)] vary from a single comb structure to cells spread along a suspension (Hansell 1981), and differences in nest design, even between closely related species, is so great that the classification scheme proposed for nest architecture by Saussure and revised by Richards and Richards (1951) cannot be applied to this group (Yoshikawa *et al.,* 1969). Regardless of the architecture, nests are always small with 5 to 80 or more cells (Pagden, 1958).

Some species of Stenogastrines construct ant guards, umbrella shaped structures on the filamentous support of the nest, to discourage attacks by ants (Pagden, 1958; Turillazzi and Pardi, 1981). These guards are constructed from material exuded as tiny droplets from the tip of the female's gaster and collected by her on the hind legs, then rapidly passed forward to the mouth parts for application to the nest support. Females also exude a material from the tip of the gaster which is used to feed young larvae, but Turillazzi and Pardi (1981) suggest that the ant guard substance is a different material.

c. Prey. Studies have revealed that female stenogastrines have well developed mandibular glands that are much larger than those in other vespids. These glands have been suggested as the source of the jellylike food mass placed on the unhatched egg, a feature unique to the stenogastrines (Spradbery, 1973). The female also provides the larvae with a similar material, which contains arthropod fragments. The fragments are probably remnants of small Diptera (especially midges), Hymenoptera, and spiders. The food material is supplied in quantities which last a day or more so this provisioning style appears intermediate between mass provisioning of solitary wasps and progressive provisioning of other vespids (Spradbery, 1975).

d. Behavior. Some stenogastrine wasps have the ability to hover in flight, an adaptation to their preferred habitat of dense, tangled vegetation where flight paths are frequently blocked by multiple obstacles. They maneuver with great agility, especially species of the genus *Parischnogaster* that "can fix themselves in the air like a pendulum, and change their flight course in any direction(s)" (Yoshikawa *et al.,* 1969). This permits them to examine vegetation while hovering and darting from surface to surface. They have also been reported to collect small insects from spider webs (Pagden, 1958; Iwata, 1976). They can hover in front of the web and extract the prey without becoming entangled. Unlike other vespids, they were never observed walking or crawling while foraging. This tendency to hover and thus to forage for prey in dense vegetation also occurs in two North American vespines, *Vespula acadica* (Sladen) and *V. consobrina* (Saussure), which also nest and forage in areas with dense vegetation and deep shade (Akre *et al.,* 1982; Reed, 1981).

Little is known of social behavior organization in the Stenogastrinae. Only a single foundress inhabits most nests, but some species regularly have more than one female per nest, usually simultaneously with one or more males. Most are probably offspring that join the nest for varying periods of time. However, the females are morphologically indistinguishable. Even though many nests contain several individuals, social interactions between nest inhabitants have been recorded for only three species (see Spradbery, 1975). Two females of *Liostenogaster varipictus* Rohwer were observed dividing prey that one had brought back to the nest, then both fed the prey to larvae. The second observation on interactions was that a male and a female *Stenogaster cilipennis* (Smith) were seen exchanging food. The type of food was not indicated.

The third account involved an unnamed species of Malayan *Parischnogaster* that is probably a eusocial species (Yoshikawa *et al.,* 1969). The paper nest consisted of a number of individual cells attached in a linear series to the underside of a twig or vine hanging at a 45° angle from the horizontal. The cells were nearly parallel with the twig, facing downward with one below the other, and ran for about 30 cm along the twig (Fig. 2). There was a total of approximately 35 cells. Six females and three males were on the nest. All were

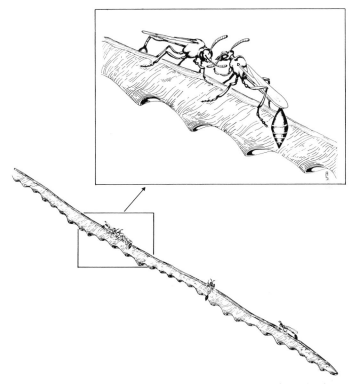

Fig. 2. A paper nest of *Parischnogaster* sp., upper drawing showing a dominant–subordinate interaction between two females on the nest. (Drawn from photographs by Yoshikawa *et al.,* 1969.)

marked with paint for purposes of individual recognition. Behavioral interactions were observed over an entire week, although intensive observations were recorded for only one 90-min period during this time (Fig. 3). Observations on other days showed similar results. Analysis of recorded interactions showed a definite linear dominance hierarchy among the females similar to that reported for *Polistes* (Pardi, 1948) and for members of the polybiine genus *Belonogaster* Saussure (Roubaud, 1916). The dominant female spent most of her time resting although she engaged in some nest construction and larval feeding. She never left the nest. Most nest construction, larval feeding, foraging, and other maintenance duties were done by the subordinates that left the

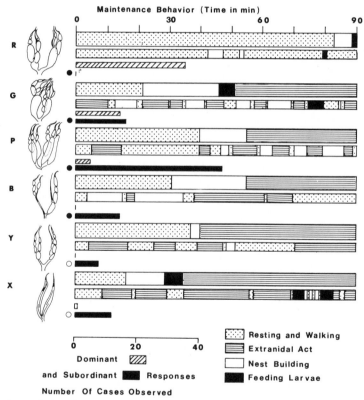

Fig. 3. Relationships in dominance order, conditions of reproductive system, and work performance among six females in a nest, *Parischnogaster* sp. Four horizontal bars given for each individual mean (from above): (A) ratios occupied by various types of maintenance behavior during 90 min. (IX 18, 9:15–10:45); (B) actual chronological sequence of these behaviors; (C) number of dominating responses (cf. Table II), and (D) that of subordinate responses (dominated and avoidance, cf. Table II) during a week. Circles given together with ovaries show spermatheca (black: inseminated; white: uninseminated). (From Yoshikawa *et al.*, 1969.)

nest for varying periods of time. There were no morphological differences between females, but an examination of the ovaries showed a definite gradation of ovariole development, with the dominant female having the greatest development. Three other females had sperm in their spermatheca while the two at the bottom of the dominance order were not mated. These two spent more than half their time during the observations away from the nest. The males were not part of the hierarchy and their interactions with the females were stated as "indifferent." Although this study was brief and many unanswered questions about the behavior of these wasps remain, this species obviously satisfies the criteria for being a eusocial species. Cooperative brood care, division of labor (including reproductive), and probably an overlap of generations (if it is assumed that some individuals were offspring from this nest) are all present in this stenogastrine.

Other species of stenogastrines tend to place their nests close together (20–30 cm), and while these wasps are definitely presocial, marking experiments on one species, *Parischnogaster striatula* Buysson, showed some females joined different nests on successive days (Yoshikawa *et al.,* 1969). Among these species with clumped nesting sites the nest inhabitants are very possibly close relatives and different females may have brood in one nest, or may even cooperate in brood rearing. Future investigations will resolve this problem.

Comparing the known behavior and nesting biology of the Stenogastrinae with members of the sphecid genus *Microstigmus* reveals many points of similarity (Table I). Although only one species in each group has been reported to exhibit social behavior, the intriguing possibility exists that investigations of other species will reveal more eusocial species or at least in-

TABLE I
Comparison of *Microstigmus* sp. (Sphecidae) and Stenogastrinae

Character	*Microstigmus* sp.	Stenogastrines
Larval provisioning	Mass	Intermediate mass– Progressive
Female tends larvae	Yes	Yes
Trophallaxis	?	?
Two or more females per nest	Frequent	Frequent
Morphological caste	No	No
Clumped-nest distribution	Some species	Some species
Nest	Mud–vegetable fiber Pedicel (petiolate) Small No. cells	Mud–vegetable fiber No pedicel Small No. cells
Nest inhabitants close relatives	Probable	Probable

termediate types of behavior that will provide a better insight into the origins of sociality in the Hymenoptera. Spradbery (1975) urges caution in looking to the stenogastrines for this purpose, at least in relation to social organization in the higher vespids, because they are unique in many ways and have only very distant ancestors in common. However, this is probably precisely the reason that these groups should be studied. These studies, in combination with studies of more closely related but still relatively primitive species, may provide insight into the selective pressures important in the origin of eusociality in wasps.

2. Polistinae

The Polistinae are divided into three tribes: Ropalidiini, Polybiini, and Polistini (Richards, 1962). Members are mostly tropical in distribution except for *Polistes* Latreille, *Mischocyttarus* Saussure, a few species of *Ropalidia* Guérin, and several species of Polybiini which also occur in temperate regions (Richards, 1971, 1978a,b; Iwata, 1976) (Figs. 4 and 5). In general, worker and queen castes of members of this entire subfamily are not externally distinguishable, but exceptions occur (Richards, 1978a).

a. Ropalidiini. *i. Taxonomy and Distribution.* The tribe Ropalidiini contains a single genus, *Ropalidia,* with ca. 126 species occurring in the Indo–

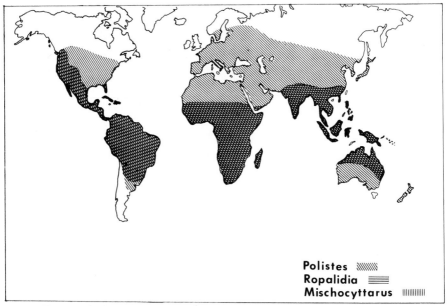

Fig. 4. Distribution of the Polistinae: genera *Polistes, Ropalidia, Mischocyttarus.*

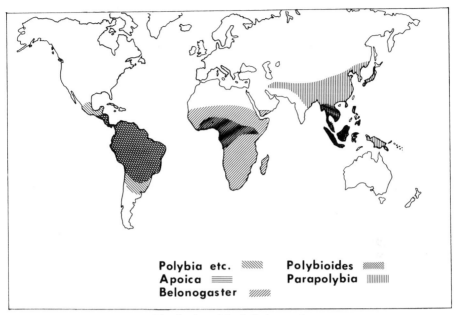

Fig. 5. Distribution of the Polistinae: other genera.

Australian and Oriental Regions (southeast Asia, India, Australia) and in Africa and Madagascar (van der Vecht, 1941, 1962; Iwata, 1976; Richards, 1971, 1978b; Yamane and Yamane, 1979). A few species extend into temperate areas with the northernmost species occurring on islands (principally Amami Oshima) just south of Japan while others are found in warm temperate Australia. Richards (1971) suggested that the genus will ultimately be divided into several genera, but presumably this will not be done until there is a detailed taxonomic study of the New Guinea species, only poorly known at present. Although Richards (1978b) recently published an excellent study on the social wasps of Australia which contained a section on the taxonomy of *Ropalidia,* including descriptions of new species, the taxonomy of this group is still not complete and awaits additional studies.

Even though the Ropalidiini are abundant in tropical areas, especially in the Indo–Malayan region, New Guinea and adjacent islands, very little is known of their biology and behavior with most accounts being based on a single species or, at most, several species with simple types of nests and therefore the most easily studied. This is perhaps at least partially attributable to the taxonomic status of the group with the attendant difficulties such as species identification and undescribed species. Efforts to compare the biology of species (van der Vecht, 1962; Richards, 1971) have been greatly hampered by the lack of available information. Richards (1978b) also reported on the biologies of many previously undescribed Australian species, added new information on some other species, and compared these data to that reported in the literature. This report clearly shows that *Ropalidia* exhibit a diversity of nest architecture and social organization previously unrecognized and unappreciated.

ii. Nests. Most species of *Ropalidia* construct paper nests that are very similar to those of *Polistes* and that are composed of a simple uncovered comb suspended by a pedicel (Richards, 1978b). However, like the stenogastrines, nest architecture in this genus is so diverse that the standard scheme of nest classification does not work well. The simplest nests are theadlike and consist of a pedicel and a single row of cells suspended vertically one below the other, containing as few as nine cells. More complex nests such as those of *R. variegata jacobsoni* (du Buysson) are composed of two rows of alternating cells with the largest nests up to 40 cm long (Yoshikawa, 1964). Next in complexity are *Polistes*-type combs, nests with an elongate, narrow comb (two or more cells wide), a convex comb, or a broader oval or nearly circular comb, with a pedicel in the center or at one end. As does *Polistes, Ropalidia* sometimes add more pedicels to support the comb as it becomes larger. Single comb nests of this type usually hang horizontally, but may be vertical or parallel to the nesting substrate as in *R. dubia* Saussure which attaches its nest to the trunks of trees (FitzGerald, 1950). Still, other species such as *R. fulvopruinosa* (Cameron) make several large, more or less circular combs, one suspended below the other and without an envelope (Richards, 1978b). *Ropalidia flavopicta* (Smith) also makes a multicombed nest with one comb below another, but is unusual in that alternate combs face in opposite directions, one up and one down. The combs are enclosed in an envelope of very thin, light brownish-grey paper (van der Vecht, 1962). The largest nests have a tendency for the combs to be interconnected in a spiral arrangement as in *R. montana* Carl where one large nest consisted of 35 interconnecting combs composed of about 60,000 cells (Carl, 1934). Other species such as *R. romandi* Cabeti and *R. kurandae* Richards also have a tendency to interconnect combs in a spiral arrangement. Many of these large, multiple combed nests are enclosed in an envelope, which may or may not have leaves folded about the combs and papered into place, but have at least some envelope and/or leaves protecting the top of the nest. One species, *R. opifex* van der Vecht, has an envelope which resembles polyethylene plastic (van der Vecht, 1962).

All species remove the meconium through the bottom of the cell, and then close the cell, presumably with a salivary secretion which hardens, so that cells which contain or have contained a pupa have a transparent window (Richards, 1978b). This was first reported for *R. tomentosa* (Gerstaecker) by FitzGerald (1940) and was later confirmed by observing colonies of *R.*

fasciata (Fab.) (van der Vecht, 1962). In addition, the nests of *R. tomentosa* have a varnished or glazed pedicel (FitzGerald, 1940) similar to that reported for *Polistes* (West Eberhard, 1969), the polybiine *Mischocyttarus* Saussure (Jeanne, 1972), and the vespines *Vespa* L. (Matsuura, 1971b) and *Dolichovespula* Rohwer (Greene *et al.,* 1976, Jeanne, 1977a). Whether the materials in the window, the pedicel, and the polyethylenelike envelope are similar is unknown.

Nests are usually constructed in some location protected from the rain such as under leaves, twigs, and eaves of structures, on trees, or in natural cavities such as in the hollow interior of dead bamboo; however, they have also been found in extremely exposed locations such as hanging from wires (van der Vecht, 1962; Yoshikawa, 1964). Many species that make simple *Polistes*-type combs construct nests close together. For example, an extreme case was reported for *R. plebeiana* Richards where closely adjacent combs in one nesting site covered 10 m² (Richards, 1978b).

Richards (1978b) summarized all available information on nests and nesting habits of the group and attempted to relate nest architecture to the current taxonomic groupings. There was a slight correlation, but as nests of more species are investigated and described, this will probably change.

iii. Prey. There is little information on the prey of species of *Ropalidia* and nothing is known of their foraging behavior. References to prey and other foods includes only caterpillars (Roubaud, 1916; FitzGerald, 1950; Iwata, 1976), *Perkinsiella saccharicidia* Kirkaldy (sugarcane leafhopper) (Williams, 1928), and the secretions of Aleyrodidae (Richards, 1978b). However, the Ropalidini, because of their large colonies or abundant small colonies, must be important predators in many areas.

iv. Behavior. Colonies of *Ropalidia* are usually small (less than 50 individuals) and the nest may be initiated by a single female (haplometrosis), by several females that cooperate (pleometrosis), or either way, even in a single species. *Ropalidia variegata* (Smith) (Davis, 1966a), *R. cincta* (Lepeletier) (Darchen, 1976), *R. fasciata* (Fab.), and *R. artifex fuscata* van der Vecht (van der Vecht, 1962) were reported to be haplometrotic. *Ropalidia marginata* (Lepeletier) (Gadgil and Mahabal, 1974), *R. cincta,* and *R. guttatipennis* (Saussure) (Roubaud, 1916) were reported to have nests established by a small number of females, while *R. timida* van der Vecht has been reported to establish nests by swarming (Pagden, 1976). These few reports indicate that the wasps are versatile and may switch from haplometrosis to pleometrosis depending on environmental and colony conditions (Iwata, 1976). Time of observations may also be a factor. As Davis (1966a) stated, nests of *R. marginata* are first established by a single female and then other, cooperating, females appear on the nest. Most colonies are annual, but colonies of some species are perennial with nests in various stages of construction at all times.

Richards (1969) reported on eight nests of *R. cincta* containing 3–200 cells, the largest nest having eight females. There is also a report on *R. timida* van der Vecht, a species which constructs somewhat larger nests of 250 cells or more, having colonies of about 50 individuals (Pagden, 1976). Other studies include descriptions of colonies of *R. variegata* (Smith) which builds adjacent combs where 47 nests represent 11 colonies and 53 combs represent 5 colonies (Davis, 1966b). *Ropalidia hova minor* Saussure (FitzGerald, 1950) and *R. plebeiana* Richards construct hundreds of combs closely adjacent to each other. There are no reports of the number of individuals in a colony or even what constitutes a colony in this case. Reports are also lacking on the number of adults in the larger nests which contain up to 60,000 cells (see Carl, 1934). The great diversity of nest structure and colony size in various species of *Ropalidia* indicates a wide variance in their social organization and behavior, but unfortunately reports are available only on species making small, open comb nests that have few individuals in a colony.

Ropalidia are generally regarded as primitively social wasps that lack obvious morphological caste differences between females. One of the few early reports (Roubaud, 1916) on social interactions between adults on the nest merely stated that more than one egg layer was present in colonies of *R. cincta* and *R. guttatipennis.* For many years this report was used as the basis for comparing *Ropalidia* to other wasps. Knowledge of colony organization was expanded when Davis (1966a), observing *R. variegata,* reported that colonies are usually founded by a single female, but when the first larvae approach maturity the foundress is joined by 7–9 females. These females serve the function of workers in cleaning and trimming cells, tending and feeding larvae, and even cutting the edges of caps to help emerging individuals. As the colony enlarges with the addition of progeny, some females leave to join other nests, or they may initiate nests of their own. Observations of another species, *R. marginata,* by Gadgil and Mahabal (1974) indicated still more complex interactions. *Ropadilia marginata* colonies are usually founded by several females and are perennial. However, these studies showed that often only one female, or at most a few, had fully functional ovaries. Most females had undeveloped or atrophied ovaries and functioned as workers. It was suggested that these workers spent more time in food gathering than the egg laying female(s), but received only a small portion of the food. This division of labor was thought to be accomplished by behavioral interactions (i.e., dominance, see Gadagkar, 1980) which resulted in nutritional castration. The aforementioned studies were brief and, with the exception of Gadgil and Mahabal (1974) and Gadagkar (1980), contain only incidental references to behavioral interactions.

The first comprehensive study of behavior among individuals within the

TABLE II

Distribution, Number of Species, and Degree of Sociality of Subfamilies of Social Wasps[a]

Subfamily	Principal (or social) genera	Estimated No. species	Distribution	Degree of sociality
Pemphredoninae	*Microstigmus*	ca. 50	Paraguay to Costa Rica	Solitary– One species social
Stenogastrinae	*Stenogaster* *Anischnogaster*	17	Papuan region	
	Eustenogaster *Holischnogaster* *Liostenogaster* *Metischnogaster* *Parischnogaster*	50	Oriental region	Subsocial– Communal– One species social
Polistinae Ropalidiini	*Ropalidia*	126	Africa, Madagascar, Iran, India, southern China, New Guinea, and Australia	Unknown– Social
Polistini	*Polistes*	203 sp., 106 ssp.	Worldwide	Social
	Sulcopolistes	3	Palearctic	Social parasites
Polybiini	*Belonogaster*	ca. 35	Tropical Africa, Arabia	
	Polybioides	6	West and Central Africa, Indonesia	
	Parapolybia	6	Iran to Japan, Korea, Indonesia	
	Polybia	54	Mexico to Argentian	
	Protopolybia	23	Tropical Central and South America	Unknown– Social
	Brachygastra	16	Southern United States to Argentina	
	Parachartergus	17	Mexico to Argentina	
	Stelopolybia	23	Mexico to Argentina	
	Mischocyttarus	202	Argentina to British Columbia, Canada	
Vespinae	*Provespa*	3	Indo–Malayan region	
	Vespa	20	Europe to southern China, Mediterranean to Ethiopia, Malaysia	Social
	Dolichovespula	14	Holarctic	
	Vespula	ca. 20	Holarctic	

[a] Modified from Richards (1971) by use of Richards (1978a,b); van der Vecht (1966); Wilson (1971).

TABLE III
Colony and Nest Size of Polybiine Wasps[a]

Principal genera	Approx. No. adults	Nest	
		No. combs	Total No. cells
Belonogaster	1–60, usually < 10	1	100–600, usually < 150
Parapolybia	4–150	1	50–1900
Polybioides	400–3000	6–7	1600–9000
Apoica	50–500	1	300–1500, most < 500
Mischocyttarus	1–30	1	?–300, most < 30
Metapolybia	20–60 +	1	?–7000, most < 500
Clypearia	40–70	1	100–300(?)
Synoeca	100–600	1	200–6000, most < 600
Polybia	20–10,000 +	5–27	1000–50,000 +
Protopolybia	3000–20,000	5–14 +	800–54,000
Chartergus	300–3000	5–12	2000–22,000 +
Brachygastra	80–15,000	2–12 +	400–50,000(?)
Angiopolybia	20–400	2–4	250–1200
Pseudopolybia	30–600	3–5	300–500 +
Stelopolybia	5000–22,000 +	7–9	15,000–100,000

[a] Principal literature sources: Richards and Richards, 1951; Richards, 1969, 1978a; Jeanne, 1972, 1975b, 1977b; Naumann, 1968, 1970; Piccioli and Pardi, 1978. Some total cell counts estimated from nest size. All colonies not at same stage of development so values given are only for general comparison purposes.

tached together by struts [*Stelopolybia vicina* (Saussure)], several combs arranged one above the other with the comb perimeter fastened to the edge of the envelope (*Polybia* Lepeletier, modified somewhat in *Brachygastra* Perty, *Chartergus* Lepeletier, *Epipona* Latreille, *Protonectarina* Ducke, *Synecoides* Ducke), or nearly spherical concentric combs (*Stelopolybia* Ducke, see Jeanne, 1973, for a detailed description of this remarkable type of nest). Many other variations exist but are not nearly so easily described.

Nests of *Polybioides* are unusual in that the bottoms of cells containing mature larvae are removed by the adults so that fecal pellets can fall to the ground or be removed. This opening is not glazed over as in *Ropalidia*. *Belonogaster* also remove cell bottoms and meconia (Marino Piccoli, 1968), but unlike *Ropalidia,* where the adults are presumed to glaze over the hole, this glaze is produced in some way by the larvae (Pardi, 1977).

Most species build nests of vegetable fibers mixed with salivary secretions, but the resulting envelope and carton of most species is coarser than in the Vespinae and Polistini as the particles are less malaxated (Richards and Richards, 1951). Two subgenera of *Polybia* (*Furnariana* Richards, *Pedothoeca* Richards) build the envelope of mud, but the comb portion is mainly paper. One species, *Polybia emaciata* Lucas, constructs both the thin envelope and the cells of the comb of mud. Since the fiber particles are not well chewed and are cemented together with (sometimes) copious amounts of oral secretions, some nests contain large transparent windows or holes (*Mischocyttarus*), or transparent micalike specks in the envelope [*Metapolybia pediculata* (Saussure)] (Richards, 1978a). A few species [*Pseudochartergus fuscatus* (Fox) and *Ps. chartergoides* (Gribodo)], have a transparent, cellophanelike envelope consisting entirely (?) of oral secretions (Jeanne, 1970), while *Protopolybia pumila* (Naumann, 1970) coat the leaves incorporated into the inside of the nest with a thin translucent material, again presumably an oral secretion.

Most Polybiini build aerial nests in a variety of locations such as on the underside of sloping tree trunks, on the sides of tree trunks, on the eaves, walls, and ceilings of structures, on or under leaves, and attached to twigs or large branches. Some that do not construct an envelope around the combs (*Stelopolybia*) frequently nest in hollow trees or other cavities. Most reported nests are 2–15 m high. Although some nests undoubtedly occur at greater heights, they are frequently difficult to see and are, for all practical purposes, inaccessible.

iii. Prey. There are few reports of the prey or food habits of the Polybiini (see Naumann, 1970; Evans and West Eberhard, 1970; Iwata, 1976). This may be attributable, at least in part, to the manner in which prey is brought to the nest. Most species return with the prey as a malaxated pellet of various sizes, infrequently as a recognizable, fairly intact unit. Another difficulty exists in *Protopolybia,* which bring in most prey in the crop completely ingested, chewed, and mixed with nectar. Available information indicates that various species prey on flies, lepidopteran caterpillars, winged ants and termites, beetles, and Homoptera. From the list given, most species appear to be opportunistic foragers and probably take readily available arthropods. All species probably forage for nectar and honeydew, a few collect pieces of ripe fruit (e.g., *Polybioides melainus* Waldo), and others scavenge for meat from carcasses or from households. *Angiopolybia pallens* (Lepeletier) and *A. obidensis* (Ducke) occasionally prey on living and dead tadpoles in foam nests in the Amazon forest (Lacey, 1979) (Fig. 6). In British Guiana, *Synoeca surinama* (L.) and *Mischocyttarus drewseni* Saussure attack honey bee hives. The former attacks and kills workers while the latter enters weak hives to plunder both honey and brood. Many species of polybiines store honey in their nests, and some species of *Brachygastra* Perty are maintained in a semidomestic state by Mexican Indians for their honey caches.

iv. Behavior. Most tropical polybiines establish nests by swarming, during which several to many queens and a number of workers leave the old nest as a group or swarm to rapidly establish a nest at a new site. However, all

Fig. 6. *Angiopolybia pallens* preying on tadpoles in a foam nest in the Amazon. (Photograph by Barbara Gibbs, in Lacey, 1979.)

modes of nest building occur among species whose nests are initiated by a single foundress (haplometrosis), nests initiated by a single female that is joined later by other females (foundress association), or nests originally started by a number of cooperating, egg-laying females (pleometrosis). *Mischocyttarus mexicanus* (Saussure) colonies are usually founded by one female in the spring, but by groups of females in the fall (Litte, 1977). It is quite possible that the method of colony founding of other polybiines may vary seasonally. *Mischocyttarus, Belonogaster,* and *Parapolybia* are reported as usually haplometrotic and nonswarming, while *Apoica* Lepeletier, *Brachygastra, Synoeca, Polybia, Angiopolybia* Araujo, *Stelopolybia, Metapolybia, Parachartergus* Ihering (Evans and West Eberhard, 1970), and *Protopolybia* (Naumann, 1970) swarm. Ten other genera are assumed to swarm, but evidence is weak or lacking (Evans and West Eberhard, 1970). However, Richards (1978a) recently stated that all South American genera except *Mischocyttarus* swarm, and all available evidence indicates that this is the case (see also Jeanne, 1980).

Information on the caste composition of swarms is meager, but males are assumed to fertilize the queens before the swarm departs and are not present in the swarm. *Apoica* may be an exception because males have been found in newly established nests (see Richards, 1978a). The number of queens in a swarm is unknown for most species. However, Naumann (1970) recorded

100–400 queens in swarms of *Protopolybia pumila,* while the workers in these swarms were primarily young wasps, rather than foragers. Similarly, swarms of *Stelopolybia areata* (Say) are 5–12% queens; the rest are workers (Jeanne, 1975b).

Swarming is essentially a tropical adaptation that provides for rapid construction of the new nest with the advantage that many workers can serve as guards to protect the developing brood (Richards and Richards, 1951). Eggs are usually laid in the cells only when the nest is nearly complete. In addition to the normal swarms, secondary swarming also occurs, usually when the nest is destroyed or raided by army ants and most of the adult population are unharmed (Richards, 1971; Young, 1979).

While externally distinct castes are generally said not to be present in the Polybiini, actually variations exist ranging from species with a morphologically distinct queen (*Stelopolybia, Protonectarina*) to those in which there is little or no external difference among colony members [*Parachartergus fraternus* (Gribodo), *Angiopolybia pallens* (Lepeletier)]; however, females differ in behavior and in ovarian development. In between these extremes are the polybiines with slight, but statistically significant, morphological differences in wing length, number of hamuli (wing hooks), etc. (Richards, 1971). In general, species having large colonies tend to have more morphologically distinct queens (Richards and Richards, 1951; Jeanne and Fagen, 1974).

There are two types of behavior associated with colony duration in the Polybiini (Richards and Richards, 1951). The first type is shown by colonies which persist for a relatively short period, usually lasting about three developmental cycles (culminating in the production of males and queens each time), for a total duration of usually less than 1 year. *Mischocyttarus, Metapolybia, Polybia occidentalis* (Oliver), *P. bistriata* (Fab.), and *Protopolybia minutissima* (Spinola) are in this group. The second type concerns colonies that are perennial and last a much longer time, up to 25 years (see Richards and Richards, 1951), although this length of time is probably exceptional. Two to six years is probably more typical. Swarms originate from these colonies annually, but a portion of the population remains with the parent nest. This group is represented by *Polybia rejecta* (Fab.), *P. scutellaris* (White), *Protopolybia pumila* (Saussure), *Epipona tatua* (Cuvier), *Synoeca surinama* (L.), and *Brachygastra lecheguana* (Latreille). Most swarmers seem to have perennial colonies while colonies initiated by a single foundress or by foundress association last 1 year or less. However, *Metapolybia* swarms and is annual, so at least a few exceptions occur (Evans and West Eberhard, 1970). Colony duration is, of course, related to colony size (Table II). Species with less populous colonies tend to be of the short duration type. Regardless of colony duration, most tropical species have colonies at various stages of development throughout the year.

All polybiine colonies are cyclical in that after the nest is established, the colony expands exponentially to a certain point when reproductives (males and queens) are produced. These reproductives mate, the males die, and the new queens (often accompanied by workers) establish nests. While colony founding can occur at any time, in tropical areas with a pronounced wet and dry season colonies are generally largest near the end of the dry season, and many new nests are established at this time.

v. Representative Genera. As previously stated, little is known of the social biology of polybiine wasps, and much of that information concerns a few species with small colonies and naked comb nests. Since there seems to be a positive correlation between nest population size and social complexity (Naumann, 1970), brief biological and behavioral sketches of *Belonogaster* and *Mischocyttarus* (small colonies), *Protopolybia,* and *Stelopolybia* (large colonies) follow as a representation of the extreme diversity of behavior present among polybiines. These are not intended to represent typical polybiines, but are used as examples solely because the most complete studies have been done on species of these genera.

BELONOGASTER

Belonogaster make simple nests consisting of a single stalked comb. The only known exceptions are nests of *B. brevipetiolatus* Saussure and *B. hildebrandtii* Saussure that do not have a pedicel, but rather are fastened directly to the substrate (du Buysson, 1909). Nests are usually hammock shaped, curving down from the attachment point and back up at the distal end (Marino Piccioli and Pardi, 1978) (Fig. 7). Most nests contain less than 150 cells, although nests of *B. griseus* (Fab.) can be as large as 600 cells. The adult population on the nest is also small, usually less than ten. The nest is usually initiated by a single foundress, although she may be joined by other females soon after nest initiation.

Much of the early information on *Belonogaster* was presented by Roubaud (summarized, 1916) on *Belonogaster junceus* (Fab.). From this information it was concluded that these wasps were primitive and quasisocial (see Wilson, 1971). The first indication that this was not correct was contained in a study by Richards (1969) on the same species. He observed individually marked wasps on 6 nests over a period of 7–26 days, and afterward collected the nests and adults for analysis. He determined that the adults were of five types: males, old queens, intermediates, workers, and young queens. Old queens had 2–8 mature eggs in their ovarioles, intermediates had much smaller eggs and their ovaries were only slightly developed, while worker ovarioles were threadlike. The queen and callows tended to sit on the face of the comb while workers usually stayed on the outer surface. It was concluded that Roubaud (1916) was

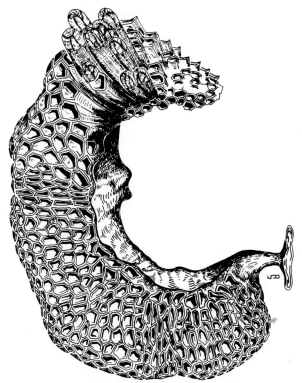

Fig. 7. Stylized drawing of mature nest of *Belonogaster griseus* showing the characteristic hammocklike shape. Cells near the petiole are not reused, but are gradually demolished. (Drawn from a photograph in Marino Piccioli and Pardi, 1978.)

not correct in assuming all females were equally capable of reproduction at the same time.

After studying *Belonogaster griseus* (Fab.), Marino Piccioli and Pardi (1970, 1978; Pardi and Piccioli, 1970; Pardi, 1977) demonstrated that *Belonogaster* are primitively eusocial wasps with behavior similar to *Mischocyttarus* and *Polistes*. A linear dominance hierarchy exists among the females with only the highest ranking individuals laying eggs. The dominant individual (queen) also practices differential oophagy by eating any eggs laid by subordinates and replacing them with her own. The dominant individuals demanded and received more food from foragers, always initiated new cells, rarely left the nest, usually positioned themselves on the face of the comb, were dominant in all encounters with other wasps, and in general exhibited royal behavior (Pardi and Piccioli, 1970). Dominant females were also the only individuals seen to twirl about the nest pedicel with their abdominal sterna appressed to its surface, probably depositing an odorous substance (perhaps an ant repellent, see Jeanne, 1975a).

Castes are not externally recognizable, although, at least in some colonies, dominant females tend to have a wider, longer thorax (Pardi and Piccioli, 1970). Inseminated females are always dominant while noninseminated females generally assume the duties of workers. A number of females of both castes are frequently on the nests. In some cases certain egg-laying, noninseminated females manage to rank quite high in the dominance hierarchy. When a colony is mature and has males, they solicit food from returning foragers and distribute it among the larvae. Even more unusual, males have been observed returning to the nest with solid food and participating in its distribution to the larvae (Pardi, 1977), similar to behavior reported for two species of *Polistes* (Hunt and Noonan, 1979).

It is not surprising that Roubaud did not record or even suggest behavior indicating a dominance hierarchy in *Belonogaster*. The studies of Piccioli and Pardi indicated that dominance behavior was rare on some nests, and it took intensive studies of marked individuals to demonstrate a hierarchy.

While it is probably safe to assume that this behavior is similar for all *Belonogaster,* only additional studies of the behavior of the remaining 33 species will verify this assumption.

MISCHOCYTTARUS

Mischocyttarus also build a nest of a single, naked comb with one pedicel. Nest size varies from 50 to 400 cells (Richards and Richards, 1951), although the mature nest of most species is usually from 100 to 200 cells. Nests are nearly always constructed in some protected location such as on the eaves or ceilings of man-made structures (many species), or on the underside of leaves.

Although *Mischocyttarus* consists of 202 species, biological and behavioral studies have been conducted only on *M. drewseni* Saussure (Jeanne, 1972; Jeanne and Castellóu Bermudez, 1980), *M. mexicanus* (Saussure) (Litte, 1977), *M. immarginatus Richards* (Gorton, 1978), and to a lesser extent, *M. flavitarsis* (Saussure) (Landolt and Akre, 1978; Litte, 1979). Other references to *Mischocyttarus* biology may be found in Jeanne (1972).

Colonies of *Mischocyttarus drewseni* were studied in the lower Amazon region of Brazil from August, 1967, to May, 1969. During this time 53 colonies were located, and hundreds of hours of observations of marked individuals were conducted to determine colony cycle and to record behavioral interactions between colony members. In this species there are no morphological differences between queens and workers although queens tend to be larger.

Most colonies are founded by a single queen, although up to eight females (assumed to be siblings) will associate to initiate a nest. The first 10–15 females emerge within 40–65 days to serve as workers, while the first males emerge about 6 weeks after the first larvae have pupated. Soon after male emergence, nonworkers (potential queens) begin to emerge and over the next 3–4 months the ratio of males and nonworkers to workers increases as fewer workers are produced. This ratio, and not the physiological senescence of the queen, is probably what finally accelerates colony decline to the point that brood cannibalism occurs and the nest is abandoned. The progeny then establish new colonies and the cycle continues, with new colonies founded at any time of the year. The entire cycle takes about 6 months although many colonies under observation were destroyed by predators or by other means before the cycle was completed. Workers foraged for various arthropod prey and during periods of peak abundance, nectar was concentrated and stored in cells of the nest.

Mischocyttarus drewseni has a linear dominance hierarchy very similar to that occurring in *Polistes*. At the top of the hierarchy is the queen (primary egg layer) and her cofoundresses. Their rank is apparently correlated to ovariole development, although this has not been investigated by dissections. The queen is the primary egg layer, eats the eggs of any egg-laying subordinates (differential oophagy), initiates the construction of most cells, and forages primarily for pulp. The workers are subordinate to the queen, extend cell walls, forage for pulp and food, distribute the food to nestmates and the brood, and also bring water to the nest. High ranking individuals also have a trophic advantage in that they solicit and receive more food from foragers and other subordinates. Dominance behavior varies in intensity from merely rushing at a subordinate that responds by crouching low on the comb in submission, head down and gaster slightly elevated, to a vigorous chewing with the mandibles by the dominant over the face and body of the subordinate. The hierarchy is not static and the dominant position is usually usurped once or even twice during a colony cycle.

This hierarchy differs from that in *Polistes* in that subordinate females are usually involved in dominance interactions primarily during the first 2 weeks of their life. This is related to ovarian development which apparently can be supressed only during a certain critical period of 2–15 days.

Females have a gland at the base of the terminal abdominal sternum that apparently secretes a material onto a tuft of hairs from which it is applied to the pedicel of the nest to repel ants and protect the brood. Females are also observed to drag the gaster across the face of the comb, but the reason for this behavior remains obscure.

The studies on *M. mexicanus* in Florida and *M. flavitarsis* in Washington indicate that these species exhibit behavior quite similar to that reported for *M. drewseni,* and dissections of queens and workers definitely indicate that rank in the dominance hierarchy is related to ovarian development.

PROTOPOLYBIA

Polybia colonies are representative of the populous, perennial colonies of polybiines occurring in the tropics that have more complex social interactions than those that occur in the small colonies of *Belonogaster* and *Mischocyttarus*. However, excellent summaries of the known biology and behavior of various *Polybia* species have already been presented by Evans and West Eberhard (1970) and Wilson (1971). Therefore, *Protopolybia* and *Stelopolybia* were chosen here to represent this important group.

Protopolybia pumila [= *P. sedula* (Saussure), Richards, 1978a] was studied in the Canal Zone, primarily Barro Colorado Island, from April, 1966, until February, 1968 (Naumann, 1970). During these 20 months, 49 colonies of *P. pumila* were located, and 19 of these were collected for nest and population analyses. Five nests were located where detailed observations could be made of inhabitants by means of an elevated platform or tower and the use of red cellophane windows placed in the nest envelope. To date, this is the only study ever made of interactions between and among nest inhabitants inside a populous, envelope-covered nest.

Protopolybia pumila is atypical of the genus in that this is the only species with large, multilayered, elaborate nests housing up to 20,000 adults. However, behaviors found in colonies of this species are probably representative of the social complexity found in populous colonies of other genera and species.

The aerial nests of *P. pumila* are constructed on the tips of flexible branches and palm fronds at heights of 3–15 m or more. The first comb is always placed directly on the substrate, and is usually quite small, less than 50 cells. Combs are then added below until the mature nest is a large, multicombed structure, covered with an envelope. The carton and envelope are closely packed plant hairs mixed with a salivary secretion. Ordinarily the combs are not attached to each other, but in some nests the combs are continuous and built into a spiral configuration, similar to a spiral staircase. Leaflets and branches are usually woven into the nest, and these, and not the envelope, serve as the major support of the combs. This permits numerous entrances on the sides and top of the nest where the envelope is discontinuous, and most wasp traffic takes place through these lateral passageways around the edges of the combs. The leaves sticking inside the nest are frequently covered with a thin layer of translucent material that is presumed to be an oral secretion. It is probably the same secretion used to make the carton. Most nests are quite large and vary in length from 8.5 to 44 cm. However, external dimensions are usually a poor indicator of size, especially if the nest is an odd shape because it is build around obstructions. The number of cells is usually more reliable, and mature nests range in size from 17,000 to 54,000 cells with the average of ca. 32,000 cells. Just as in several wasp groups already mentioned, variability among nests of this species is so great that de Saussure's classification scheme is applied only with difficulty, and even this requires some modification.

The seasonal cycle of *P. pumila* is fairly simple if several colonies which do not fit the general scheme are excluded. Colonies founded in the early dry season (December–January) grow rapidly into large male and queen producing colonies, and several swarms issue from these colonies at the end of the dry season in April–May. These swarms contain old queens and a number of inseminated young queens. The colonies founded by these swarms (wet season colonies) grow to a smaller size (about 10,000 cells) and remain fairly stable in size while the young queens that were in the founding swarm mature. The general population level appears to decline somewhat during the wet season either through smaller colonies, fewer colonies, or perhaps both. At the end of the wet season the nest population increases and the colony, with the young queens now mature, divides into several swarms which found colonies late in the wet season. These new colonies become the highly productive, dry season colonies. Thus, 2–5 swarms occur during the course of 1 year.

All colonies of *P. pumila* contain a number of queens (Table IV). These queens are morphologically distinct from the workers by their larger size, darker color, reduced markings, and by the greater width of the second abdominal tergum. This facilitated sampling of nest populations at various times which showed that 0.6–12% of the population are queens. Males are produced only in dry season colonies from January to July, with the peak male emergence in May. Few males take part in swarms. Instead they inseminate new queens at their own, or possibly other, colonies prior to swarming. The composition of swarm populations is uncertain since very few swarms have been observed, much less collected and analyzed. However, evidence indicates *P. pumila* swarms vary in size from 1,000 to 3,000 wasps, and consist mostly of workers, with 100–300 queens. Larger swarms of 5,000–8,000 may occur at the end of the dry season when colonies are at their largest. The causes

TABLE IV

Population Composition of Selected *P. pumila* Colonies[a]

		Population			Population				
:ason	Collection date	Queens (%)	Workers (%)	Males (%)	Queens (No.)	Workers (No.)	Males (No.)	Total No.	No. cells in nest
ry	Jan 16	1.53	98.50	0.00	113	7269	0	7382	16,600
	Feb 13	1.50	98.08	0.42	249	16251	70	16,570	—
	March 10	0.61	81.56	17.83	120	16104	3521	19,745	48,500
	May 16	0.45	69.88	29.68	39	6075	2580	8694	17,500
'et	Dec 6	3.48	96.52	0.00	36	998	0	1034	4300

[a] Modified from Naumann, 1970.

of swarming remain largely obscure although it has been suggested that the new queens stimulate colony division (swarming) in some unknown way at the end of the dry season. Another possibility is that the ratio of queens to workers increases and induces swarming.

A swarm constructing a nest at the end of the wet season–early dry season enters into an expansion phase characterized by rapid cell construction, a high oviposition rate, a rapid increase in worker population, and an increase in worker activities such as foraging, ventilating, and nest construction. Later, males and new queens are produced. The expanding nest lacks an envelope at the bottom, the area of rapid construction. As many as 40 queens may be active on the bottom comb, checking cells and laying eggs. Oviposition rate appears to determine expansion rate, and one nest expanded about 400 cells/day for at least 2 months. In addition to new cells, 600 callows emerged daily, so nearly 1,000 cells were available for oviposition each day.

After males and new queens emerge, the colony enters the postexpansion and swarming phase. This phase is characterized by queen activity on the outer surface of the nest, and changes in worker behavior. Workers cease construction activities, fight among themselves, chase and bite males, and cannibalize brood. Just prior to swarming the population tends to cluster above the nest and all normal nest activities cease. Especially obvious are "buzzing runs" performed by workers on the nest surface, which increase in frequency until up to 25 workers make runs at one time. The runs start with the worker walking rapidly, wings straight up, and climax in longer runs with wings rapidly vibrating. Runs generally terminate in flight. These runs tend to break up groups of young workers and probably stimulate emigration to a new site. Species of *Angiopolybia, Leipomeles* Moebius, *Polybia* (Jeanne, 1981), and *Stelopolybia* also make similar buzzing runs while dragging their gaster on the substrate at the old nest site and along the path to and at the new nest site. Evidence indicates that a pheromone is deposited, probably from sternal glands, which is used to lay a trail to the new site to enable the entire swarm to emigrate rapidly (Naumann, 1975; see also Chadab and Rettenmeyer, 1979). Information on swarming and emigration behavior is not available for nests established in the wet season.

Workers forage for arthropod prey, pulp for nest construction, water, and nectar. Most prey is returned to the nest completely ingested and is carried in the crop, although some solid prey is brought to the nest in various stages of dismemberment. Identified prey include various Homoptera, adult and larval beetles, caterpillars, adult and larval flies, bees, and small wasps. Depending on supply and demand, nectar from various floral sources is fed to nestmates and larvae, or concentrated as honey and stored in empty peripheral and egg cells. Workers collect water for evaporation and fan to control nest temperature when it becomes too warm, and also rid the nest of excess water immediately after heavy rains. The water is sucked into the crop and dumped outside the nest.

Most foraging appears to be done by older workers that remember foraging sites and return to them repeatedly during the course of a day. Foraging for both food and pulp is greatest in the morning, as is done in yellow jacket colonies (Potter, 1964). There is a definite division of labor among workers, and the behavior of a particular worker tends to be uniform during any one day. Data on whether various duties change over time are not available. Foragers often display a "departure dance" after giving prey or nectar to nestmates and before leaving to forage again. This dance does not have a regular pattern, but the worker runs about excitedly and is licked and antennated by other wasps. The dance suggests that some sort of communication occurs, but no evidence has been found to support this idea. It does not increase the number of foragers present at a given foraging site, and may just function to excite nestmates to depart and to forage.

Interactions between and among workers inside the nest are extremely diverse. Pulp loads are shared with nestmates by foragers (sharing pulp loads is an unusual behavior rarely seen in other social wasps), as are loads of ingested prey–nectar mixtures. Surprisingly, solid prey is less often solicited. As might be expected, trophallaxis between nest inhabitants is common. It has been suggested that dominance interactions occur between workers engaged in trophallaxis as has been reported for *Vespula* Thomson by Montagner (1966), but only a limited number of observations have been made on this behavior.

Queens aggressively solicit food from foragers to the extent of grasping them by the thorax with their front legs while biting until the foragers offer food. Stored honey is also used by nonforaging queens as an energy source.

Most workers rest on or above the nest, or slowly walk over the surface of the combs, essentially doing nothing. However, the prime function of these workers is colony defense against both invertebrate and vertebrate predators. Upon stimulation they may assume the alarm position, body immobile, wings folded and down, abdomen held straight up with the string extended, or they may give an "alarm buzz," a short, audible buzz of the wings with the abdomen usually in contact with the nest carton, causing it to vibrate. A buzz is sometimes given repeatedly, with a rapid pumping of the abdomen between buzzes, or, if stimulation increases, the buzz may develop into an "alarm dance" with the worker running in an ever widening circle, abdomen held high and wings buzzing intermittently. These defensive behaviors tend to threaten the intruder and/or alert nestmates. In addition, there is strong evidence for an alarm pheromone released by stinging workers which stimulates other workers to sting.

The behavior of *P. pumila* indicates that polybiines with large colonies are not as primitive as might be inferred from behaviors reported for species with small colonies, and future studies on interactions among adult wasps, on possible pheromones, and on other possible communication among the wasps may show them to be quite advanced.

STELOPOLYBIA

The most complete information on *Stelopolybia* biology was provided by Jeanne (1973, 1975b, 1975c) and Jeanne and Fagen (1974) for *Stelopolybia areata* in Mexico, although Jeanne (1970) also reported on a nest of *S. testacea* (Fab.) from Brazil.

Stelopolybia areata nests differ from the nests of closely related species in that the nest has an envelope, and is frequently built in exposed locations, usually at the tops of trees. Seven nests were collected from heights that varied from 2 to 25 m. Mature nests are large, 45–48 cm long × 34–36 cm diameter, and contain about 8–9 combs. The combs are concentric spheres, interconnected by rampways to form a continuous surface spiralling outward, with an entire nest containing as many as 100,000 cells. The adult population of these nests are among the largest of all social wasps; one colony had 22,000

Stelopolybia areata has morphologically distinct queens; larger in size, and with a much larger first gastral tergum (petiole) than workers. Queens composed 5–12% of the adult population of four nests collected in January–February.

New nests are constructed in the fall by swarms of several thousand adults, including 5–12% queens, issuing from a parent colony. However, a resident population probably remains behind, so this colony continues on in the old nest. From February to July the worker population in the new nests increases greatly as the founding queens gradually disappear. By July, males are produced and the first new queens emerge in August. Several swarms may then split from this colony to establish new nests and to start the cycle anew.

Although this species is remarkably nonaggressive and workers do not readily sting, there are indications that stinging workers release an alarm or attractant pheromone that has a sweetish odor like Ivory soap. No information was given indicating that this odor excited other workers or stimulated them to sting. *Stelopolybia areata* also appears to have a trail pheromone used to guide the swarm to the new nest site. Workers land on conspicuous objects along the route, and drag their gasters on the substrate, presumably marking the area with pheromone. This behavior is very similar to that reported for several other polybiines that also apparently use trail pheromone (Naumann, 1975; Jeanne, 1981).

No information is available on behavior inside the nest, but the large colony size, swarming behavior, complex nest design, and possible presence of several different types of pheromones suggests that the social behavior of this species is at least as complex as that of *P. pumila.*

c. Polistini. *i. Taxonomy and Distribution.* The tribe Polistini is represented by only two genera: *Polistes,* with 203 species and 106 subspecies divided among 11 subgenera, and *Sulcopolistes* Blüthgen with only three species, all social parasites (Richards, 1973, 1978a,b). Older taxonomic reports of a somewhat general nature on this group that are still quite useful include Yoshikawa (1962a) and Guiglia (1972), while regional distribution and taxonomy have been covered by Blüthgen and Gusenleitner (1970), Bohart and Bechtel (1957), Buckell and Spencer (1950), Chandler (1964), Guiglia (1971), Kemper and Döhring (1967), Krispyn and Hermann (1977), Løken (1964), Pekkarinen (1973), Snelling (1954, 1970, 1974), Soika (1975), van der Vecht (1971), and Yamane and Yamane (1979). In addition, a key to Costa Rican species by Roy Snelling was included in a biological analysis by Nelson (1971). However, even with the above studies, and many others of more limited scope (see Wilson, 1971), the taxonomy of the Polistini is still in a constant state of flux and has many problems. Some ambiguously defined species have been found to include at least several species (Wade and Nelson, 1978; MacLean *et al.,* 1978), and many color forms or subspecies await study to determine if they might also be distinct species. At the present time there is no adequate taxonomic key for North American species, and with the possible exception of the Polistini of Europe, much taxonomic investigation still remains to define species limits. Most taxonomic studies have relied heavily on adult morphology and color patterns, but a few studies (Nelson, 1969; Wade and Nelson, 1978) have shown that larval morphology is also very useful for separating and identifying species. Several current taxonomic studies also include the determination of chromosome numbers which appears to vary among the species (Hung *et al.,* 1981).

The Polistini occur in all regions of the world except the coldest parts of the continents and New Zealand (Richards, 1971; Iwata, 1976). *Polistes humilis* (Fab.) now occurs in New Zealand; it was accidentally introduced from Australia (Ferro, 1976), and recent reports indicate that two other exotic *Polistes* (*P. chinensis antennalis* Pérez and *P. olivaceus* De Geer) have probably established residency (Harris, 1979). No *Polistes* occur naturally in Great Britain either, although *P. gallicus* (L.) (Richards, 1971) and *P. metricus* (Say) (Cooter, 1976) have been introduced. Apparently they are unable to survive for more than one season. From a practical standpoint, few *Polistes* occur above 50° north latitude in North America, nor above 65° north latitude in Europe (Sweden and Norway, which are warmed by ocean currents), but even considerably below these latitudes (e.g., the northern states of the United

States) most *Polistes* colonies are small and not particularly successful most years. In contrast, the genus is well represented in southern areas of the United States (e.g., Texas, Louisiana) where more species occur and colonies are relatively large and abundant. It has been stated that in Europe and North America colonies of *Polistes* outnumber colonies of all other social wasps combined (Wilson, 1971). Actually, the Vespinae (e.g., hornets, yellow-jackets) are dominant in these regions, but their nests are mostly subterranean and therefore not nearly as noticeable as nests of *Polistes*.

The biology of Polistini has been studied quite extensively in comparison with other social wasps. Indeed, there are more published reports of *Polistes* than of any other group of social wasps. The biology of the group has been summarized by Evans and West Eberhard (1970), Wilson (1971), Spradbery (1973), Ebeling (1975), and Iwata (1976). Excluding most of the older reports, the more important studies of the biology of individual species are by Bohm and Stockhammer (1977), Corn (1972), Garcia (1974), Gibo (1972, 1978), Gillaspy (1971, 1973), Hermann and Dirks (1975), Hermann *et al.* (1974, 1975), Hunt and Gamboa (1978), Ibrahim (1974), Jeanne (1979), Matsuura

Fig. 8. Mature nest of *Polistes aurifer* attached to roof of shed in Pullman, Washington. Several peripheral cells contain stored honey.

(1970), West Eberhard (1969), S. Yamane (1969, 1971, 1972), Yamane and Kawamichi (1975), Yamane and Okazawa (1977), Sk. Yamone (1973), Yoshikawa (1962b, 1963a,b), Prattle and Gervet (1980), Miyano (1980), Suzuki (1980, 1981), Strassman (1981), Lester and Selander (1981), and Dew and Michener (1981).

ii. Nests.   The nest of most *Polistes* is a single exposed comb, usually supported by a single pedicel (Richards, 1978a) (Fig. 8). All nests are constructed from dry, woody fibers scraped from weathered wood and other plant sources. The range of nest architecture is considerable, and even nests of the same species can vary greatly (Iwata, 1976). This is frequently dependent on the nesting site. Some species (e.g., *P. goeldii* Ducke) construct a nest of a series of cells, one below the other, while other species construct a vertical comb (*P. infuscatus* Lepeletier, *P. exclamans* Viereck, *P. instabilis* Saussure), or the more familiar horizontal comb [*P. fuscatus* (Fab.), *P. carnifex* (Fab.), *P. apachus* Saussure, *P. metricus, P. major* Palisot de Beauvois]. Only a few species have been reported to construct multiple combs (Jeanne, 1979). The single pedicel may be central (*P. carnifex*), eccentric (*P. metricus*), or attached peripherally [*P. annularis* (L.), *P. exclamans*] (see Reed and Vinson, 1979a). A few species such as *P. apachus, P. carolina* (L.), and *P. metricus* (Say) may have nests with multiple pedicels, while *P. peruvianus* Bequaert constructs pedicels that are often branched. This characteristic also appears to vary with the nest site.

Polistes nests are usually constructed in protected, but quite visible locations. While many species construct nests on and in vegetation, most studied species appear to prefer man-made structures. For example, in Costa Rica, *P. erythrocephalus* Latreille constructs nests on the eaves and roofs of buildings, under bridges, on other man-made structures, and on the trunks of trees. Nests in trees are nearly always located on large solitary trees in otherwise cleared areas; neither wasps nor their nests have been seen in the forest (Nelson, 1971). In urban Texas most nests of *P. metricus* and *P. exclamans* are located on buildings and other man-made structures in fairly well lit locations. However, *P. carolina,* and occasionally *P. metricus,* nest in concealed locations such as attics, holes in trees, preformed cavities covered by grass on the soil surface, and inside clothesline posts made of pipe (Reed and Vinson, 1979a). Nests in these concealed locations frequently have multiple pedicels.

Polistes colonies are generally small. Among the largest nests reported are those of *P. annularis* (L.) (1886 cells, Nelson, 1968), *P. erythrocephalus* (nearly 1000 cells, Nelson, 1971), and *P. flavus* Cresson (more than 1500 cells, Nelson and Crowell, 1970) (see also Table V), although larger nests undoubtedly occur. However, most nests probably contain fewer than 400 cells, and most colonies rear 200 or fewer individuals to adults (even assuming that

TABLE V
Comparison of Nest Size of Several Species of *Polistes*

Species	Locality	No. nests	No. cells largest nest	Avg. No. cells/nest	Reference
Polistes annularis	United States	101	1886	393	Nelson, 1966, 1968
P. fuscatus	United States	2	306	180	Nelson, 1966, 1968
P. metricus	United States	136	479	92	Nelson, 1966, 1968; see also Bohm and Stockhammer, 1977
P. exclamans	United States	145	569	152	Nelson, 1966, 1968
P. carolina	United States	19	191	98	Reed, 1978
P. biglumis (L.)	Japan	28	152	100	Yamane and Kawamichi, 1975
P. chinensis antennalis Pérez	Northern Japan	12	99	57	Yamane, 1972
P. c. antennalis	Japan	15	1106	138	Iwata, 1976
P. gigas Kirby	Taiwan	8	126	60	Matsuura, 1970
P. hebraeus Saussure	India	1	1192	—	Kundu, 1965
P. humilis	New Zealand	15	1290	140	Cumber, 1951
P. peruvianus Bequaert	Peru	20	—	112	Garcia, 1974
P. rothneyi iwatai van der Vecht	Japan	5	137	74	Yamane, 1972
P. spilophorus Schletterer	West Africa	8	218	138	Richards, 1969
P. tepidus malayanus Cameron	New Guinea	4	662	273	Yamane and Okazawa, 1977

some cells of the nest are used 2–3 times to rear adults). Adult population (excluding males) of a colony at its peak rarely exceeds 150, but most colonies are considerably smaller. Surprisingly, tropical nests are generally not larger than nests in temperate regions (Evans and West Eberhard, 1970).

iii. Prey. The principal prey of these wasps is probably caterpillars (Rabb, 1960; Iwata, 1976; Dew and Michener, 1978), and most studies of prey have been involved with investigating *Polistes* are possible biological control agents (Rabb and Lawson, 1957; Lawson *et al.,* 1961; Kirkton, 1968, 1970; Gillaspy, 1971; Nakasuji *et al.,* 1976; Hasui, 1977; Yamasaki *et al.,* 1978). These projects were concerned with agriculturally important prey such as caterpillars of tobacco and tomato hornworm, tobacco cutworm, and cotton bollworm. *Polistes* also penetrate the silk nests of fall webworm to prey on the larvae (Schaefer, 1977). Diptera are usually regarded as moving and flying too swiftly for the slower *Polistes,* but Garcia (1974) observed *P. peruvianus* preying on various dipterans. *Polistes* obviously capture a number of slow-moving insects besides caterpillars, but these have received little or no investigation. They also feed on the honeydew of many homopterans, the sweet juices of many fruits, and the floral and extrafloral nectar of many plants (see Heithaus, 1979). Most species store honey in various amounts in cells of the nest (see Strassman, 1979). The amount of honey stored varies with the species of *Polistes,* and on the abundance of nectar in the area.

iv. Behavior.   In the cold temperate areas of the world only inseminated females (potential queens) overwinter. In the spring nests are initiated by a single foundress queen (West Eberhard, 1969; Iwata, 1976) which is then joined by other females that have not established a nest of their own. Depending on locality, nests are initiated in April–June, produce reproductives in July–September, and then abandon the nest. About this time, males and queens mate, the males usually die soon after (an exception occurs in *P. annularis* where males overwinter and mate in the spring), and only the inseminated females overwinter. Tropical *Polistes* do not hibernate, and nests are initiated by females leaving the parental nest at any time of the year. These nests are frequently constructed by multiple foundress queens. While these colonies are called pleometrotic, *Polistes* seem essentially haplometrotic because when there are several foundresses one eventually becomes the effective queen and egg layer, while over a period of time the other foundresses desert, become idle, or serve as workers (Richards, 1978a). The length of a colony cycle of several tropical species is 6–7 months as compared to only 4–5 months for temperate species, and it has been suggested that the ultimate colony decline depends on the reproductive longevity of the queen (Evans and West Eberhard, 1970).

Caste differences in *Polistes* are extremely small (Eickwort, 1969), especially in tropical species, and caste is more a matter of behavior than of structure (Richards, 1978a).

While it has been stated that colony foundation in the tropics may be by swarming (Richards and Richards, 1951), only three instances were found in the literature that might provide evidence for this behavior: an abnormal swarm which occurred when the nest was destroyed (Rau, 1933); a discussion of "pseudoswarms" (Rau, 1941); and a report of about 200 males and females of *P. humilis* on a nest of 400 cells containing no brood which could possibly have been constructed by these individuals (Cumber, 1951). It is probable that most species do not establish nests by swarming.

The social biology of several species of *Polistes* has been investigated in great detail with a select list of the more important papers in this area including Pardi (1948), Deleurance (1955), Morimoto (1960), Yoshikawa (1963a,b), West Eberhard (1969), Evans and West Eberhard (1970), Corn

(1972), Hermann and Dirks (1975), Hermann *et al.* (1975), Maher (1976), Perna *et al.* (1977), Turillazzi and Pardi (1977), and Gamboa *et al.* (1978).

One of the most significant behaviors exhibited by *Polistes,* a dominance hierarchy among female nestmates, was first observed in *P. gallicus* (L.) by Pardi (several papers, largely summarized in Pardi, 1948) and expanded by Deleurance (1955). These studies have served as the basis for the investigations of dominance interactions in other species of *Polistes* as well as in other genera of social wasps (e.g., *Belonogaster*). The hierarchy is established soon after a number of inseminated females join a single foundress that has already initiated a nest. Through a series of behavioral interactions, one female, often the foundress, becomes dominant with respect to receiving the most food, initiating cell construction, laying most of the eggs, and occupying the comb face. In contrast, the subordinates do most of the foraging for food and fiber, lay fewer eggs, lengthen cells, and occupy mostly the outer edges of the comb. They are subordinate to the dominant individual (i.e., queen) in all interactions despite being inseminated and fully functional as egg layers. Eventually the ovaries of these individuals regress, and they become idle. They usually desert the nest after the first workers emerge. Since excellent summaries of earlier literature by Evans and West Eberhard (1970) and Wilson (1971) are already available, the dominance interactions of *P. annularis* in Georgia (Hermann and Dirks, 1975) are described in the following discussion.

After the nesting cycle is completed in the fall, inseminated females (potential queens) of *P. annularis* show strong tendencies to aggregate into groups of closely related individuals (i.e., siblings). The establishment of dominance hierarchies begins in these groups during a prehibernation period in late fall. Females attack each other and sometimes become entangled and fall to the ground ("falling fights") while grasping each other with their legs while attempting to bite and sting. A few individuals are actually stung, with serious consequences, but most encounters end with no bodily harm to either combatant. These females then hibernate over the winter in protected locations (usually as a group), mate in early spring (Hermann *et al.,* 1974), and these aggressive interactions continue through the posthibernation period of early spring. About this time up to 8–10 females (cofoundresses) initiate a nest, and the hierarchies established earlier apparently determine duties (division of labor) among the cofoundresses. Additional females sometimes join the colony after nest initiation and, of course, their position in the hierarcy is established by aggressive behavior soon after their arrival. Subordinate cofoundresses that are vigorously attacked by the dominant female often leave the colony, probably to establish small nests of their own. These nests usually contain only a few cells, and most are eventually abandoned. Dominance interactions continue among the remaining cofoundresses, and as a more stable hierarchy evolves, subordinates essentially function as a worker caste (see also

Gamboa *et al.,* 1978). This differs from *P. erythrocephalus* (= *canadensis erythrocephalus,* see Richards, 1978a) where subordinate females become idle residents of the nest (West Eberhard, 1969). The establishment of a stable hierarchy takes up to 2.5 months, until the first workers emerge. During this time aggressive interactions continue, although the frequency and intensity of the encounters decrease. Even when a stable hierarchy is finally established, the queen still rushes at and threatens nestmates, which respond by crouching down on the comb, but more aggressive encounters rarely occur. The initial aggressive interactions to establish dominance tend to interfere with nest building so nests with multiple foundresses (> 3) are not as efficient (in terms of number of cells constructed) as might be assumed. In nests with only 2–3 cofoundresses, hierarchies are established earlier, and efficiency loss is not as apparent. At first all females lay eggs and practice oophagy. This also contributes to the inefficiency of the colony, but as a single female becomes dominant, her ovaries increase in size and she becomes the sole egg layer, while the ovaries of the subordinate females atrophy. Later, as the first workers emerge, the cofoundresses leave the nest and do not return. Workers assume the duties associated with the nest, but they also soon begin to tail wag (abdominal wagging, see also Gamboa and Dew, 1981), and to "smear"; behavior that is normally restricted to the queen and cofoundresses. The queen stays with the nest through August, but leaves the nest (possibly dies) long before cold weather begins. About the same time or shortly after the queen leaves the colony (August–September), males and new queens begin emerging.

The dominant behavior of the queen of *P. annularis* is similar to that reported for other species (e.g., *P. metricus, P. fuscatus*). She collects little food and fiber, rarely leaves the nest, spends most of her time on the comb face, receives a greater share of the food, and is dominant in all interactions with other nest inhabitants. She also tail wags (abdominal wagging of other reports), or vibrates her gaster laterally, a behavior not well understood at this time (Gamboa and Dew, 1981). The gaster is in contact with the comb during the behavior and vibrates the comb loudly enough that it can be heard several feet away (West Eberhard, 1969). This behavior may stimulate greater activity by nestmates, and/or it may be a clear signal of dominance by the queen. Similar behavior occurs in *Metapolybia* ("shaking dance," West Eberhard, 1978), honey bees (Fletcher, 1975), yellowjackets ("gastral vibration," Greene *et al.,* 1976, 1978), and other species of *Polistes* (West Eberhard, 1969; Esch, 1971; Fluno, 1972; Maher, 1976; Gamboa *et al.,* 1978). Another behavior restricted to the queen or cofoundresses is smearing, an application of a repellent substance to the top of the nest, the nest pedicel, and the surrounding substrate (Hermann and Dirks, 1975; Turillazzi and Ugolini, 1978). The repellent material is produced by glands of the sixth and seventh abdominal sterna and is applied by means of a sternal brush (Post and Jeanne, 1980;

Delfino *et al.*, 1979). The material is effective against formicine ants (Post and Jeanne, 1981) and serves the same function as the material reported for *Mischocyttarus* by Jeanne (1970).

Male *P. annularis* are ordinarily at the bottom of the hierarchy and soon leave the nest. They overwinter in aggregations with the new females (potential queens), become active in the spring, mate, and die soon after (Hermann *et al.*, 1974). This sequence is similar to that reported for *P. erythrocephalus* which has at least some nonmated females initiating nests; however, they are soon inseminated (West Eberhard, 1969). This differs from some other species of *Polistes* (e.g., *P. metricus* because males mate soon after emergence and die within a short time (Bohm and Stockhammer, 1977).

Males of *P. exclamans* (Lin, 1972) and *P. erythrocephalus* (West Eberhard, 1969; Alcock, 1978; Alcock *et al.*, 1978) maintain small territories they defend against intruders, especially conspecific males. The defended area, about 0.5–1.0 m diameter, usually consists of a convenient perch on a prominent piece of vegetation, rock, or wall near the nest or hibernation sites. Males show constancy to their territory and some return to the same site for 5 days (Alcock, 1978). Defense usually consists of a short flight to chase the intruder, but grappling fights occur with conspecific males in which both combatants fall to the ground. They sometimes continue to fight, but most break apart quickly. The presumed function of these territories is to uniformly space the males and thus ensure successful mating.

Since colony founding by associations is considered intermediate between founding by swarms and colony founding by a single queen (Wilson, 1971), much recent research on *Polistes* biology has been concentrated on investigating foundress associations with regard to kin selection theory (Hamilton, 1964, 1972) and parental manipulation theory (Alexander, 1974) in an attempt to better understand the evolution of social behavior, and especially to gain insight into the evolution of the worker caste (see Chapter 2, Volume I of this series). Recent reports on this subject, specifically as it relates to *Polistes* colonies, are by Lin and Michener (1972), Gibo (1974, 1978), Metcalf and Whitt (1977a,b), Gamboa (1978), Noonan (1978), West Eberhard (1978), and Lester and Selander (1981).

Obviously much remains to be learned about these primitively eusocial wasps. All investigated species have relatively small colonies and a somewhat limited repertoire of behavioral interactions between colony members. Physical dominance manifestations, rather than pheromones, are apparently responsible for the queen's control over egg laying and ovarian development by subordinates, although abdominal wagging could possibly be the behavioral act associated with the deposition of a pheromone on the comb face. No species has been reported to have an alarm pheromone. However, the exocrine gland system of *Polistes* has not been examined in detail, and investigations may reveal or suggest glands as possible pheromone sources. For example, males of *P. aurifer* Saussure (= *fuscatus aurifer,* see MacLean *et al.,* 1978) have a large mandibular gland which may be a site for the possible production of an aggregating or mating pheromone (Landolt and Akre, 1979). These unknown areas, plus the fact that few of the 203 species of *Polistes* have been studied at all, leaves all conclusions about the "primitive" eusociality of *Polistes* as tentative. They are obviously not as behaviorally complex as honey bees or large colonies of social wasps such as *Protopolybia,* but they probably have a more complex social behavior than presently realized.

3. Vespinae

a. Taxonomy and Distribution. The subfamily Vespinae is essentially a tropical Asian group of wasps that has extensively invaded the temperate regions (Richards, 1971) (Fig. 9). The subfamily has only 4 genera: *Provespa* Ashmead, *Vespa* L., *Dolichovespula* Rohwer, and *Vespula* Thomson. While the taxonomy of vespines has been relatively well studied, problems still remain, especially at the species level. Therefore, the number of species should be regarded as tentative. There are presently ca. 20 species of *Vespa,* 3 of *Provespa* (van der Vecht, 1957, 1959), ca. 18 of *Dolichovespula,* and 20 or more species of *Vespula* (Edwards, 1980). *Provespa* occur only in southeast Asia, *Vespa* has both tropical and temperate species, and *Dolichovespula* and

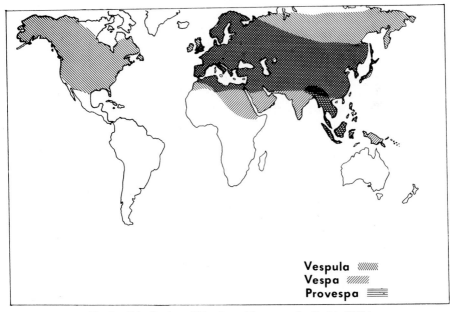

Fig. 9. Distribution of Vespinae. (From van der Vecht, 1967.)

Vespula are temperate species, with the possible exception of *V. squamosa* (Drury) (Miller, 1961) and an unnamed species occurring in Mexico. Unfortunately, there is no recent taxonomic treatise on the vespines (however, see Archer, 1981a,b; Eck, 1980, 1981; Yamane *et al.*, 1980), and Richards (1978a) excluded this group from his revision.

b. Biology. All vespine colonies are initiated by a single queen (see discussion in van der Vecht, 1957), and all build paper-carton nests. The nests consist of a number of nearly round combs attached one below the other, usually covered with a multilayered envelope. Some species construct subterranean nests, some nest above ground in aerial locations, and others use either location. Colonies vary in size from less than 100 individuals to several thousand. The Vespinae are the only wasps known to construct special cells, usually on the last several combs of the nest, which are decidedly larger and used to rear queens (Richards, 1971; Spradbery, 1973).

Most vespines have large, easily recognizable queens, and a distinct, smaller, worker caste, However, colonies of *Vespa tropica* L. and *V. analis* Fab. contain many individuals that are larger than workers, but smaller than most queens, making caste separation difficult or impossible (van der Vecht, 1957). A similar situation apparently occurs in some colonies of *Dolichovespula maculata* (L.).

Summaries of vespine biology have been prepared by Richards (1971), Spradbery (1973), and Edwards (1980).

c. *Provespa.* *Provespa* occur exclusively in the tropical areas of the Oriental region, from Burma southeastward through the Indochina and Malayan areas to the islands of Indonesia (van der Vecht, 1936, 1957, 1959). Based on morphology, *Provespa* is considered the most primitive genus in the Vespinae. However, castes are distinct and queens and workers are readily separated by size differences, by pubescence [*P. anomala* Saussure, *P. barthelemyi* (Buysson)], and also by the shape of the gaster (*P. nocturna* van der Vecht) (van der Vecht, 1957).

Virtually nothing is known about the biology of these wasps. All three species are nocturnal, have uniformly brown bodies, and have greatly enlarged ocelli. Only three or four nests have been found and all were aerial (see van der Vecht, 1957; Pagden, 1958). One of these nests, probably a nest of *P. anomala,* was 12 cm in diameter, spheroidal, and had a single entrance. It was located about 1 m above the ground in a tree and contained the queen and 130 workers. Unfortunately, no other nest descriptions are available.

Reports on behavior are equally meager. The only known prey of these wasps are blue-bottle flies (Diptera: Calliphoridae), and even this observation was made under unusual circumstances. It was made by a prisoner in a

Japanese camp during World War II who watched the wasps at night attacking flies around the latrines and subsequently relayed this information to van der Vecht (1957). Since all members are nocturnal, workers are not found actively foraging during the day. However, they apparently fly quite readily to defend the nest if disturbed (Pagden, 1958). The only other behavioral observation of record is that all species are attracted to artificial light, sometimes in great numbers, and for this reason can be a nuisance in homes. Most of the specimens used for taxonomic studies have evidently been collected at lights.

d. *Vespa* **(Hornets).** *i. Taxonomy and Distribution.* The genus *Vespa* contains 20 species and ca. 58 subspecies and most of them occur in eastern Asia and the adjacent archipelago. In the western Palearctic region only two species, *V. crabro* L. and *V. orientalis* L., are widely distributed (see Guiglia, 1972). *Vespa* does not occur in Africa south of the Sahara desert and did not occur in North America until 1840–1860 when *V. crabro* was introduced. The greatest number of species and the greatest diversity in characters are found in the eastern Himalayas–southern China which is regarded as the evolutionary center of the genus. From this center members of the group migrated outward to colonize the adjacent islands (van der Vecht, 1957). The taxonomy of *Vespa* is fairly stable (van der Vecht, 1959; Yamane, 1974) and although it is unlikely that new species will be discovered, the distribution of some subspecies is still unknown (van der Vecht, 1957).

ii. Biology. Among the Oriental species, *V. affinis* L., *V. analis* Fab., *V. tropica* (L.), and *V. velutina* Lepeletier occur in populous areas where the forest has been cleared, and nearly all available information on biology and habits concern these species. However, *V. tropica* and *V. affinis* have often been confused in the past and have been referred to in older literature as "*V. cincta.*" Thus, it is extremely difficult to separate the species and their respective biologies in these reports. *Vespa basalis* Smith, *V. mocsaryana* Buysson, *V. multimaculata* Pérez, and *V. bellicosa* Saussure are forest-inhabiting species. All are rare, and there is little information on their habits (van der Vecht, 1957). Recently, *V. analis insularis* Dalla Torre, *V. crabro flavofasciata* Cameron, *V. mandarinia latilineata* Cameron, *V. mongolica* André (= *xanthoptera* Cameron, *simillima* Smith), and *V. tropica pulchra* Buysson have been studied in Japan (Iwata, 1971; Matsuura, 1969, 1971a,b, 1973a,b, 1974, 1977; Matsuura and Sakagami, 1973; Yamane, 1974, 1977; Yamane and Kamijo, 1976; Yamane and Makino, 1977). Nesting data and a few behavioral observations have also been collected of *V. affinis affinis* (L.), *V. analis parallela* André, *V. basalis,* and *V. velutina flavitarsus* Sonan in Taiwan (Matsuura, 1973c).

The habits of *V. crabro* in the western Palearctic region have been detailed by several researchers in general reports on vespids (Løken, 1964; Kemper and Döhring, 1967; Guiglia, 1972; Pekkarinen, 1973; Spradbery, 1973), while the biology of this species in North America has been summarized by Akre *et al.* (1981). *Vespa orientalis* has been intensively studied in Israel by Ishay and coworkers. Unfortunately, while reports on specific aspects of behavior and physiology of this species abound, there is still little available literature on the general biology and nesting habits of this species (see Rivnay and Bytinski-Salz, 1949; Ishay *et al.*, 1968; Wafa *et al.*, 1968; Wafa and Sharkawi, 1972). The biologies of all other species, subspecies, and varieties are unknown except for a few casual observations.

All *Vespa* are haplometrotic. Colonies in temperate zones are annual with both inseminated and noninseminated queens (Matsuura, 1969) overwintering in a diapause state (Matsuura, 1974). *Vespa* in tropical regions are cyclic, with colonies in all stages of development at any given time of the year. In West Java *Vespa analis* Fab. and *V. velutina* Lepeletier show this type of periodicity (van der Vecht, 1957). There are no reports of perennial colonies.

All temperate *Vespa* appear to have a remarkably similar cycle (Matsuura, 1969, 1971b; Yamane, 1974; Wafa *et al.*, 1968). The overwintering queens emerge during the first warm days of spring from late March to May. They require food after the long diapause period, and begin to forage on flowers for nectar. The Japanese species, however, forage for tree sap from *Quercus* L. (oak), and establish an inter- and intraspecific dominance hierarchy at the sap flows. The larger species and individuals are dominant (Matsuura, 1969). Nonmated queens continue to visit the *Quercus* until July, but they never establish a nest, and eventually die. During May–early June the inseminated or mated queens establish a nest. Initially the first few cells of the nest are suspended from a single pedicel which is strengthened by "painting" a lustrous substance (oral secretion?) on its surface. The queen usually expands the nest to 35–50 cells, and raises 5–14 larvae to pupation (Matsuura, 1971b, 1973a). Until the first eggs hatch, the queen continues to rely almost entirely on tree sap for food. Afterward, she hunts insects and spiders to feed the brood. The envelope, which totally encloses the queen nest, is partly removed by the workers that emerge in June–early July and begin to expand the nest. Soon after the emergence of the first workers, the queen no longer leaves the nest to forage, but continues with activities inside the nest such as cell enlargement, larval feeding, and thermo-regulation. After the workers emerge, queens of *V. mongolica, V. analis insularis,* and *V. tropica pulchra* continue to initiate cells, but queens of *V. crabro flavofasciata* do not. In nearly all species the queen gradually ceases activities inside the nest except egg laying, but queens of *V. tropica pulchra* continue these activities (e.g., cell enlargement, feeding prey to larvae) throughout their entire life span (Matsuura, 1974). By August–early September reproductive cells are constructed and the first males and new queens usually emerge by mid-September. These leave the nest in October–November, and in *V. mandarinia* the males wait near the nest entrance to mate with the queens as they leave (Matsuura and Sakagami, 1973). The new queens then seek a sheltered place to hibernate for the winter.

iii. Nests. The nests of tropical *Vespa* are very incompletely known. Depending on species and locality, these wasps construct nests on the branches of trees or bushes at various heights above the ground (most *V. affinis,* most *V. analis,* most *V. basalis,* a few *V. crabro,* most *V. luctuosa* Saussure, most *V. mongolica,* a few *V. tropica,* most *V. velutina*), on the eaves of houses (some *V. affinis,* some *V. analis,* a few *V. tropica*), or in protected locations such as cavities in trees (most *V. crabro,* some *V. mandarinia,* a few *V. orientalis*), wall voids or attics of houses (many *V. crabro,* many *V. orientalis,* many *V. tropica*), or underground (a few *V. basalis,* many *V. crabro,* most *V. mandarinia,* a few *V. mongolica,* most *V. orientalis,* most *V. tropica,* a few *V. velutina*) (Duncan, 1939; van der Vecht, 1957; Matsuura, 1971a, 1973c; Chan, 1972; Guiglia, 1972; Kojima and Yamane, 1980). Subterranean nests are frequently constructed in rodent burrows or similar cavities and have entrance tunnels of 3–420 cm (Matsuura, 1971a). Nests of all species are very similar, varying only slightly in shape, but queen nests of *V. analis insularis* are somewhat unique in having a long tubular entrance (funnel) much like *Dolichovespula maculata* (L.) (Matsuura, 1971b). This entrance tube is cut off by the workers shortly after they emerge. Nests of *V. crabro* (Fig. 10) in protected cavities in Europe and North America have little envelope, but nests of *V. crabro flavofasciata* in Japan have a complete envelope regardless of nest location. Nests of *V. mandarinia* also have a poorly developed or incomplete envelope, but all other species apparently enclose the nest almost completely (Matsuura and Sakagami, 1973). All species of *Vespa* create a scalloped envelope, with the scallops enclosing trapped air pockets, and some tropical species construct aerial nests with a conical roof, probably an adaptation to withstand heavy rains (van der Vecht, 1957). Typically the nests have only one entrance hole. The envelope and carton (comb material) are usually composed of dead plant fibers collected from various sources, including loose powder from rotten wood. However, a few species use other materials in addition to plant fibers. *Vespa mongolica* uses tree resin (van der Vecht, 1957), while *V. orientalis* sometimes incorporates clay (Bequaert, 1936), or sand and pebbles into the envelope (Ishay, 1964). Some species have also been reported as collecting fibers from living plants. *Vespa crabro* attacks and girdles lilac, birch, and rhododendron (Duncan, 1939), and *V. basalis* workers damage the bark of young *Eucalyptus* trees (van der Vecht, 1957). These reports are probably erroneous, especially in the case of *V. crabro,* because the wasps girdle the young bushes to obtain sap, not fiber, and are never seen carrying fiber away (see Akre *et al.,* 1981).

Mature nest size of *Vespa* spp. varies from about 600 to more than 14,000

Fig. 10.   Mature nest of *Vespa crabro* built in playhouse in Con-
necticut. (Photograph by C. W. Rettenmeyer.)

cells (10 combs, *V. mongolica*). The number of combs have been recorded for
nests of the following seven species: *V. affinis* (12–13), *V. analis* (8–9), *V.
crabro* (3–5), *V. mandarinia* (4–8), *V. mongolica* (9–11), *V. tropica* (3–4), and
V. velutina (11) (Iwata, 1976). *Vespa crabro* in Europe builds nests of 6–9
combs (Spradbery, 1973; Archer, 1981), and nests of *V. orientalis* in India
consist of 5 or more combs (Maxwell-Lefroy and Howlett, 1909). The
smallest *Vespa* nest is probably that of *V. dybowskii* André, a facultative social parasite, which has mature nests of 3 combs of less than 20 cm diameter
and weighs less than 1 kg (Sakagami and Fukushima, 1957a). Since *Vespa* are
large wasps, their nests are quite large even with relatively few cells. For example, a nest of *V. affinis* measured 109 cm long, one of *V. analis* 80 cm, and two
of *V. velutina* measured 75 cm and 122 cm (van der Vecht, 1957). Nests of *V.*

crabro are up to 60 cm long (Spradbery, 1973). Total number of cells and adult inhabitants of a mature nest would be a much more reliable indicator of colony size than comb numbers or external nest dimensions. Unfortunately, this information is not available for most species. In general, *Vespa* colonies are usually less populous than colonies of certain species of *Vespula* (yellow-jackets). A mature nest of *V. affinis affinis* (L.) had 6178 cells in 5 combs of about 50 cm diameter. The entire adult population consisted of 1095 workers and the queen (Matsuura, 1973c), whereas many mature colonies of yellow-jackets have physically much smaller nests, but these consist of 10,000–15,000 cells, and the colonies have 2000–5000 workers.

iv. Prey. *Vespa* are powerful, agile wasps, and many hunt active insects as prey. Prey includes spiders, butterflies, moths, caterpillars, flies, bees, and wasps (van der Vecht, 1957; Iwata, 1976). However, most records of prey are based on rather casual observations and/or deal with only a few species. For example, *V. analis* and *V. crabro* attack cicadas in addition to other arthropods (Matsuura, 1977). *Vespa velutina* prefers to hunt fast flying Diptera such as stable flies and blow flies, although workers also capture honey bees. In fact, a number of species appear to prey mostly on social bees and wasps. *Vespa tropica* has been seen attacking colonies of *Ropalidia, Parapolybia,* and stenogastrines (van der Vecht, 1957), and in Japan *V. tropica* preys nearly exclusively on *Polistes* (Sakagami and Fukushima, 1957b). *Vespa crabro* also takes some *Polistes* larvae as prey, and *V. mongolica* workers have been seen removing *Vespula* workers from spider webs (Iwata, 1971). Various species of *Vespa* prey heavily on honey bees, and are a problem to beekeepers in all areas where these wasps occur. *Vespa crabro* and *V. analis* cause some problems in beehives, while *V. orientalis, V. mongolica,* and *V. mandarinia* cause serious problems by killing adult bees and removing all the brood. They can literally devastate entire apiaries in a short period of time (see Akre and Davis, 1978).

Although it has been stated that *Vespa* rarely visit flowers for nectar (van der Vecht, 1957), Yamane (1974) collected a number of species of *Vespa* from various flowers, and Ishay *et al.* (1968) stated that *V. orientalis* frequently visits flowers. Most species probably forage for nectar from flowers, as well as for tree sap, honeydew (Yamane and Kamijo, 1976; Ishay *et al.,* 1968), and other sources of carbohydrates. Indeed, species of *Vespa* are reported to cause extensive damage to ripe fruits of papaya, guava, peach, fig, grapes, apricots, and apples (van der Vecht, 1957). *V. orientalis* is also attracted to sweetmeat (a sugary confection).

A few species of *Vespa* scavenge for meat to feed their larvae. This is especially true of *V. orientalis* workers that forage for small bits of fish or ground meat (Ishay *et al.,* 1968). Apparently this habit is not as well developed in *Vespa* as it is in yellowjackets. However, further investigations may reveal that other species also exhibit this behavior.

The hornets quickly kill prey with a bite of their powerful mandibles on the head or thorax, and then usually fly to a nearby perch to trim the prey to a convenient size before returning to the nest. Typically, the wings, legs, and head are removed and discarded, while either the thorax or abdomen is brought to the nest. The only known exception is *V. tropica pulchra* which apparently carries all food as liquid in the crop (Matsuura, 1974). Hornets have been reported to use the sting to kill prey (van der Vecht, 1957; Ishay *et al.,* 1968), but this is considered extremely unlikely. Yellowjackets are much smaller than species of *Vespa,* and they do not normally use their sting for this purpose.

v. Behavior. Detailed studies of foraging behavior of hornets are limited to observations of workers of *V. crabro flavofasciata* (Matsuura, 1973b). One thousand thirty-three foraging trips by workers from an observation colony were recorded, and it was found that 57% were for fluid, 13% for prey, 17% for fiber for nest construction, 7% for water, and on 5% of the trips the workers returned with nothing. Some workers appear to specialize in collecting food and water, and all workers are fairly constant to a particular activity. The specific task a worker performs is not associated with physical size, and no particular trend has been observed in respect to foraging activity and age. Most foraging trips for food and pulp are made in the morning, while more water is brought in during the heat of the day when the nest temperature increases. This is similar to the observations by Iwata (1971) of foraging by workers of *V. mongolica* (= *xanthoptera*). They were most active early in the morning before sunrise, and foraging decreased slightly by noon. The workers then began to bring in water. Although *Vespa binghami* Buysson is the only truly nocturnal *Vespa* (van der Vecht, 1957), workers of *V. crabro* frequently fly (possibly to forage) on moonlit nights (see Akre *et al.,* 1981).

Few observations of activities inside the nest of *V. crabro flavofasciata* exist, but these observations indicate that hornet behavior is similar in many aspects to that of their smaller relatives, the yellowjackets. Matsuura (1968, 1974) has found that foragers returning to the nest with pulp or water never share their load with their nestmates. However, all foragers with prey are solicited immediately on their return and share their prey with one or more workers. After malaxating the prey thoroughly, all then feed those large larvae which scratch the walls of their cells indicating that they require food. Small larvae are fed only liquid. All workers, and in early nests, the queen, engage in fanning when the nest temperature rises too high. This also elicites more workers to bring in loads of water, and these are usually spread on the silken caps of cells containing pupae. The evaporation of the water, in combination with the fanning, holds the nest temperature within reasonable limits. In these early nests there is little interaction between the queen and her workers, and even when the first workers emerge, the queen ignores them. As the colony grows larger, trophallaxis between nestmates increases, but the queen never behaves dominantly or aggressively when she is engaged in trophallaxis with a worker. She solicites prey or liquid from foragers like any worker, and is sometimes ignored, just like any worker. The workers are not attracted to the queen in these early nests, but this apparently changes as the colony becomes larger. In the populous colonies of August the workers are extremely attracted to the queen and form a "royal court" about her while licking and nibbling at her body surfaces.

The most extensive studies of behavior inside the nest have been conducted on colonies of *V. orientalis* by Ishay and coworkers in Israel (Ishay *et al.,* 1968). They designed nest boxes for hornets that permit the viewer to sit in a darkened room to observe behavior inside the nest while the colony has access to the outside and apparently behaves quite normally. It was found that the egg to adult period was 29–42 days (Ishay, 1976), and that workers live 12–64 days. No worker left the nest until at least 2 days after eclosion, and none left the nest after they were 35–38 days old, but they still remained active inside the nest (Ishay *et al.,* 1968). Activities observed were similar to those reported for *V. crabro.* The colony became active at 7:30–8:00 AM when the nest temperature reached 27°C, and the first foraging of the day was always for pulp for nest construction. This was followed by a period of foraging for prey, and then for carbohydrate food. Water was brought into the nest at all times, and was used in combination with fanning to reduce nest temperature when it became greater than 29°C.

Much of the workers' activities were associated with feeding and caring for the larvae which were visited frequently, up to 95 times per hour. Larvae were apparently warmed by the workers that covered them with their bodies if the nest temperature dropped below 26°C. It was suggested that one reason for the great attraction of the larvae to the adult wasps was that the adults did not have proteases and were required to obtain protein nourishment from the larvae. The adult wasps captured prey, malaxated it, fed it to the larvae, and the larvae used their proteases to convert the protein into materials readily useable by the adults (Ishay *et al.,* 1968; Ikan *et al.,* 1968). However, investigations by Spradbery (1973) and by Grogan and Hunt (1977) on *Vespula* and Kayes (1978) on *Polistes* showed the adults do have proteases and this has suggested some experimental errors in the original *Vespa* studies.

Trophallaxis between adult wasps in the nest is frequent and older workers appear to exhibit dominance in demanding to be fed by younger workers (Ishay and Ikan, 1968) similar to the "dominance hierarchy" among yellowjacket workers reported by Montagner (1966). This behavior has not been reported for *V. crabro.*

Other investigations indicate that the wasps communicate by sound (Ishay *et al.,* 1974; Ishay, 1975, 1978). In the morning, as the colony first becomes active, the workers tap their abdomens on the comb (awakening tap or dance) to arouse the larvae which then indicate their hunger by scraping on their cell walls with their mandibles. Once active, the larvae continue to scrape the cell walls throughout the day as they become hungry. This sound can be heard several feet away by an observer, but it is probably the vibrations traveling through the comb which stimulate the workers to attend the larvae. Workers also tap their abdomens against the comb as they stand in a circle around the queen, and it was proposed that this sound may encourage the queen to continue to lay eggs. While the explanations of the functions of these sounds are thus far somewhat tentative, it appears that sound plays a major role in colony communication.

Quite early in his studies Ishay (1965) found that the queen was highly attractive to workers, and these workers avidly licked and nibbled at her head and body. He postulated that the queen secreted a pheromone or queen substance which was similar in function to that of the honey bee. Later, Ikan et al. (1969) isolated a material, δ-n-hexadecalactone, primarily from the heads of queens of V. orientalis, that was proposed as the queen pheromone. Experiments showed that this material stimulated construction of queen cells by workers at the end of the season, even in the absence of the queen, and inhibited ovarian development by workers (Ishay, 1973a). Recently, Coke and Richon (1976) were able to synthesize this lactone which will permit controlled experiments to study the role(s) of this material in normal colony life. Since the queen pheromone was discovered, a number of additional pheromones have been proposed for V. orientalis. Ishay (1972, 1973b) postulated a pheromone in pupae that elicits the workers to warm them when the temperature falls too low, and Saslavasky et al. (1973) showed workers have an alarm pheromone. When disturbed, workers come out of the nest and spray venom. The alarm pheromone in the venom excites the hornets further, and stimulates the stinging response. Other proposed pheromones include a wasp assembly pheromone, a male wasp assembly pheromone, a building depressor pheromone, and a building initiating pheromone (Ishay and Perna, 1979). The wasp assembly pheromone, produced by workers and fertilized queens, stimulates the wasps to assemble into groups, either while on the nest under natural conditions, or in a particular spot inside a nest box under experimental conditions. Males do not respond to this material, but have a similar pheromone of their own. The building depressor pheromone is released by the queen each time she oviposits and tends to regulate the rate of cell construction to her oviposition rate. Thus, workers are unlikely to have excess cells in which they might oviposit. The building initiating pheromone, on the other hand, is a material released by the workers, either from tarsal glands or from van der Vecht's gland (sixth sternal gland), that impregnates the substrate and stimulates workers to build at a particular spot. Investigations are continuing on all these materials.

or no reports are available on other Asian (especially Chinese, Indian, Siberian) species.

In temperate areas yellowjacket colonies are primarily annual and newly produced queens are the only members to survive the winter. These queens overwinter in sheltered locations such as under loose tree bark or in decaying stumps (see Yamane and Kanda, 1979) and emerge from hibernation during the first warm days of spring, usually in April or May. Soon after emerging, they begin to feed on flowers and other sources of nectar, and will also catch and malaxate arthropod prey. Spring queens have small ovaries but they begin to develop rapidly as reproductive diapause is terminated. Many queens are seen during this period seeking nest sites, alighting under the eaves of houses or entering cracks and crevices. After selecting a suitable location, the queen gathers plant fibers to construct the first cells of the nest. This fiber is usually gathered from weathered wood but may be collected from decayed wood, or even living plants. The queen's nest ultimately consists of 20–45 cells covered by a paper envelope. The queen lays eggs in the cells as they are constructed and forages for nectar and anthropod prey to feed the developing larvae. In about 30 days the first 5–7 workers emerge and assume such duties as cell and envelope building, foraging for prey and nectar, and nest sanitation (Akre *et al.*, 1976). After a short period of additional foraging, the queen no longer leaves the nest and her primary function is oviposition.

Workers are much smaller than the queen, and under normal colony conditions they do not lay eggs (with the possible exception of *Dolichovespula maculata, Vespula acadica* and *V. consobrina*). Until the second brood of workers emerges colony growth is slow but becomes exponential by midsummer as successive broods of workers emerge. At this time the general, multitiered architecture of the vespine nest is evident with horizontally arranged combs comprised of ventrally open cells, typically enclosed in a paper envelope. Later in the season the workers start building the larger reproductive cells in which both males and queens are produced. Males are also reared in worker cells. The colony enters a declining phase shortly thereafter during which workers pull larvae from the comb and feed them to other larvae or discard them. When new queens and males emerge, they leave the nest and mate, and then the inseminated queens hibernate. Large numbers of males are often seen at this time swarming on hill tops, or near shrubs and other vegetation. There is some indication that mating may enhance successful overwintering, at least in *Vespula* (MacDonald *et al.*, 1974). The cycle is repeated the following spring (Fig. 11). More detailed explanations of this cycle are given by Duncan (1939, MacDonald *et al.* (1974), and Spradbery (1973).

iii. Prey. Yellowjackets do not store honey as do the bees and some of the other vespids. They feed meat (i.e., arthropods, especially insects) and prob-

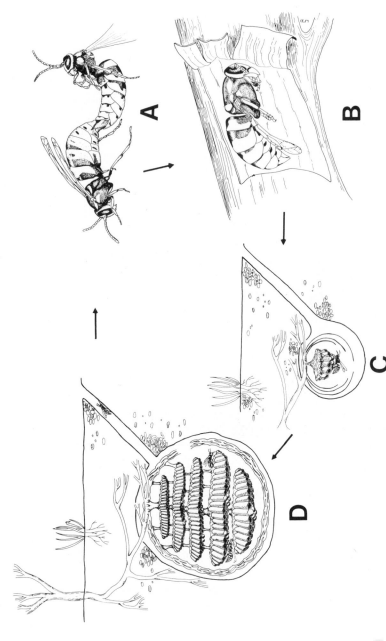

Fig. 11. Typical yellowjacket life cycle (*Vespula pensylvanica*): (A) mating of new queen and a male in September–October, (B) fertilized queen in diapause under bark during winter months, (C) queen nest beneath soil surface, usually in abandoned rodent burrow, and (D) nest at peak of colony development. This nest is rarely reused, it gradually decomposes and disintegrates over the winter.

ably nectar and honeydew to their larvae. The adults feed on juices while malaxating prey in the nest (Fig. 12). They also feed on nectar and larval secretions. As is the case for most social wasps, trophallaxis is a prominent activity in the colony.

Yellowjackets prey on a wide variety of insects, spiders, other arthropods, and even mollusks to feed their larvae. All species of *Dolichovespula* usually attack only live prey although there are a few records of workers scavenging from animal carcasses (Greene *et al.*, 1976). *Vespula* spp. are frequently divided into the *V. vulgaris* species group (*Paravespula* Blüthgen of European

Fig. 12. Worker of *Dolichovespula maculata* malaxating prey (housefly) while hanging from the edge of the comb. Prey is usually divided with one or two workers and subsequently fed to larvae.

workers) and the *V. rufa* species group (*Vespula* of European workers) (Table VII) (Bequaert, 1931; MacDonald *et al.,* 1974, 1976). *Vespula rufa* group species forage only for live prey, whereas members of the *V. vulgaris* group will also scavenge meat from animal carcasses, picnic tables, garbage cans, and many other locations which makes them pestiferous to man (MacDonald *et al.,* 1976; Akre *et al.,* 1981).

iv. Nests. *Vespula* nests are usually subterranean (Fig. 13), although some species build their nests in hollow logs, trees, or in the attics, between the walls, or rarely under the eaves of houses (Duncan, 1939; Ebeling, 1975; Green *et al.,* 1970; Kemper and Döhring, 1967; Spradbery, 1973). Nest size varies from about 300 to 120,000 cells, although most nests have 2000–6000 cells and are 8–15 cm diameter. The largest nests are those of atypically perennial colonies, which do not die out in the winter but continue on for an additional season with multiple queens. For example, the nest of one such *V. vulgaris* colony in California was nearly 4 ft long (Duncan, 1939); a *V. squamosa* (Drury) nest in Florida was 9 ft tall (Tissot and Robinson, 1954) (see Fig. 14). However, the record for any vespine species clearly belongs to *V. germanica* (Fab.) in New Zealand, where perennial colonies are not uncommon (Thomas, 1960). One nest was nearly 15 ft tall, probably contained several million cells in about 180 combs, and weighed an estimated 1000 lb. The total adult populations of these nests were not determined, but must have numbered at least 30,000 or more. For comparison, most annual nests of *V. vulgaris* group species contain 1000–4000 workers.

Dolichovespula nests are usually aerial (Fig. 15) although all species probably nest underground on occasion and at least two species [the Nearctic *D. arenaria* (Fab.) and the Palearctic *D. sylvestris* (Scopoli)] may do this frequently (Spradbery, 1973; Greene *et al.,* 1976; Archer, 1977a; Roush and Akre, 1978a). Most nests are small (ca. 300–1500 cells), but those of vigorous *D. arenaria* colonies may contain up to about 4300 cells and those of *D. maculata* up to about 3500.

TABLE VII

Characteristics of Species Groups of the Subgenus *Vespula*[a]

Parameter	*Vespula vulgaris* species group	*V. rufa* species group
Colony size	500–5000 ♀ at peak	75–400 ♀ at peak
Nest	3500–15,000 cells	500–2500 cells
Foraging	Predators and scavengers for protein and carbohydrates	Predators, will scavenge for carbohydrates
Decline	Late Sept–Nov–Dec	Aug–early Sept

[a] Modified from MacDonald *et al.,* 1976.

Fig. 13. Large subterranean nest of *Vespula pensylvanica* on Washington State University golf course, September, 1977. Nest contained nearly 4000 workers.

Fig. 14. A perennial nest of *Vespula squamosa* from 13 miles WNW Gainesville, Florida, February 2, 1977, with 120,000 cells in 14 combs. (Nest courtesy of J. Sharp, USDA, Gainesville.)

Fig. 15. A nest of *Dolichovespula maculata* in flowering crabapple tree, Pullman, Washington, August, 1979. A worker is attaching fiber, seen as a dark (moist) strip, to the envelope.

Members of the *V. rufa* group have small colonies of 75–400 workers, nests of 500–2500 cells, and are strictly predaceous on live arthropods (Table VII). Most nests are subterranean or are located in decaying logs or stumps (Mac-Donald *et al.,* 1974, 1975b; Akre *et al.,* 1982). The nest of members of this group differs from that of the *V. vulgaris* group by having broad supporting buttresses (Fig. 16), a strong, flexible envelope paper, and a single comb of worker cells (MacDonald *et al.,* 1974). The nests of several species are also quite filthy and unkempt with remains of prey and dead workers incorporated into the envelope. Yellowjackets of the *V. rufa* group are usually considered highly beneficial because of their predation on a number of economically important insects (MacDonald *et al.,* 1976).

Members of the *V. vulgaris* group have colonies of up to 5000 workers, nests with 15,000 cells or more, and scavenge for flesh in addition to preying on live arthropods. Most colonies construct subterranean nests, but many are also located in wall voids and attics of buildings. Members of this group have

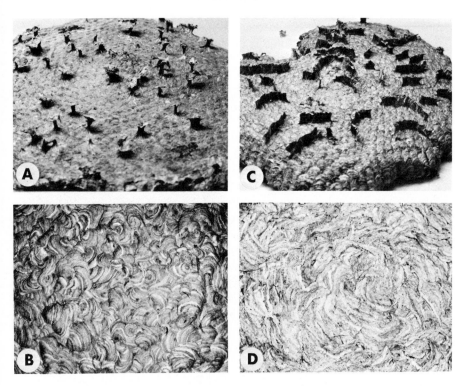

Fig. 16. Comparison of paper and nest suspensoria of *Vespula pensylvanica* (*V. vulgaris* group species) and *V. atropilosa* (*V. rufa* group species). *Vespula pensylvanica* has cordlike suspensoria for comb supports (A) and scalloped envelope paper (B), while *V. atropilosa* has buttresslike suspensoria on the uppermost comb (C) and laminar envelope paper (D).

a longer colony duration than members of the *V. rufa* group (Table VII). Because these yellowjackets will scavenge flesh, they are pestiferous to man.

v. Behavior. The foraging behavior of several species of yellowjackets has been investigated extensively with regard to light, temperature, weather, and daily cycle (see Spradbery, 1973; Edwards, 1980). Foraging for arthropod prey is largely visual, but those workers scavenging for protein also respond to odors (Gaul, 1952). Yellowjackets hunt independently and are unable to communicate food sources to other members of the colony (Kalmus, 1954; Kemper, 1961).

Potter (1964) studied the foraging behavior of workers of *V. vulgaris* in England. It was found that workers start to forage about 1 hour before sunrise and do not cease until 1 hour after dark. However, foraging reaches a peak in the early morning, declines slightly during the middle of the day, and reaches another smaller peak in late afternoon or evening. Similar to *Vespa,* peak foraging in yellowjackets primarily involves obtaining liquid food. Foraging is highly dependent on light intensity, but the wasps will forage whenever light is above a minimum threshold.

Few studies have been conducted on the foraging distances of yellow-jackets. Arnold (1966) determined that workers of *V. rufa* (L.) and *D. sylvestris* forage within 180–275 m of the nest, and Rogers (1972) found that most *V. pensylvanica* (Saussure) and *V. vulgaris* workers forage within 400 m of the nest and do not appear to have any direction preferences. Akre *et al.* (1975) studied the foraging distances of *V. pensylvanica* workers by use of small metal tags and a magnetic recovery system (Fig. 17). They recovered 80% of the workers foraging within 340 m of the nest. Therefore, these studies indicate that most foraging occurs quite close to the nest, although workers are capable of much longer flights (ca. 1000 m, see Akre *et al.,* 1973, 1975).

Foraging behavior has also been investigated in *V. germanica* (Free, 1970) and *V. maculifrons* (Buysson) (Lord *et al.,* 1977). Archer (1977b) investigated the loads carried by workers of *V. vulgaris* and found that flesh loads could be up to 50% or more of the worker's weight, while even heavier soil particles were moved when the wasps were excavating.

Few detailed observations had been made of yellowjackets inside their nests until Potter (1964), Montagner (1966), and Akre *et al.* (1976) designed and constructed glass-bottomed nest boxes for this purpose. One of the most ob-vious and constant behaviors observed was trophallaxis between and among adults (Fig. 18), and between adults and larvae. Males engaged in trophallaxis with workers and, rarely, with one another, but never fed larvae, workers, or the queen. Trophallaxis with the queen was nearly always one way, with the queen the recipient, either from workers (rare) or large larvae (usually). Workers engaged in trophallaxis with all colony members and frequently engaged in "mauling"; a peculiar behavior that involves vigorous chewing by one individual on another, while the recipient remains motionless. Mon-tagner (1966) considered this type of behavior to be a manifestation of dominance with the more dominant individuals demanding food by biting subordinates. This dominance was not correlated with ovarian development. However, Akre *et al.* (1976) presented evidence showing that the dominant in-dividuals in mauling encounters are seldom fed by the subordinate, and dissection of workers indicated that this behavior is definitely not related to ovarian development (Akre *et al.,* 1982). Therefore, at least in this respect, this behavior is not similar to dominance behavior as typically exhibited by *Polistes,* and investigations are needed to clarify its functions in the colony. Workers returning to the colony with foreign chemical odors such as ether are mauled, so perhaps there is a relationship between mauling and colony odor. Whatever its function, this behavior occurs in all yellowjacket species that

Fig. 17. (A) Anesthesized worker in holder showing color-coded steel label which was glued to the gaster. (B) Box with row of magnets which was placed over the entrance to subterranean colony of yellowjackets. The magnets recovered the labels from returning workers so foraging distances could be determined.

Fig. 18. Workers of *Dolichovespula maculata* in trophallaxis.

have been studied including *V. acadica* (Sladen), *V. atropilosa* (Sladen), *V. consobrina* (Saussure), *V. germanica, V. pensylvanica, V. vulgaris, D. arenaria,* and *D. maculata.*

Ishay and Brown (1975) analyzed sounds produced by larvae of *V. germanica,* and Es'Kov (1977) did similar work with larvae of *D. sylvestris* and *V. vulgaris.* The larvae scrape the sides of their cell walls when hungry, as in *Vespa* colonies, and thus stimulate the adults to feed them. Larvae of *V. pensylvanica* and *V. atropilosa* twitch in unison to stimulate the adults, but were not seen to actually scrape the cell walls, although under certain conditions they do so, and produce an audible sound (Akre *et al.,* 1976). *Dolichovespula arenaria* (Greene *et al.,* 1976) and *D. maculata* queens and workers also probably produce a sound for communication with the larvae by means of a rapid dorso-ventral gastral vibration, while moving on the comb. Workers especially display this behavior between feeding visits to larvae. The vibration always ceases while a larva is fed and presumably this behavior functions in some way to let the larvae know food is available. This behavior has not been previously reported for *Vespula*, but occurs in *V. atropilosa, V. vulgaris,* and *V. consobrina* (see Akre *et al.*, 1982).

Yellowjacket queens are not attractive to workers as in *Vespa* (Akre *et al.,* 1976; Greene *et al.,* 1976, with the exception of *Dolichovespula maculata,* Greene, 1979). Indeed, there is little physical contact between the queen and workers. Even trophallaxis between the queen and workers rarely occurs except in declining colonies. In *V. vulgaris* colonies the workers avoid the queen and give her a "space" of 0.5 cm or more (Fig. 19). Obviously the queen does not maintain control over the workers by physical dominance, and Potter (1964) postulated that an air-borne queen pheromone may perform this function; however, Landolt *et al.* (1977) investigated this possibility and instead

Fig. 19. Regardless of the number of workers on a comb, the queen of *Vespula vulgaris* usually has a space (ca. 1 cm diameter) around her which is void of workers. Workers entering this area quickly detect the queen and leave. This queen space probably serves to prevent worker interference during oviposition and when in trophallaxis with larvae.

postulated the existence of a nonvolatile queen pheromone passed throughout the colony by trophallaxis or by some other, unknown means. This material controls ovarian development in workers, and workers removed from the presence of the queen develop and lay eggs within 6–8 days (see Akre and Reed, 1982). Yellowjacket queens have a number of exocrine glands (Landolt and Akre, 1979), and these are currently being investigated as possible sites for pheromone production (Fig. 20).

Other pheromones reported to occur in yellowjackets are an alarm pheromone (Maschwitz, 1964, 1966; Edwards, 1980) and a footprint pheromone in *V. vulgaris* (Butler *et al.*, 1969), and thermoregulatory pheromones produced by *V. germanica* larvae to stimulate adults to warm them (Ishay, 1972). In addition, noradrenaline has been found in envelope paper of *V. germanica* (Lecomte *et al.*, 1976) and *V. vulgaris* (Bourdon *et al.*, 1975) although no social function for this substance has been postulated. Undoubtedly more pheromones (or possible pheromones) will be reported as investigations advance.

Studies of division of labor among yellowjacket workers (Potter, 1964; Akre *et al.*, 1976) indicate that a typical worker is first a nurse, then a collector of pulp and liquid, then a forager of prey, and later once again a collector of liquid (Fig. 21). Older workers in *V. vulgaris* become guards (Potter, 1964), while those of *V. atropilosa* and *V. pensylvanica* (Akre *et al.*, 1976) stay in the nest and are relatively inactive. Division of labor is not sharply defined and workers perform a number of different activities during the same day.

Another behavior apparently common among yellowjackets is inter- and intraspecific competition for nests. Akre *et al.* (1976) and MacDonald and Matthews (1975) reported that dead queens are often found in entrance tun-

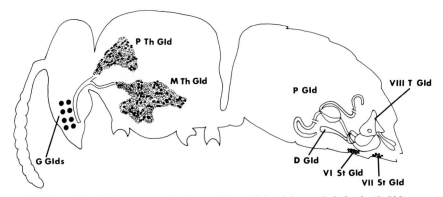

Fig. 20. Yellowjacket queen, showing positions of the eight gnathal glands (G Gld), prothoracic and mesothoracic sections of the thoracic gland (P Th Gld and M Th Gld), poison gland (P Gld), Dufour's gland (D Gld), sixth sternal gland (VI St Gld), seventh sternal gland (VII St Gld), and the eighth tergal gland (VIII T Gld). (From Landolt and Akre, 1979.)

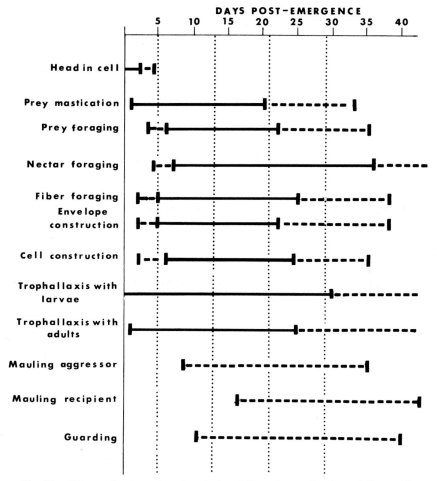

Fig. 21. History of early workers in colonies of *Vespula pensylvanica* and *V. atropilosa:* range in postemergent activity correlated with age; (—) range in age-related activity typical of most early workers; (---) the maximum range for unusually precocious or long-lived workers. (From Akre *et al.,* 1976.)

nels of subterranean nests. Since queens are observed seeking nesting sites after many nests are already founded, it may be assumed that these queens are probably killed attempting to usurp the nests. It is now known that *V. squamosa* frequently usurps the nests of *V. maculifrons* (Buysson) (Mac-Donald and Matthews, 1981), and *V. pensylvanica* sometimes usurps the nests of *V. vulgaris* (Akre *et al.,* 1977; Roush and Akre, 1978a). Intraspecific competition for nests is probably also frequent and intense, but so far no detailed studies have been conducted on this behavior.

Other behavior in the colony such as nest sanitation, malaxation of prey, and thermoregulation is similar to that reported for other vespines.

Greene *et al.* (1976) presented data on the behavior of *D. arenaria* indicating that members of this genus have a comparatively weak dichotomy between queen and worker castes. The foundress malaxates prey throughout the season, new queens malaxate prey, trim caps of cells from which callows have emerged, and are mauled by workers; the foundress also exhibits differential oophagy, all behavioral traits unreported for *Vespula*. Preliminary studies show similar behavior occurs in *D. maculata* (Greene, 1979). This indicates the social organization of *Dolichovespula* spp. may be somewhat more primitive than that of *Vespula* (see Greene, 1979). On the other hand, the intracolony behavior of *Vespa* and *Vespula* show a number of similarities (with the exception of the royal court of *Vespa*). However, it should be remembered that the social behavior within the colony has been investigated in only a few vespines including two species of *Vespa* (*V. crabro, V. orientalis*), three species of the *V. vulgaris* species group (*V. germanica, V. pensylvanica, V. vulgaris*), and one member of the *V. rufa* species group (*V. atropilosa*). Currently, investigations are underway on other members of the *V. rufa* group (e.g., *V. acadica, V. consobrina*).

III. PHEROMONES

Table VIII summarizes pheromones that have been suggested as occurring in the social wasps. This summary does not include materials rubbed on the petiole or other parts of the nest by species of *Mischocyttarus, Polistes, Belonogaster,* and several other genera. There is no evidence that these materials act as pheromones, although in *Polistes* (Hermann and Dirks, 1974; Turillazzi and Ugolini, 1978) and *Mischocyttarus* (Jeanne, 1972) this material acts as an allomone to repel ants.

In general, wasps that have small colonies do not have pheromones such as alarm pheromones, since they rely on vision and substrate or nest vibration to alert them to possible dangers. Reports of alarm pheromones in *Polistes* (Rau, 1939) and *Mischocyttarus* (Jeanne, 1972) were stated tentatively, and are based on meager evidence. In addition, in most species of wasps with small colonies, the queen maintains control over other females by physical aggression (e.g., dominance, dominance hierarchy) and differential oophagy (e.g., eating subordinates' eggs). As long as the colony remains small (usually less than 100 adults), this method of control seems to work well. However, in colonies with more individuals, the possibility of a single female (the queen) recognizing potential egg layers and controlling them by overt aggression, while still performing her duties as the principal egg layer, rapidly diminishes.

TABLE VIII
Possible Pheromones of Social Wasps

Wasp	Pheromone	Source	Reference
Mischocyttarus drewseni	Alarm	Gland opening into sting chamber	Jeanne, 1972
Polistes variatus (= *fuscatus variatus*)	Alarm	?	Rau, 1939
Polistes aurifer (= *fuscatus aurifer*)	Aggregating	Male mandibular gland	Landolt and Akre, 1978
Various Polybiini (16 genera)	?	Gland on fifth gastral sternum (= sixth abdominal) of female (Richard's gland, West Eberhard, 1977)	Richards, 1971, 1978a
Angiopolybia pallens *Leipomeles dorsata* *Polybia catillifex* *Stelopolybia myrmecophila*	Trail, directs swarm to new nest site	Gland on fifth gastral sternum (?)	Naumann, 1975
Stelopolybia areata	Trail, directs swarm to new nest site	Gland on fifth gastral sternum (?)	Jeanne, 1975c
Stelopolybia areata	Alarm or attractant	Released during stinging (gland associated with sting apparatus)	Jeanne, 1973
Metapolybia	Queen recognition	?	West Eberhard, 1977
Protopolybia pumila	Alarm	Gland associated with sting apparatus	Naumann, 1970
Vespula germanica *Vespula vulgaris*	Alarm	Venom	Maschwitz, 1964, 1966
Vespula vulgaris	Footprint (nest marking)	?	Butler *et al.*, 1969
Vespula germanica *Vespula vulgaris* *Dolichovespula saxonica* *Dolichovespula media* *Vespa crabro* *Vespa orientalis*	Pupal brooding (stimulates warming by adults)	Pupae	Ishay, 1972
Vespula atropilosa	Female sex attractant	?	MacDonald *et al.*, 1974
Dolichovespula sylvestris	Female sex attractant	?	Sandeman, 1938
Vespula germanica	Female sex attractant	?	Thomas, 1960

(*cont.*)

TABLE VIII (*Continued*)

Wasp	Pheromone	Source	Reference
Vespula vulgaris	Queen phero-mone (volatile) (inhibits ♀ ovary develop-ment)	?	Potter, 1964
Vespula atropilosa *Vespula pensylvanica* *Vespula vulgaris*	Queen pheromone (nonvolatile)	Glands of head (?)	Landolt *et al.,* 1977 (also experimental data, Akre)
Dolichovespula arctica	Queen pheromone	Dufour's gland or gland on seventh sternum	Greene *et al.,* 1978
Vespa orientalis	Queen pheromone	Glands of head	Ishay, 1965; Ikan *et al.,* 1969
Vespa orientalis	Alarm pheromone	Venom	Saslavasky *et al.,* 1973
Vespa, Vespula *Dolichovespula* *Polistes*	Wasp assembly	Fertilized ♀, ♀	Ishay and Perna, 1979
Vespa, Vespula *Dolichovespula* *Polistes*	♂ Wasp assembly	?	Ishay and Perna, 1979
Vespa, Vespula *Dolichovespula*	Building de-pressor	?	Ishay and Perna, 1979
Vespa orientalis *Polistes foederatus*	Building ini-tiating	Tarsal glands or sixth sternal gland of gaster	Ishay and Perna, 1979

The pecking order in chickens breaks down as the flock becomes too large; surely the same principle must apply to wasps, with strife replacing order if the colony becomes too large. If pheromones exist in species with small colonies, especially alarm pheromones, they must play a minor role in normal colony activities.

Most pheromones have been reported for wasps with populous colonies such as certain species of Polybiini and the Vespinae. These wasps (at least the Vespinae) appear to have replaced dominance hierarchies and differential oophagy with a chemical means of queen control (i.e., queen pheromone that inhibits ovarian development of workers), and several also appear to have alarm pheromones to alert the colony to protect and defend the nest. Observational evidence also strongly indicates that several species of polybiines use a pheromone to lead the colony to a new nest site during swarming (Naumann, 1975; Jeanne, 1975c; Jeanne 1981).

Only one pheromone, the queen pheromone of the Oriental hornet, *Vespa orientalis,* has been identified (Ikan *et al.,* 1969) and synthesized (Coke and Richon, 1976). This pheromone, δ-*n*-hexadecalactone, inhibits ovarian development of workers, and has also been suggested to have a number of other functions such as monogynia, caste differentiation, and the initiation of reproductive cells (Ishay, 1973a).

A queen pheromone has also been postulated for yellowjackets, based on experiments using colonies of *Vespula atropilosa* and *V. pensylvanica* (Landolt *et al.,* 1977; Akre and Reed, 1982). Again, this material was suggested to be a controlling agent of ovarian development in workers. In colonies of these two species and in most other yellowjackets, the queen and workers have very limited physical interaction, and there is no indication of any behavior suggesting control of workers by overt dominance. In vespine species which have large colonies (i.e., 3000 or more workers) and a nest of multiple combs (e.g., *V. pensylvanica*), it would be impossible for the queen to patrol the various combs to prevent even a fraction of the workers from laying eggs. However, in *Dolichovespula arenaria* and *D. maculata,* species with much smaller colonies (i.e., usually less than 500 workers), the queen may maintain control by means of a queen pheromone and differential oophagy (see Greene *et al.,* 1976). Current experiments using *V. pensylvanica* suggest that the queen pheromone may be produced in glands in the head, although a number of other exocrine glands exist in yellowjackets (see Landolt and Akre, 1979) (Fig. 20).

Although *Vespula germanica* and *V. vulgaris* have been reported as having alarm pheromones (Maschwitz, 1964, 1966; Edwards, 1980), these reports should be considered to be very tentative, at least until more definitive experiments are conducted. Tests in Pullman, Washington, on colonies of *V. atropilosa, V. pensylvanica,* and *V. vulgaris* did not reveal any behavior that indicated the presence of any alarm substance. These experiments consisted of holding workers in front of the nest entrance with forceps, crushing workers and presenting them to the colony, extracting various body parts in solvents, applying them to filter paper and presenting them to the colony, and other combinations of the above. However, the slightest vibration of the substrate or nest elicits instant alarm and defense behavior. If an alarm pheromone exists in these wasps, its role in colony defense must be very minor. It surely does not cause a reaction in workers comparable to that in army ants [i.e., *Eciton burchelli* (Westwood)], where crushing the head of one major worker stimulates the column into a frenzy of agitation and attack.

Mating or sex pheromones are also suggested in yellowjackets by the behavior of queens and males during mating. For example, queens of *Vespula atropilosa,* even while in copulation with one male, were highly attractive to other males (MacDonald *et al.,* 1974), just as males of *D. sylvestris* balled up in clumps of 15 or more on a mating pair (Sandeman, 1938). However, *V. pen-*

sylvanica queens did not show this high degree of attraction during mating, indicating at least some species may not have sex pheromones.

Other suggested pheromones, such as wasp aggregating pheromone and pupal warming pheromones, have not been thoroughly investigated. Instead of pheromones eliciting the behaviors observed, they might also be explained by sight, colony or species odor, or by a multitude of other factors. Furthermore, instead of a single pheromone for each observed behavior, it is equally possible that one, or at most 2–3 pheromones serve a number of different functions or elicit different types of behavior at different concentrations or titers.

Yellowjackets and hornets have as many as 14 exocrine glands or gland systems, and functions are known for only a few (Landolt and Akre, 1979). Some seem to be associated with food and pulp mastication and other typical colony funtions, while functions of the remaining glands are totally unknown. Transmission electron microscope studies have shown that several of these glands may be capable of protein synthesis, and these are currently under investigation as pheromone production sites.

IV. OTHER COMMUNICATION

Social wasps, especially the more highly evolved species, communicate by means of physical stance (i.e., dominance), sound, or pheromones. Wasps, specifically yellowjackets, hunt independently and were reported to be unable to communicate the location of food sources to other members of the colony (Kalmus, 1954; Kemper, 1961). However, Maschwitz *et al.* (1974) reported that *Vespula germanica* and *V. vulgaris* can communicate information about a food source. Workers were trained to a scented food source, then recruits were counted at test and control stations. The number of individuals visiting the test stations was significantly higher and it was concluded the wasps were communicating the location of the food source. These results are not supported by extensive testing of a related species, *V. pensylvanica*, in Washington state. All colonies in a certain area had been located and were marked with fluorescent dyes poured into the entrance tunnels. This identified any worker as being from a particular colony. Then bait stations (fish) were set up and the first worker from a colony was marked with paint on the thorax as it alighted to take some flesh. Once a worker from a particular colony was marked at a bait station, all workers from other colonies were killed as they appeared so there would be less chance of error. Although these same workers returned to the station repeatedly over several weeks, there was no significant increase in the number of workers from that particular colony visiting the station. These tests were repeated a number of times and it was concluded that workers do not communicate the food location to other colony members. Of course, these are different species, and the tests are not quite comparable, but communication of this type really needs more investigation before any definitive conclusions can be drawn. If, indeed, communication of location of a food source exists, this raises the important question of how a forager communicates this information to other potential foragers. No waggle dance has been observed in vespines, but a gastral vibration does occur in some vespine species which is apparently related only to larval feeding (with the exception of *Dolichovespula maculata*).

V. SOCIAL PARASITISM

In true social parasitism, the parasite has no worker caste of its own, and usurps the colony of another wasp species. The workers of the host species then rear the parasite's brood (queens and males).

Taylor (1939) presented four stages he considered important in the development of social parasitism in wasps, and these were modified slightly and discussed by Spradbery (1973).

1. *Stage 1. Intraspecific, facultative temporary parasitism*

In this initial evolutionary step a queen invades a colony of the same species and kills the resident queen, or sometimes takes over a nest where the queen has died. The parasitism is brief, since the new queen's progeny soon take over the nest. Intraspecific competition for nests is probably much more common than realized, although there are few observations to support this assumption (Akre *et al.*, 1977). However, dead queens (3–4) are frequently found in the entrance tunnels of *V. pensylvanica* and *V. atropilosa* in Washington, and it is assumed that these queens are killed attempting to usurp the nest. MacDonald and Matthews (see Akre *et al.*, 1981) found a similar situation in colonies of *V. maculifrons* where nests were frequently usurped by conspecific queens (MacDonald and Matthews, 1981). To test this assumption, Akre *et al.* (1976) introduced queens into established nests to observe the behavior and found that the resident queen usually vigorously attacks and drives off or kills the introduced queen. Furthermore, queens introduced into orphan nests frequently adopt the colony with little aggression exhibited by any workers that may be present.

2. *Stage 2. Interspecific, facultative temporary parasitism*

This stage is identical to stage 1, except that two closely (presumably) related species are involved. Sakagami and Fukushima (1957a) discovered that *Vespa dybowskii* André, a species that frequently builds its own nest and has a worker caste, also usurps the nests of *V. crabro* and *V. mongolica*. A similar situation occurs between *V. pensylvanica* and *V. vulgaris,* with the *V.*

pensylvanica queen usurping the *V. vulgaris* colony when it is quite small, usually less than 50 workers (Akre *et al.,* 1977; Roush and Akre, 1978a). These two species are both in the *V. vulgaris* species group and are probably closely related, although they differ greatly in size and in the use of materials for nest construction. *V. pensylvanica* has larger workers and makes a relatively durable gray envelope and carton, whereas *V. vulgaris* uses decayed wood for its envelope and carton, giving it a tan-brown color. Thus when a colony is usurped, the carton of the comb gradually changes from tan to gray and leaves unmistakable evidence that the nest was once a *V. vulgaris* nest. MacDonald and Matthews (1975, 1981; see also Akre *et al.,* 1981) found that *V. squamosa* frequently usurped the nests of *V. maculifrons,* and it has also been reported to usurp a nest of *V. vidua* (Saussure) (Taylor, 1939). However, *V. squamosa* is probably not closely related to either of its hosts. This violates Emery's rule, a generalization concerning social parasites which states that a social parasite tends to more closely resemble its host morphologically than any other freeliving species and thus both probably arose from the same ancestral stock (Wilson, 1971).

TABLE IX
Social Parasites and Their Hosts

Parasite	Host(s)	Reference
Sulcopolistes atrimandibularis (Zimmermann)	*Polistes bimaculatus* Fourcroy *Polistes omissus* Weyrauch	Scheven, 1958
Sulcopolistes sulcifer (Zimmermann)	Polistes gallicus L. *Polistes nimpha* Christ	Scheven, 1958
Sulcopolistes semenowi (Morawitz)	*Polistes gallicus*	Scheven, 1958
Dolichovespula adulterina (Buysson)	*Dolichovespula norwegica* (Fab.) *Dolichovespula saxonica* (Fab.)	Weyrauch, 1937; Guiglia, 1972
Dolichovespula omissa (Bischoff)	*Dolichovespula sylvestris*	Weyrauch, 1937; Guiglia, 1972
Dolichovespula arctica	*Dolichovespula arenaria,* *Dolichovespula norvegicoides*	Greene *et al.,* 1978
Vespula austriaca	*Vespula rufa* (Eurasia) *Vespula acadica* (North America)	See Reed *et al.,* 1979
Vespula squamosa	*Vespula vidua* *Vespula maculifrons*	Taylor, 1939 MacDonald and Matthews, 1975; see also Akre *et al.,* 1981
Vespa dybowski	*Vespa crabro* *Vespa mongolica* (= *xanthoptera* Cameron)	Sakagami and Fukushima, 1957a

3. *Stage 3. Interspecific, obligatory temporary parasitism*

This was considered a transitional stage and has not been documented in social wasps. In this stage, the queen loses the ability to initiate her own nest and is forced to usurp the nest of a host species. The parasitic stage ends when the workers of the parasite emerge and gradually replace all the host workers.

4. *Stage 4. Interspecific, obligatory permanent parasitism*

In this final and common stage of inquilism, the worker caste is lost and only males and queens exist. The female parasite must usurp the nest of a closely related host species and depends on host workers to tend and feed her brood. All species listed in Table IX (excluding *Vespula squamosa* and *Vespa dybowskii*) belong to this category.

Social parasitism has not been recorded in the Polybiinae, except for the vague account by Zikan (1949) of 23 parasitic species of *Mischocyttarus*. Very little behavioral data is presented in this monograph, and the parasitic status of these species is based almost entirely on morphological characteristics. Richards (1978a) questioned the validity of Zikan's claims and stated that no evidence exists to prove they are parasites, but rather these species are really only variants of their supposed hosts. Only future research will reveal the existance of any true polybiine social parasites.

Three socially parasitic species of *Polistes* have been studied in Europe: *P. sulcifer* Zimmermann, *P. semenowi* F. Morawitz, and *P. atrimandibularis* Zimmermann (Scheven, 1958). All three are obligate, permanent parasites. The parasites closely resemble their hosts species but minor differences in head shape and coloration as well as their more recurved mandibles have led some researchers to place these three species in the genus *Sulcopolistes* (Kemper and Döhring, 1967; Richards, 1978a). Scheven (1958) gave a detailed account of the usurpation process of these inquilines. A nest of the host species (Table IX) is invaded in early June when the nest is small and before the emergence of the first workers. The parasite fights violently with the α-female (queen) and may even bite and "sting" her into submission. The inquiline apparently wins by physical superiority and greater endurance, not with the use of her unique mandibles. Only rarely is the host queen killed; usually she remains on the nest with the parasite, or she may disappear, as is the case when *P. sulcifer* invades the nest. The resident queen apparently is not harmed physically but assumes a rank subordinate to the inquiline. The parasite continues to dominate the host queen with vigorous attennation while attempting to mount on her dorsum. The host responds by crouching low and assuming other submissive postures. The parasite always receives liquid food and never donates during trophallaxis with the host queen. Emerging workers also become submissive to the inquiline.

Social parasitism in *Polistes* species in other regions has not been confirmed, but three species (see Table 19–3 in Wilson, 1971) may be possible candidates based on morphology and scanty field data. Thorough investigations of species in these areas are needed to prove the existence of other parasitic paper wasps.

The first reported, and one of the best known cases of true social parasitism, was that of *Vespula austriaca* (Panzer) in nests of *V. rufa* in Ireland (Robson, 1898). Evidence indicated that the *V. austriaca* queen killed the host queen when usurping the nest or shortly thereafter, and then workers of the host reared a number of the parasite's brood to adults. Since that time relatively few parasitized colonies have been found, indicating that this parasite is quite rare. However, frequent net collections of *V. austriaca* queens indicate that this species is active about 1 month later than queens of the host, *V. rufa,* presumably an adaptation to invade already established nests (Evans, 1903). Reed *et al.* (1979) recently found the North American host of this parasite to be *V. acadica* (Sladen) and summarized information regarding its occurrence in the nests of *V. rufa* in Eurasia (Fig. 22).

A more common social parasite, *Dolichovespula arctica* (Rohwer), occurs in nests of *D. arenaria* (Fab.) and *D. norvegicoides* (Sladen). Although most parasites are assumed to be active later than queens of the host species, *D. arctica* is active at the same time as *D. arenaria* queens in Washington and sometimes usurps the nest of *D. arenaria* when the colony consists of only the host queen, one or two workers, and 35–40 cells with brood. Evans (1975), Jeanne (1977c), and Greene *et al.* (1978) studied the biology and behavior of this species and reported a number of previously unknown behaviors. The host queen is usually not immediately killed by the parasite, but is dominated by her during a period of coexistence, usually until a number of workers

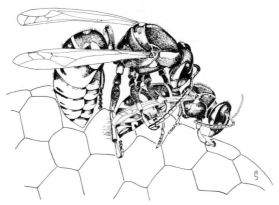

Fig. 22.   A female social parasite, *Vespula austriaca,* mauling a *Vespula acadica* worker. Approximately one-half of these maulings ended in a forced trophallaxis between the social parasite and the host worker. (Drawn from a photograph by R. D. Akre.)

(> 15) emerge. Then the parasite kills the queen, and becomes very aggressive toward the host workers rearing her brood. This aggressiveness toward the workers is probably related to their ovarian development, and sometimes erupts into fighting during which the *D. arctica* female kills a number of workers, and may be killed herself. More commonly, the *D. arctica* leaves the nest, and the host workers continue to rear the *D. arctica* brood to adult males and females.

While still in the nest the *D. arctica* was more active than any other colony member, fed larvae, and participated in some nest construction. Workers were attracted to the inquiline and frequently nibbled and antennated its body (Fig. 23). This and other behaviors suggested that the parasite produced a compound(s) which attracted and pacified the host workers to some degree (Greene *et al.,* 1978). This is quite plausible, since a pacifying chemical has now been isolated for another yellowjacket species (Franke *et al.,* 1978). *Dolichovespula arctica* has a Dufour's gland that is much larger than that of freeliving vespines (Jeanne, 1977c) and this, or the seventh sternal gland of the abdomen, may be the source of this material(s).

An examination of either *V. austriaca* or *D. arctica* females shows that these parasites are well adapted for combat. The cuticle is presumed to be thicker, and abdominal sclerites fit more closely together, and they have a curved sting which makes it easier to penetrate the intersegmental membranes of the host (Spradberry, 1973).

Taylor's scheme for the development of social parasitism seems quite logical and is supported by available information. Since there appears to be frequent conspecific fighting for nests, this presents ample opportunity for

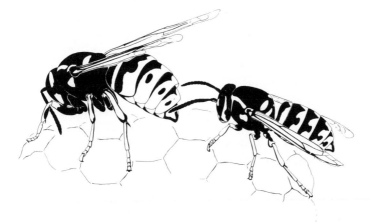

Fig. 23.   Antennation of *Dolichovespula arctica* by a worker of *D. arenaria.* This attraction of workers to the inquiline occurred during the period the social parasite invaded the nest and throughout its occupation of the colony. (From Greene *et al.,* 1978.)

the development of parasitism, and most interspecific, facultative parasitism seems to occur in areas of overlap of species distribution as was more or less predicted by Richards (1971). This is definitely the case where *V. pensylvanica* queens usurp the nests of *V. vulgaris*. The evolution of permanent, obligatory parasitism from this facultative stage (i.e., how the worker caste is lost) is still uncertain but future studies may shed some light on this problem. It is possible that under certain experimental conditions, one of the obligate social parasites may produce a worker caste (revert to stage 3).

VI. CYCLICAL POPULATIONS

Population dynamics of wasp populations were discussed by Richards and Richards (1951) and Richards (1971). Population dynamics involve the growth of individual colonies through the season and their reproductive success (i.e., the number of colonies initiated by individuals from the original colony). There is little information available on these parameters for most social wasps; however, Lövgren (1958) constructed a mathematical model for predicting the growth and productivity of a wasp colony based on information given by Richards and Richards (1951). However, the parameter estimates used to make the model were incorrect or inaccurate. A more accurate model of wasp colony growth was done by Spradbery (1971) using data from colonies of *Vespula germanica* and *V. vulgaris* (see also Archer, 1981d). In the Vespinae each new queen is potentially capable of initiating a colony and the number of queens produced can be determined by examining the large reproductive cells in the nest. Spradbery compared Lövgren's model of wasp population growth to his own observations. Using Spradbery's parameters, the model was fairly accurate in predicting seasonal growth and ultimate production of the colony. The optimal reproductive strategy in annual yellowjacket colonies depends on producing workers only to a point in time that will permit the maximum production of new queens. This depends on temperature conditions. Production of workers must cease and queen production start in time to allow the queens to emerge before the onset of cold weather.

There is little additional information available to discuss the population dynamics of any species of social wasp. However, there is some information on an intriguing problem: the tremendous rise and fall of yellowjacket populations in successive years. This sometimes results in outbreaks or years of incredible abundance, especially common in the Pacific northwest of the United States. Beirne (1944) and Fox-Wilson (1946) discussed the phenomenon of cycles of yellowjacket abundance and scarcity from year to year and emphasized the importance of climatic factors. Beirne suggested that heavy rainfall in April, May, and June decreased wasp populations, but rainfall alone did not account for all observations, so he suggested that some additional factor, possibly disease, might also be operating. Archer (1973, 1979) hypothesized that this additional factor was the wasps themselves, and proposed that characteristics of queens, which change from generation to generation, affect population levels. Aggressive queens form colonies that produce few new queens, while nonaggressive, fertile queens form colonies that produce many queens. The balance of these two types during a particular year determines wasp abundance.

Analyses of nearly 400 colonies and nests of *Vespula* spp. (mostly *V. pensylvanica*) in Washington, from 1971–1978, did not suggest that these two types of queens were affecting yellowjacket populations, but rather that tremendous yellowjacket abundance during certain years was related directly to rainfall and temperatures in April, May, and early June (see Long *et al.*, 1979; Akre and Reed, 1981; see also Madden, 1981). Excess rainfall and low temperatures during these months caused many colonies to fail. During early June (sometimes May) the colony is at a critical stage composed of the queen and the first 5–6 workers. The queen is foraging less and less, and if two or three of the first workers are lost because of excess moisture or cold, the colony never grows very large even if it survives. Conversely, dry, warm weather during April, May, and June tends to favor survival of colonies and results in outbreak populations as experienced in the Pacific northwest during 1973, 1979, and to a lesser extent in 1977 (see MacDonald *et al.*, 1976; Akre *et al.*, 1981). Nest analyses also indicated that during years of peak abundance colonies are smaller, and this is probably attributable to competition for food among colonies. During 1973, workers were seen bringing in many yellowjackets (mostly conspecifics, *V. pensylvanica*) as prey. It was not known if the workers were actually killing each other for prey, or whether these workers were being found dead and scavenged from car radiators and similar sites. Other prey was scarce because of the drought conditions.

Disease may also be a factor that influences yellowjacket populations in certain areas. During 1977, populations of *Vespula vulgaris* were at outbreak levels in northern Idaho until mid-August, when yellowjackets became quite scarce. Several colonies which had been located and marked earlier were checked, and many nests were completely gone, probably eaten by rodents. A few formerly thriving colonies were represented by empty nests, or at most 2–3 workers and no brood. The colonies had been very active during the first part of August until cold, rainy weather moved into the area and persisted for ca. 1 week. The rapid population decline after the weather cleared is indicative of some pathogen killing the brood and causing colony demise. Unfortunately, no brood was present in any of the remaining nests (possibly taken by ants) to allow investigation for possible pathogens. The entire sequence of the population crash took about 3 weeks, and no explanation, other than a possible pathogen, was suggested by field observations.

These fluctuations in yellowjacket populations, so common and pronounced in the Pacific northwest, occur to some extent in England and Europe, but are less common in other parts of North America, if they occur at all. Since weather conditions do not fluctuate quite so widely in these areas, weather is probably not responsible for population fluctuations observed in these areas, and other, unknown factors must be in operation. One possibility is inter- and intraspecific competition for nests (MacDonald and Matthews, 1981).

VII. NEST ASSOCIATES

The nests of social wasps often contain a variety of associated fauna, such as scavengers, parasites, predators, and commensals. A number of these associates, mostly arthropods, were discussed by Evans and West Eberhard (1970), and a detailed list of associates was made and discussed for European Vespinae (*Vespula, Dolichovespula, Vespa*) by Spradbery (1973). In addition, Richards (1978a) briefly listed associates found in nests of Polistinae which included cockroaches, moths, parasitic Hymenoptera, Strepsiptera, Diptera, and mantispids. This list included studies by Nelson (1966, 1968) on parasites and symbionts in nests of *Polistes* wasps. Most references reported the occurrence of various associates, but few studied the impact, if any, of the associates on the wasp colonies. Since these sources are readily available, the discussion that follows concentrates on associates of *Polistes* and yellowjackets of North America, with the latter discussion dealing particularly with the species occurring in the Pacific Northwest.

In North America *Polistes* brood are adversely affected primarily by *Chalcoela iphitalis* Walker and *C. pegasalis* Walker (Lepidoptera: Pyralidae), *Pachysomoides stupidus* (Cresson) and *P. fulvus* (Cresson) (Hymenoptera: Ichneumonidae), and *Elasmus polistis* (Burks) (Hymenoptera: Eulophidae). *Chalcoela iphitalis* is the most apparent and common parasite in Illinois, Kansas, and Texas (Nelson, 1968). The pyralid larva consumes wasp larvae and pupae and even tunnels horizontally to consume wasp brood in adjacent cells. Infestation rates of this parasite in *Polistes* nests can be very high. For example, Nelson (1968) found 44% of the nests collected during a 2 year study in southern Illinois contained this moth. This parasite often consumes much of the brood with some nests containing *C. iphitalis* cocoons in 90–95% of the cells. Reed and Vinson (1979b) also found this pyralid to be the most common (31 of 81 nests) and destructive (55.6% parasitization of the brood) parasite of *Polistes exclamans* in Texas. *Pachysomoides* spp. were not as common nor as destructive as the pyralids in *Polistes* nests in southern Illinois, and were not as abundant as *E. polistis* in Texas. *Elasmus polistes* is commonly encountered in nests of *P. exclamans* and consumes an average of 30% of the wasp brood in 21 of 81 collected nests (Reed and Vinson, 1979b). Undoubtedly, *C. iphitalis* and *E. polistis* reduce the number of new queens and males produced by *Polistes* colonies and must be important selective agents on their hosts. Heavy parasitism by these or similar parasites may select for *Polistes* that build nests in locations difficult for the parasites to locate, and may possibly prevent some populations from successfully nesting close to their parental nest site as is common in temperate *Polistes* (see Starr, 1976, 1978). Indeed, a predaceous moth has been suggested as a selective force on the peculiar habit of multiple comb building in *Polistes canadensis* despite a very low brood infestation rate (Jeanne, 1979).

Wilson (1971) explained the relative scarcity of nest associates in colonies of social wasps on the basis of their compact, tightly sealed nests in arboreal conditions. These nests are easy to keep clean since all refuse is merely taken out the entrance and dropped, and this provides fewer opportunities for scavengers and other potential colony associates. However, subterranean *Vespula* nests provide an ideal habitat for many arthropods including soil dwellers, mycetophagous species, saprophagous Diptera larvae, and brood predators and parasites. Most of these associates live outside the nest envelope, in the refuse piles below, and notably absent are associates with obvious morphological or behavioral adaptations for living with their hosts, so common in myrmecophiles (MacDonald *et al.*, 1974) and termitophiles.

A study of the associates of *Vespula pensylvanica* and *V. atropilosa* showed that five species were commonly encountered in nest cavities or nests: *Dendrophaonia querceti* (Bouche) (Diptera: Muscidae); *Fannia* spp. (Diptera: Muscidae); *Cryptophagus pilosus* (Gyllenhof) (Coleoptera: Cryptophagidae); *Sphecophaga vesparum burra* (Cresson) (Hymenoptera: Ichneumonidae); and *Triphleba lugubris* (Meigen) (Diptera: Phoridae) (MacDonald *et al.*, 1975a). *Sphecophaga* (Fig. 24), a pupal parasitoid in over 80% of the *V. atropilosa* nests, appears to adversely affect incipient colonies, while *Triphleba* probably influences colony decline of *V. pensylvanica* late in the season by destroying some developing queen pupae and lowering colony productivity. All other associates are found infrequently or do not adversely affect colony development.

Further studies of *Vespula consobrina* (Saussure), *V. acadica*, *V. vulgaris*, *Dolichovespula arenaria,* and *D. maculata* indicate that these species have similar nest associates. *Sphecophaga* infests the nests of all these species and cause severe damage to young colonies of *V. acadica* and *V. consobrina* where it lowers the worker force to a level that could cause small colonies to fail. In

Fig. 24.   A female parasitoid, *Sphecophaga vesparum burra,* on cells of a *Vespula atropilosa* nest.

mature colonies, *Sphecophaga* parasitism of reproductive cells probably also lowers the production of new queens. In 1978, *Melittobia acasta* (Walker) (Hymenoptera: Eulophidae) was found in colonies of *V. acadica* (Reed *et al.,* 1979). It is not known how common these parasitoids are in colonies of this species, but observations of the adverse effects of *Melittobia* spp. on bees and other Hymenoptera indicate that it could contribute to colony failure (Erickson and Medenwald, 1979). In 1979, the trigonalid *Baregonalos canadensis* (Harrington) was found in three nests of *Vespula vulgaris* and one of *V. acadica* collected in northern Idaho (Carmean *et al.,* 1981). These rare insects were recorded previously only from nests of *Vespula pensylvanica* (British Columbia) and *Dolichovespula arenaria* (California) (Stage and Slobodchikoff, 1962). Another trigonalid, *B. huisuni* Sk. and S. Yamane has been collected from three nests of *Vespula karenkona* Sonan in Taiwan (Yamane and Yamane, 1975).

Few nest associates of yellowjackets seem to have any serious effects on colonies, but are mentioned only as interesting addenda to biological studies of the wasps. In one respect this is unfortunate since pestiferous yellowjackets frequently require control and an effective biological control agent would be ideal for this purpose. However, there have been few serious considerations of this possibility, probably because none have shown any promise in this regard. However, Poinar and Ennik (1972) discussed the possibility of using a nematoda, *Neoplectana carpocapsae* Weiser, as a control measure for pestiferous yellowjackets in California, and *Sphecophaga vesparum burra* has been imported into New Zealand for use against *Vespula germanica*. However, there is little hope that these agents will reduce populations of yellowjackets to acceptable levels, if they have any effect at all. These and other considerations led MacDonald *et al.* (1976) to state that no effective biological control agent exists for yellowjackets and none is foreseen.

VIII. COMMUNAL DEFENSE

The well-known associations of ants, wasps, and birds nesting together in trees in South America (Myers, 1929, 1935; Richards, 1978a) have received relatively little study. Since the ants (most commonly species of *Azteca, Pseudomyrmex, Oecophylla*) are the most aggressive and numerous inhabitants, they are probably the first to establish themselves, and the social wasps and the birds establish their nests later. Nearly any animal traversing or settling on the tree is vigorously attacked by the ants, so the associated wasps and birds must either escape notice by the ants (highly unlikely) or avoid their attacks in some unknown manner. Although little information is available, the obvious assumption is that all gain from the association, perhaps by combining forces to repel vertebrate or invertebrate predators. Other invertebrates may also gain some advantage by associating with wasp colonies. For example, at least two species of katydids usually roost in close proximity to aggressive wasp colonies and are probably protected from predators by the wasps (Downhower and Wilson, 1973; Richards, 1978a). Different species of social wasps also sometimes associate together as shown by *Mischocyttarus immarginatus* Richards which usually nests in the vicinity of nests of the more aggressive *Polybia occidentalis* (Oliver) (Windsor, 1972) (Fig. 25). Jeanne (1978) has also reported that colonies of the fierce *Polybia rejecta* (Fab.) nest in close proximity to each other in a single tree with up to 23 nests found in a $3 \times 3 \times 4$ m area. Mutual defense against vertebrate predators was again suggested as a possible reason for these nesting associations.

Although the associations of ants and social wasps can present a mutual, and probably formidable, defense against possible vertebrate predators,

Fig. 25.   Nest of *Polybia occidentalis* (lower) and *Mischocyttarus* ? *emarginatus* (upper) on same palm frond, Palo Verde, Comelco, Costa Rica. (From color slide by G. W. Frankie.)

reports of vertebrates preying on the wasps are scarce (see Windsor, 1972; Gibo and Metcalf, 1978; Cobb, 1979). However, it could be that few vertebrates attack wasp nests, especially those species with populous colonies, or that this association is quite effective in detering possible attacks. Since subterranean yellowjacket nests in North America are subject to destruction by skunks, racoons, coyotes, and bears (Akre *et al.*, 1981), perhaps the associations do help deter attacks. However, invertebrate predators, especially army ants, do not attack *Azteca* (and aggressive species of ants in several other genera), and nesting in close proximity to the ants protects the wasps from one of their principal enemies in the Neotropics. This also raises an interesting question. Doryline ants are common in the Indo–Malayan area and must prey on the tropical vespines in that area. Why have no similar associations of ants and vespines been reported in this area? Perhaps the associations are not as noticeable as large nests of *Azteca* and polybiines hanging together, or perhaps they do not exist.

IX. PHYLOGENY

The possible steps in the evolution of social behavior in wasps have been discussed by Evans (1958), Evans and West Eberhard (1970), Richards (1971), Wilson (1971), Spradbery (1973) and a number of other researchers, and the entire subject has been reviewed and updated by Starr (Vol. 1, Chap. 2, this series). Nothing further need be added to these excellent discussions. However, there are new data and new considerations that should be examined in regard to the phylogeny of social wasps.

It is generally agreed that a solitary scolioidlike ancestor (very primitive wasp) gave rise to the present day members of the superfamilies Vespoidea and Sphecoidae (Evans, 1958); the two groups containing all the known social wasps. The Vespidae, as revised by Richards (1962), contains the Steno-gastrinae, Polistinae, and Vespinae. Van der Vecht (1977a) enumerated the various ways in which the stenogastrines differ from the Polistinae and Vespinae and suggested that they evolved from a solitary cell-building ancestor closely related to Zethinae (Eumenidae). It was also suggested that these differences are great enough that the Stenogastrinae should be made a separate family or a subfamily of Eumenidae. The Vespinae are more like solitary *Eumenes* (Eumenini) than any of the Polistinae, and probably branched off earlier from some common ancestor (Richards, 1971). This is also supported by morphological studies of the larvae, since Vespinae, especially *Vespa*, have large, robust mandibles not much different from eumenids, while the mandibles of larval polistines are slender (Yamane, 1976). Richards also concluded that *Polistes* are probably as near to the ancestral Polistinae as any living wasp, and that it is possible that both the Polybiini and the Ropalidiini may have arisen from an ancestral *Polistes,* even though he considers both groups as being monophyletic. This evolutionary sequence is schematically represented in Fig. 26.

Distributional data indicate that the social wasps (excluding Sphecidae) originated in the Indo–Malayan Region, the only area where all the sub-families and tribes occur together, and then radiated over the world (van der Vecht, 1957; Richards, 1971). Diversification probably began at the end of the Mesozoic Era even though the earliest fossils (probably Vespinae) are early Tertiary. The report of a vespid from the middle Eocene Epoch (Wilson, 1977) shows that forms nearly identical to present day yellowjacket workers in wing venation and size were present in North America, so diversification may have started even prior to that time. The Polybiini probably entered America over the Bering Straits region when a bridge of land existed in that area, and the climate was much warmer. They obviously arrived early as evidenced by the great diversification of the 23 genera constituting the present day poly-

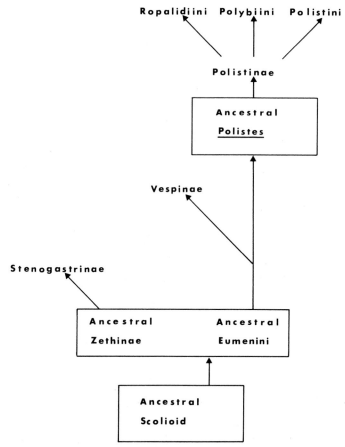

Fig. 26. Possible phylogeny of Vespidae.

biine fauna. Van der Vecht (1965) is of the opinion that *Polistes* arrived later and occupies a more subordinate position in this area than in the Old World tropics. The Vespinae occur in tropical areas only in the Indo–Malayan area where they originated, and are the dominant social wasps in the north temperate regions. They probably also arrived in North America via the Bering Straits region.

Yamane (1976) proposed a phyletic sequence for the Vespinae based on a study of morphological characteristics of larvae. There is general agreement that the ancestral yellowjacket probably evolved from the true hornets, *Vespa*, but the evolutionary sequence from this point onward is problematical. Yamane suggested a sequence in which the *Vespula vulgaris* species group (*Paravespula* of European workers) was considered closely related to true hornets, at the base of the yellowjacket lineage, with *Dolichovespula* being at the apex as the most recently evolved genus. However, Greene (1979) recently proposed an alternative yellowjacket phylogeny, based primarily on behavior, stating that *Dolichovespula* is probably the most ancestral yellowjacket group. This conclusion is based on his studies of *D. arenaria* (Greene *et al.,* 1976) and *D. maculata.* His studies have shown that the behaviors of these *Dolichovespula* are more simple (probably more primitive) than that observed in colonies of highly advanced *V. vulgaris* group species. *Dolichovespula* foundresses malaxate prey throughout much of the season, new queens also exhibit this behavior, and even trim cell caps after adult eclosion, behaviors reported for only *Vespula rufa* group species (Akre *et al.,* 1982). Perhaps more importantly, reproductive dominance in *Vespula* seems to be completely pheromonal as there is little interaction between the queen and her workers, and oophagy by the queen does not occur (Akre *et al.,* 1976). Conversely, the queen of *D. maculata* competes with egg-laying workers for some cells and also chases and mauls these workers. Other evidence supporting the view that *Dolichovespula* are ancestral is based on morphology. *Dolichovespula* is similar to *Vespa* in several structural aspects including male genitalia, antennal tyloides of males, vertical carinae on sides of pronotum, collar processes of mature larval spiracles, and fusion of the posterior pair of abdominal ganglia.

Since behavior is likely to evolve much faster than morphology, I agree with Greene that the *Dolichovespula* are more primitive than *Vespula.*

Current studies of two *V. rufa* group species, *Vespula acadica* and *V. consobrina,* show that some behaviors similar to *Dolichovespula* occur in these two species. New queens chew prey in colonies of *V. acadica* and they also trim cap remnants of cells where individuals have emerged. They are very inefficient at trimming caps, but sometimes manage to remove several small pieces. In addition, a queen of *V. acadica,* on encountering an ovipositing worker, bit its head and thorax and physically pulled it from the cell. Queens of *V. consobrina* chase and maul workers (as in *D. maculata*), and although oophagy by the queen was never observed, workers were laying eggs while the queen was still functional as an egg layer. Gastral vibration is also commonly exhibited by the queen and workers in *V. consobrina* colonies. In addition, *Dolichovespula* and *V. rufa* group species have small colonies. To be sure, most *Dolichovespula* are aerial nesters, but at least *D. arenaria* frequently nests in subterranean sites in cold, windy areas in the mountains of western North America, and workers excavate soil to allow for nest expansion.

These similarities in behavior and colony size suggest that a primitive *Dolichovespula* was probably the ancestral stock for a *V. rufa* group species. If, indeed, they had a common ancestor or if a *V. rufa* group species evolved from a *Dolichovespula,* it would be an easy step from the partly subterranean nesting habit to a fully subterranean nesting site. Since members of the *V. vulgaris* species group have larger colonies, a more advanced social organization (i.e., little physical contact between the queen and workers), they probably evolved later in the sequence.

Obviously, this suggested phylogeny is very tentative and it is hoped that it will stimulate additional studies of yellowjacket behavior and morphology and a re-examination of yellowjacket phylogeny.

X. SOCIAL WASPS AND MAN

Social wasps are probably among the most beneficial as well as the most injurious insects in the world. This conflicting statement obviously requires clarification; a somewhat difficult task with little concrete scientific data to substantiate the actual degree to which the wasps are beneficial or detrimental. Some aspects of their benefits to humans have been discussed by Evans and West Eberhard (1970) and Spradbery (1973). These include the collection of honey from certain polybiines that is used commercially in Mexico, the use of wasp larvae as human food and fish bait, and a few scattered references to the role of certain species of wasps as biological control agents in protecting crops used as food (see also Machado, 1977a,b). Most references to the latter are casual and to the effect that the local inhabitants have noticed or consider these wasps a major factor in controlling the insects damaging their crops. A few references in the sources cited deal with successful biological control of particular pests by wasps. Similarly, *Polistes* are generally recognized as beneficial insects that prey on caterpillars in home and commercial gardens in the United States. However, with the possible exceptions of studies by Fye (1972), Hasui (1977), Nakasuji *et al.* (1976), and Rabb and Lawson (1957), there are no detailed reports on the extent of benefits provided by the wasps. The role of yellowjackets as biological control agents for crop pests is also largely undocumented although they are known to prey on a number of economically important insects such as *Lygus* bugs, several species of Homoptera (including spittle bugs), caterpillars (MacDonald *et al.*, 1974), fall webworm larvae (Morris, 1972), codling moth adults (White *et al.*, 1969; Howell and Davis, 1972), and, under caged conditions, pear psylla (see Akre *et al.*, 1981).

However, this leaves totally unknown what may be considered the most beneficial aspect of social wasps, their role as natural biological control agents. Colonies of many species of polybiines and vespines are large, and the arthropod prey required to feed the developing brood must number into the thousands each day. For example, yellowjackets are the dominant social wasps occurring in temperate forests, and are probably second only to ants (especially *Formica* spp.) in protecting the forest from defoliators. Observations of colonies of *Vespula vulgaris, V. pensylvanica, V. acadica, Dolichovespula arenaria* and *D. maculata* in northeastern Oregon for 3 years (Roush and Akre, 1978a) indicated that these yellowjackets were a dominant forest predator using caterpillars, various bugs, beetle larvae, adult moths and other insects as prey. No detailed study was made of prey preferences, nor the possible impact of these wasps as biocontrol agents, but in general colonies were abundant and must have accounted for a considerable and perhaps significant reduction of several forest arthropods. However, the actual role these predators play in this system is unknown and will remain so until studies of this possible function are undertaken, or until the yellowjackets are removed, and their role becomes apparent. The latter has probably already occurred several times. Spraying pesticides over large areas of forest to control pests is not a very satisfactory solution, and often leads to reoccurring outbreaks of the same or different pests. Since many chemicals used in the past (especially DDT) have been extremely toxic to ants and yellowjackets, it is probable that these pesticides killed the predators and parasites responsible for holding the pest populations in check. It has now been realized that yellowjackets are important predators in forests, and steps have been taken to achieve control of pests without unduly affecting their populations (see Roush and Akre, 1978b).

Benefits from yellowjacket predation also occur in prairie areas of Washington where the western yellowjacket, *V. pensylvanica,* preys heavily on grasshoppers (*Melanoplus* spp.), and may help keep populations of these pests low. Excavations of colonies have revealed up to 3–5 cm of grasshopper legs (unused as food, especially tibia and tarsus) in the cavity beneath the nest. Undoubtedly other vespines and polistines play similar roles in natural control, and it is hoped that these will be investigated in future studies.

Conversely, the economic losses and injuries caused by social wasps directly affects people and, at least in some areas (United States, Europe), has received considerably more attention. Nearly all serious problems are directly attributable to the Vespinae, primarily *Vespa* spp. and *Vespula* spp. of the *V. vulgaris* species group. Injurious aspects of social wasps have been discussed by Spradbery (1973) and Edwards (1980), and the pest status of the Vespidae has been reviewed by Akre and Davis (1978). Briefly, certain species of social wasps cause problems to beekeepers throughout the world by killing adult bees, brood, and by robbing the hives of honey (especially *Vespa*, some *Vespula*) (see also DeJong, 1979). They also attack a number of soft fruits such as grapes, pears, peaches, apricots, rendering them unfit for human consumption, and, of course, they cause problems to man and his animals by stinging, frequently causing allergic reactions or sometimes death. Human deaths caused by wasps are not well documented, but Spradbery (1973) lists 70 human deaths caused by wasps (possibly yellowjackets) from 1949–1969 in England and Wales, and Akre *et al.* (1981) summarized human deaths attributable to yellowjackets in the United States. Data are few and fragmentary, and since deaths caused by stings mimic the symptoms of heart attack, many deaths are probably incorrectly reported. However, fatalities must be considerably more than the widely quoted figure of 20–50 per year.

Other economic losses to yellowjackets, at least in the United States, stem directly from the increased emphasis on outdoor recreational activities such as camping, back packing, and hiking. This brings people into more frequent contact with yellowjackets, with attendant problems, especially during years of yellowjacket abundance. For example, during the 1973 outbreak in the Pacific Northwest, private, county, state, and federal parks and recreation area were nearly deserted. Bitter complaints about yellowjackets were received by all agencies in charge of recreational areas, and budgets were increased to institute limited control programs. No figures are available, but monetary losses to private resorts were probably considerable. Similar economic losses occurred in 1979, another year of peak yellowjacket abundance in the Pacific Northwest.

Homeowner concern about yellowjacket problems has also increased and Davis (1978) reviewed the subject of yellowjackets in relation to urban environments. Perhaps one of the more serious problems encountered by homeowners deals with species that construct their nests in or on houses and/or other man-made structures. The aerial nesters (*Dolichovespula*) are usually easily eliminated by homeowners, but species constructing nests in wall voids or attics such as *V. vulgaris, V. pensylvanica,* and *V. maculifrons* are not as easily controlled. Furthermore, since all of these species have large nests and populous colonies, control of mature colonies is sometimes difficult and hazardous. Of special concern in this regard is the German yellowjacket, *Vespula germanica.* During the last 10 years this introduced species has become firmly established in the northeastern United States and adjacent areas of Canada, and is moving rapidly westward, having reached Michigan by the mid-1970s. Nearly all nests of this species in the United States have been inside structures (Akre *et al.,* 1981; MacDonald *et al.,* 1980).

With the increasing awareness of the losses, injury, and concern caused by yellowjackets in the United States, it would seem that the problems would have received considerable study and that solutions would be readily available, but this is not the case. MacDonald *et al.* (1976) pointed out that no effective and selective abatement techniques are available for pestiferous yellowjackets. The most intensive study of control methods for these wasps has been in California, where a program using poisoned meat as a bait was used to achieve a reduction in populations of *V. pensylvanica* (see Wagner and Reierson, 1969). However, the behavior of this species seems to vary in different areas. *Vespula pensylvanica* in southern California rely heavily on scavenging protein, especially late in the season, and therefore are very responsive to baits. In other areas, such as Washington, Oregon, Idaho, and British Columbia colonies do not scavenge so heavily, and this type of control is usually unsuccessful. Satisfactory abatement techniques will only be found through intensive biological and behavioral studies of pest species but few studies of this type are underway at the present time.

XI. CONCLUSIONS

Even with the recent surge in studies of social wasps, they still remain the least known of all the social insects. The studies reporting that *Microstigmus comes* (Sphecidae) and a *Parishnogaster* sp. (Stenogastrinae) apparently fulfill all the requirements of eusocial behavior were much too brief and fragmentary to obtain enough information for comparison with the known behaviors of other social wasps. Additional studies of closely related species in these groups should be an exciting and productive area of endeavor. However, these reports also raise questions as to how many and to what extent other presumably solitary wasps exhibit social behavior. Investigations of other groups, especially among the Sphecidae and Stenogastrinae, may reveal behaviors that could significantly advance our knowledge and insight into how sociality may have first arisen.

In the remaining social wasps most behavioral studies have been of species having small colonies and uncovered nests, while only a few representative species with populous colonies, and large, covered nests have been investigated. It is hoped that the information available on these species reflects a true rendition of the variability of biology and behavior exhibited by most social wasps, but much remains to be done. Most species have not been studied at all, while information on nests and life histories are lacking for many others. Communication, especially through the use of pheromones, remains largely unexplored.

At least one trend is suggested concerning social behavior in wasps. As colonies become more populous, social complexity increases, just as in other social insects. All studies of species having small colonies have shown that the queen (dominant female) maintains control of the colony through physical dominance, and conflicts with nestmates result in the establishment of a dominance hierarchy. In larger colonies dominance hierarchies are probably impossible to maintain, since there are too many individuals, and control of the colony by the queen is apparently through the use of pheromones (e.g., *Vespa, Vespula*). Studies of additional species may reveal that intermediate stages exist between physical and chemical dominance and show more clearly how the chemical system operates. Another possible key to understanding queen control is suggested by the study of social parasites. They are presumably species closely related to their host, and should have similar behaviors and, perhaps, control mechanisms. Studies of *Dolichovespula arctica* and its host *D. arenaria* have shown that the parasite female cannot maintain total control over egg laying by workers as does the host queen. A comparison of differences in behavior, exocrine glands, and possible gland secretions between parasite and host queens may help indicate how the queen is able to maintain this control. While the possibility of queen pheromones in large colony polybiines has not been investigated, this would also seem a fertile area for research, since at least trail pheromones (during swarming) are utilized by several species, and behavior reported for these species also indicates other pheromones are probably present.

Obviously, studies of social wasps are still, in many cases, at initial stages, with even the most basic biological information totally lacking for many species. It is hoped that investigators will start to use the enormous number of publications on ants and bees for comparative purposes, especially with regard to behavior, to help solve or gain insight into their own areas of endeavor.

Studies of and interest in social wasps increased dramatically during the 1970s. This is not surprising considering the role they serve or can serve to investigate questions at the forefront of current biological thought (e.g., kin selection, evolution of the worker caste, communication, population dynamics). Since sociality in wasps varies from primitively social to the complexity present in the vespines and certain polybiines, at the apex of sociality in wasps, a natural laboratory, containing infinite gradations and variability in social behavior is present for study and experimentation.

ACKNOWLEDGMENTS

H. C. Reed, A. Greene, and P. J. Landolt reviewed the manuscript and made many helpful suggestions for its improvement. A. Greene is also thanked for permission to use unpublished information on the behaviors of *Dolichovespula maculata* from his current studies. D. Carmean and S. Woods are thanked for checking literature references and for checking the many editions of the manuscript for errors. A. Stanford drew many of the illustrations used in the text. Much of the research reported on yellowjackets was supported by National Science Foundation Grant BNS 761400 to Washington State University.

REFERENCES

Akre, R. D., and Davis, H. G. (1978). Biology and pest-status of venomous wasps. *Annu. Rev. Entomol.* **23**, 19–42.

Akre, R. D., and Reed, H. C. (1981). Population cycles of yellowjackets (Hymenoptera: Vespidae) in the Pacific Northwest. *Environ. Entomol.* **10**, 267–274.

Akre, R. D., and Reed, H. C. (1982). Evidence for a queen pheromone in *Vespula* (submitted for publication).

Akre, R. D., Hill, W. B., and MacDonald, J. F. (1973). Magnetic recovery of ferrous labels in a capture-recapture system for yellowjackets. *J. Econ. Entomol.* **66,** 903–905.

Akre, R. D., Hill, W. B., MacDonald, J. F., and Garnett, W. B. (1975). Foraging distances of *Vespula pensylvanica* workers (Hymenoptera: Vespidae). *J. Kans. Entomol. Soc.* **48,** 12–16.

Akre, R. D., Garnett, W. B., MacDonald, J. F., Greene, A., and Landolt, P. J. (1976). Behavior and colony development of *Vespula pensylvanica* and *V. atropilosa* (Hymenoptera: Vespidae). *J. Kans. Entomol. Soc.* **49,** 63–84.

Akre, R. D., Roush, C. F., and Landolt, P. J. (1977). A *Vespula pensylvanica/Vespula vulgaris* nest (Hymenoptera: Vespidae). *Environ. Entomol.* **6,** 525–526.

Akre, R. D., Greene, A., MacDonald, J. F., Landolt, P. J., and Davis, H. G. (1981). The yellowjackets of America north of Mexico. *U.S., Dep. Agric., Agric. Handb.* **552** 102p.

Akre, R. D., Reed, H. C., Landolt, P. J. (1982). Nesting biology and behavior of the blackjacket, *Vespula consobrina* (Hymenoptera: Vespidae). *J. Kans. Ent. Soc.* (in press).

Alcock, J. (1978). Notes on male-locating behavior of some bees and wasps of Arizona. *Pan-Pac. Entomol.* **54,** 215–225.

Alcock, J., Barrows, R. M., Gordh, G., Hubbard, L. J., Kirkendall, L., Pyle, D. W., Ponder, T. L., and Zalom, F. G. (1978). The ecology and evoluation of male reproductive behavior in bees and wasps. *Zool. J. Linn. Soc.* **64,** 293–326.

Alexander, R. D. (1974). The evolution of social behavior. *Annu. Rev. Ecol. Syst.* **5,** 325–383.

Archer, M. E. (1973). The population ecology of *Vespula vulgaris* (L.), (Vespidae: Hymenoptera) and the computer simulation of colonial seasonal development. *Proc. Int. Cong.—Int. Union Study Soc. Insects, 7th, 1973* pp. 2–10.

Archer, M. E. (1977a). Tree wasp workers (Hym.: Vespidae) excavating soil from underground nests. *Entomol. Mon. Mag.* **112,** 88.

Archer, M. E. (1977b). The weights of forager loads of *Paravespula vulgaris* (Linn.) (Hymenoptera: Vespidae) and the relationship of load weight to forager size. *Insectes Soc.* **24,** 95–102.

Archer, M. E. (1979). A simulation model for the colonial development of *Paravespula vulgaris* (Linnaeus) and *Dolichovespula sylvestris* (Scopoli). Ph.D. Dissertation, Hull University, England.

Archer, M. E. (1981a). Taxonomy of the *sylvestris* group (Hymenoptera: Vespidae, *Dolichovespula*) with introduction of a new name and notes on distribution. *Ent. Scand.* **12,** 187–193.

Archer, M. E. (1981b). The Euro-Asia species of the *Vespula rufa* group (Hymenoptera, Vespidae) with descriptions of two new species and one new subspecies. *Kontyû* **49,** 54–64.

Archer, M. E. (1981c). Numerical characteristics of nests of *Vespa crabro* L. (Hym., Vespidae). *Entomol. Mon. Mag.* **116,** 117–122.

Archer, M. E. (1981d). A simulation model for the colonial development of *Paravespula vulgaris* (Linnaeus) and *Dolichovespula sylvestris* (Scopoli) (Hymenoptera: Vespidae). *Melanderia* **36,** 1–59.

Arnold, T. S. (1966). Biology of social wasps: Comparative ecology of the British species of social wasps belonging to the family Vespidae. M.Sc. Thesis, University of London.

Batra, S. W. T. (1980). Sexual behavior and pheromones of the European hornet, *Vespa crabro germana* (Hymenoptera: Vespidae). *J. Kansas Ent. Soc.* **53,** 461–469.

Beirne, B. P. (1944). The cause of the occasional abundance or scarcity of wasps (*Vespula* spp.) (Hymenoptera: Vespidae). *Entomol. Mon. Mag.* **80,** 121–124.

Bequaert, J. (1918). A revision of the Vespidae of the Belgian Congo based on the collection of the American Museum Congo Expedition, with a list of Ethiopian Diplopterous wasps. *Bull. Am. Mus. Nat. Hist.* **39,** 1–384.

Bequaert, J. (1928). A study of certain types of *Diplopterous* wasps in the collection of the British Museum. *Ann. Mag. Nat. Hist.* [10] 2, 138-138-176.

Bequaert, J. (1931). A tentative synopsis of the hornets and yellowjackets of America. *Entomol. Am.* 12, 71-138.

Bequaert, J. (1936). The common oriental Hornets, *Vespa tropica* and *Vespa affinis,* and their color forms. *Treubia* 15, 329-351.

Blüthgen, P., and Gusenleitner, J. (1970). Über Faltenwespen aus dem Iran (Hym.: Diploptera). *Stuttg. Beitr. Naturkd.* 223, 1-13.

Bohart, R. M., and Bechtel, R. C. (1957). The social wasps of California. *Bull. Calif. Insect Sur.* 4, 73-101.

Bohart, R. M., and Menke, A. S. (1976). "Sphecid Wasps of the World." Univ. of California Press, Berkeley.

Bohm, M. K., and Stockhammer, K. A. (1977). The nesting cycle of a paper wasp, *Polistes metricus* (Hymenoptera: Vespidae). *J. Kans. Entomol. Soc.* 50, 275-286.

Bourdon, V., Lecomte, J., Leclercq, M., and Leclercq, J. (1975). Présence de Noradrenaline conjuguée dans l'enveloppe du nid de *"Vespula vulgaris Linné."Bull. Soc. R. Sci. Liege* 44, 474-476.

Buckell, E. R., and Spencer, G. T. (1950). The social wasps (Vespidae) of British Columbia. *Proc. Entomol. Soc. B. C.* 46, 33-40.

Butler, C. G., Fletcher, D. J. C., and Walter, D. (1969). Nest-entrance marking with pheromones by the honeybee *Apis mellifera* L. and by a wasp *Vespula vulgaris* L. *Anim. Behav.* 17, 142-147.

Carl, J. (1934). *Ropalidia montana* n. sp. et son nid. Un type nouveau d'architecture vespienne. *Rev. Suisse Zool.* 41, 675-691.

Carmean, D., Akre, R. D., Zack, R. S., and Reed, H. C. (1981). Notes on the yellowjacket parasite *Bareogonalis canadensis* (Hymenoptera: Trigonalidae). *Ent. News*

Chadab, R., and Rettenmeyer, C. W. (1979). Observations on swarm emigrations and dragging behavior by social wasps (Hymenoptera: Vespidae). *Psyche* 86, 347-352.

Chan, K. L. (1972). The hornets of Singapore: Their identification, biology, and control. *Singapore Med. J.* 13, 178-187.

Chandler, L. (1964). The social wasps in Indiana (Hymenoptera: Vespidae). *Proc. Indiana Acad. Sci.* 74, 197-204.

Cobb, F. K. (1979). Honey Buzzard at wasps' nest. *Br. Birds* 72, 59-64.

Coke, J. L., and Richon, A. B. (1976). Synthesis of optically active α-n-hexadecalactone, the proposed pheromone from *Vespa orientalis. J. Org. Chem.* 41, 3516-3517.

Cooter, J. (1976). *Polistes metricus* (Say) (Hym., Vespidae) accidentally imported into Britain. *Entomol. Mon. Mag.* 112, 122.

Corn, M. L. (1972). Notes of the biology of *Polistes carnifex* (Hymenoptera, Vespidae) in Costa Rica and Columbia. *Psyche* 79, 150-157.

Cumber, R. A. (1951). Some observations on the biology of the Australian wasp *Polistes humilis* Fabr. (Hymenoptera: Vespidae) in North Auckland (New Zealand), with special reference to the nature of the worker caste. *Proc. R. Entomol. Soc. London, Ser. A* 26, 11-16.

Darchen, R. (1976). *Ropalidia cincta,* Guêpe sociale de la savane de lamto (Cote-D'Ivoire) (Hym. Vespidae). *Ann. Soc. Entomol. Fr.* [N.S.] 12, 579-601.

Davis, H. G. (1978). Yellowjacket wasps in urban environments. *In* "Perspectives in Urban Entomology" (G. W. Frankie and C. S. Koehler, eds.), XV Int. Congr. Entomol., pp. 163-185. Academic Press, New York.

Davis, T. A. (1966a). Observations on *Ropalidia variegata* (Smith) (Hymenoptera: Vespidae). *Entomol. News* 77, 271-277.

Davis, T. A. (1966b). Nest-structure of a social wasp varying with siting of leaves. *Nature (London)* 210, 966-967.

DeJong, D. (1979). Social wasps, enemies of honey bees. *Am. Bee J.* 119, 505-507, 529.

Deleurance, E. P. (1955). Contribution à l'étude biologique des *Polistes* (Hyménoptères Vespides). II. Le cycle évolutif du Couvain. *Insectes Soc.* **2**, 285–302.

Delfino, G., Marino Piccioli, M. T., and Calloni, C. (1979). Fine structure of the glands of vander Vecht's organ in *Polistes gallicus* (L.) (Hymenoptera Vespidae) *Monit. Zool. Ital.* [N.S.] **13**, 221–247.

Dew, H. E., and Michener, C. D. (1978). Foraging flights of two species of *Polistes* wasps (Hymenoptera: Vespidae). *J. Kans. Entomol. Soc.* **51**, 380–385.

Dew, H. E., and Michener, C. D. (1981). Division of labor among workers of *Polistes metriculus* (Hymenoptera: Vespidae): Laboratory foraging activities. *Insectes Soc.* **28**, 87–101.

Downhower, J. F., and Wilson, D. E. (1973). Wasps as a defense mechanism of katydids. *Am. Midl. Nat.* **89**, 451–455.

Duncan, C. D. (1939). A contribution to the biology of North American vespine wasps. *Stanford Univ. Publ., Univ. Ser., Biol. Sci.* **8**, 1–272.

du Buysson, R. (1904). Monographie des guêpes ou *Vespa* (suite). *Ann. Soc. Entomol. Fr.* **73**, 485–556, 565–634.

du Buysson, R. (1909). Monographie des vespides du genre *Belonogaster*. *Ann. Soc. Entomol. Fr.* **78**, 197–270.

Ebeling, W. (1975). "Urban Entomology." University of California, Div. Agric. Sci., Berkeley.

Eck, R. (1980). *Dolichovespula loekenae* n. sp., eine neue soziale Faltenwespe aus Skandinavien. Reichenbachia **18**, 213–217.

Eck, R. (1981). Zur verbreitung und Variabilität von *Dolichovespula norwegica* (Hymenoptera, Vespidae). *Entomol. Abh.* **44**, 133–152.

Edwards, R. (1976). The world distribution pattern of the German wasp, *Paravespula germanica* (Hymenoptera: Vespidae). *Ent. Germanica* **3**, 269–271.

Edwards, R. (1980). "Social Wasps Their Biology and Control." Rentokil, Sussex, England 398p.

Eickwort, K. R. (1969). Separation of the castes of *Polistes exclamans* and notes on its biology (Hym.: Vespidae). *Insectes Soc.* **16**, 67–72.

Erickson, E. H., and Medenwald, R. (1979). Parasitism of queen honeybee pupae by *Melittobia acasta. J. Apic. Res.* **18**, 73–76.

Esch, H. (1971). Wagging movements in the wasp *Polistes versicolor vulgaris* Bequaert. *Z. Vergl. Physiol.* **72**, 221–225.

Es'Kov, E. K. (1977). The structure of acoustic signals in the larvae of the wasp *Dolichovespula sylvestris* and the hornet *Vespa crabro. Zh. Evol. Biokhim. Fiziol.* **13**, 371–375.

Evans, H. E. (1958). The evolution of social life in wasps. *Proc. Int. Contr. Entomol., 10th, 1956* Vol. 2, pp. 449–457.

Evans, H. E. (1966). The behavior patterns of solitary wasps. *Annu. Rev. Entomol.* **11**, 123–154.

Evans, H. E. (1975). Social parasitism of a common yellowjacket. *Insect World Dig.* **2**, 6–13.

Evans, H. E., and West Eberhard, M. (1970). "The Wasps." Univ. of Michigan Press, Ann Arbor.

Evans, W. (1903). *Vespa austriaca,* Panzer, in North Durham. *Vespa austriaca* and *V. rufa* in Scotland. *Entomol. Mon. Mag.* **39**, 299–300.

Ferro, D. N., ed. (1976). "New Zealand Insect Pests." Lincoln Univ. College Agric., Canterbury.

FitzGerald, D. W. (1940). Vespidae from Mafia Island, East Africa (Hym.). *Proc. R. Entomol. Soc. London, Ser. A* **15**, 33–35.

FitzGerald, D. V. (1950). Notes on the genus *Ropalidia* (Hymenoptera: Vespidae) from Madagascar. *Proc. R. Entomol. Soc. London, Ser. A* **25**, 81–86.

Fletcher, D. J. C. (1975). Significance of dorsoventral abdominal vibration among honey-bees (*Apis mellifera* L.). *Nature (London)* **256**, 721–723.

Fluno, J. A. (1972). Grooming in *Polistes exclamans* (Viereck), a forerunner of communications in social Hymenoptera. *J. Wash. Acad. Sci.* **62**, 332–333.

Fox-Wilson, G. (1946). Factors affecting populations of social wasps, *Vespula* species, in England (Hymenoptera). *Proc. R. Entomol. Soc. London, Ser. A* **21**, 17–27.

Francke, W., Hindorf, G., and Reith, W. (1978). Methyl-1,6-dioxaspiro [4,5] decanes as odors of *Paravespula vulgaris* (L.). *Angew. Chem., Int. Ed. Engl.* **17**, 862.

Free, J. B. (1970). The behavior of wasps (*Vespula germanica* L. and *V. vulgaris* L.) when foraging. *Insectes Soc.* **17**, 11–20.

Fye, R. E. (1972). Manipulation of *Polistes exclamans arizonensis*. *Environ. Entomol.* **1**, 55–57.

Gadagkar, R. (1980). Dominance hierarchy and division of labor in the social wasp, *Ropalidia marginata* (Lep.) (Hymenoptera: Vespidae). *Curr. Sci.* **49**, 772–775.

Gadgil, M., and Mahabal, A. (1974). Caste differentiation in the parer [sic] wasp *Ropalidia marginata* (Lep.). *Curr. Sci.* **43**, 482.

Gamboa, G. J. (1978). Intraspecific defense: Advantage of social cooperation among paper wasp foundresses. *Science* **199**, 1463–1465.

Gamboa, G. J., and Dew, H. E. (1981). Intracolonial communication by body oscillations in the paper wasp, *Polistes metricus*. *Insectes Soc.* **28**, 13–26.

Gamboa, G. J., Heacock, B. D., and Wiltjer, S. L. (1978). Division of labor and subordinate longevity in foundress associations of the paper wasp, *Polistes metricus* (Hymenoptera: Vespidae). *J. Kans. Entomol. Soc.* **51**, 343–352.

Garcia, A. R. (1974). Observations of *Polistes peruvianus* (Hymenoptera: Vespidae) in the environs of Lima. *Biota* **10**, 11–27.

Gaul, A. T. (1952). Additions to Vespine biology. X. Foraging and chemotaxis. *Bull. Brooklyn Entomol. Soc.* **47**, 138–140.

Gibo, D. L. (1972). An introduced population of social wasps, *Polistes apachus,* that has persisted for 10 years (Hymenoptera: Vespidae). *Bull. South. Calif. Acad. Sci.* **71**, 53.

Gibo, D. L. (1974). A laboratory study on the selective advantage of foundress associations in *Polistes fuscatus* (Hymenoptera: Vespidae). *Can. Entomol.* **106**, 101–106.

Gibo, D. L. (1978). The selective advantage of foundress associations in *Polistes fuscatus* (Hymenoptera: Vespidae): A field study of the effects of predation on productivity. *Can. Entomol.* **110**, 519–540.

Gibo, D. L., and Metcalf, R. A. (1978). Early survival of *Polistes apachus* (Hymenoptera: Vespidae) colonies in California: A field study of an introduced species. *Can. Entomol.* **110**, 1339–1343.

Gillaspy, J. E. (1971). Paper-nest wasps (*Polistes*): Observations and study methods. *Ann. Entomol. Soc. Am.* **64**, 1357–1361.

Gillaspy, J. E. (1973). Behavioral observations on paper-nest wasps (genus *Polistes;* family Vespidae; order Hymenoptera). *Am. Midl. Nat.* **90**, 1–12.

Gorton, R. E. (1978). Observations on the nesting behavior of *Mischocyttarus immarginatus* (Rich.) (Vespidae: Hymenoptera) in a dry forest in Costa Rica. *Insectes Soc.* **25**, 197–204.

Green, S. G., Heckman, R. A., Benton, A. W., and Coon, B. F. (1970). An unusual nest location for *Vespula maculifrons* (Hymenoptera: Vespidae). *Ann. Entomol. Soc. Am.* **63**, 1197–1198.

Greene, A. (1979). Behavioral characters as indicators of yellowjacket phylogeny (Hymenoptera: Vespidae). *Ann. Entomol. Soc. Am.* **72**, 614–619.

Greene, A., Akre, R. D., and Landolt, P. J. (1976). The aerial yellowjacket, *Dolichovespula arenaria* (Fab.): Nesting biology, reproductive production, and behavior (Hymenoptera: Vespidae). *Melanderia* **26**, 1–34.

Greene, A., Akre, R. D., and Landolt, P. J. (1978). Behavior of the yellowjacket social parasite, *Dolichovespula arctica* (Rohwer) (Hymenoptera: Vespidae). *Melanderia* **29**, 1–28.

Grogan, D. E., and Hunt, J. H. (1977). Digestive proteases of two species of wasps of the genus *Vespula. Insect Biochem.* **7**, 191–196.

Guiglia, D. (1971). A concise history of Vespidae systematics in the Old World. *In* "Entomological Essays to Commemorate the Retirement of Professor K. Yasumatsu" (S. Asahina *et al.*, eds.), pp. 113–117. Hokuryukan, Tokyo.

Guiglia, D. (1972). "Les guêpes sociales (Hymenoptera: Vespidae) d'Europe occidentale et septentrionale Faune de l'Europe et du bassin Mediterranéen," No. 6. Masson, Paris.

Hamilton, W. D. (1964). The genetical evolution of social behaviour. I and II. *J. Theor. Biol.* **7**, 1–52.

Hamilton, W. D. (1972). Altruism and related phenomena, mainly in social insects. *Ann. Rev. Syst. Ecol.* **3**, 193–232.

Hansell, M. H. (1981). Nest construction in the subsocial wasp *Parischnogaster mellyi* (Saussure) Stenogastrinae (Hymenoptera). *Insecta Soc.* **28**, 208–216.

Harris, A. C. (1979). Occurrence and nesting of the yellow oriental paper wasp, *Polistes olivaceus* (Hymenoptera: Vespidae) in New Zealand. *New Zealand Entomol.* **7**, 41–44.

Hasui, H. (1977). On the seasonal variability of survivorship curves and life tables of *Pieris rapae* Boisduval (Lepidoptera: Pieridae). *Jpn. J. Ecol.* **27**, 75–82 (in Japanese, English summary).

Heithaus, E. R. (1979). Community structure of Neotropical flower visiting bees and wasps: Diversity and phenology. *Ecology* **60**, 190–202.

Hermann, H. R., and Dirks, T. F. (1974). Sternal glands in polistine wasps: Morphology and associated behavior. *J. Ga. Entomol. Soc.* **9**, 1–8.

Hermann, H. R., and Dirks, T. F. (1975). Biology of *Polistes annularis* (Hymenoptera: Vespidae) I. Spring behavior. *Psyche* **82**, 97–108.

Hermann, H. R., Gerling, D., and Dirks, T. F. (1974). The cohibernation and mating activity of five polistine wasp species (Hymenoptera: Vespidae: Polistinae). *J. Ga. Entomol. Soc.* **9**, 203–204.

Hermann, H. R., Barron, R., and Dalton, L. (1975). Spring behavior of *Polistes exclamans* (Hymenoptera: Vespidae: Polistinae). *Entomol. News* **86**, 173–178.

Howell, J. F., and Davis, H. G. (1972). Protecting codling moths captured in sex-attractant traps from predaceous yellowjackets. *Environ. Entomol.* **1**, 122–123.

Hung, A. C. F., Reed, H. C., and Vinson, S. B. (1981). Chromosomes of four species of *Polistes* wasps (Hymenoptera: Vespidae). *Caryologia* **34**, 225–230.

Hunt, J. H., and Gamboa, G. J. (1978). Joint nest use by two paper wasp species. *Insectes Soc.* **25**, 373–374.

Hunt, J. H., and Noonan, K. C. (1979). Larval feeding by male *Polistes fuscatus* and *Polistes metricus* (Hymenoptera: Vespidae). *Insectes Soc.* **26**, 247–251.

Ibrahim, S. H. (1974). Some ecological and biological aspects of *Polistes foederata. Agric. Res. Rev.* **52**, 115–120.

Ikan, R., Bergmann, E. D., Ishay, J., and Gitter, S. (1968). Proteolytic enzyme activity in the various colony members of the Oriental hornet, *Vespa orientalis* F. *Life Sci.* **7**(Part 2), 929–934.

Ikan, R., Gottlieb, R., Bergmann, E. D., and Ishay, J. (1969). The pheromone of the queen of the Oriental hornet, *Vespa orientalis. J. Insect Physiol.* **15**, 1709–1712.

Ishay, J. (1964). Observations sur la biologie de la guêpe Orientale *Vespa orientalis* F. *Insectes Soc.* **11**, 193–206.

Ishay, J. (1965). Observations and experiments on colonies of the Oriental wasp. *Int. Beekeep. Jubilee Congr. 20th, 1965* pp. 140–145.

Ishay, J. (1972). Thermoregulatory pheromones in wasps. *Experientia* **28**, 1185–1187.

Ishay, J. (1973a). The influence of cooling and queen pheromone on cell building and nest architecture by *Vespa orientalis* (Vespinae, Hymenoptera). *Insectes Soc.* **20**, 243–252.

Ishay, J. (1973b). Thermoregulation by social wasps: Behavior and pheromones. *Trans. N. Y. Acad. Sci.* [2] **36**, 447–462.

Ishay, J. (1975). Frequencies of the sounds produced by the Oriental hornet, *Vespa orientalis.* *J. Insect Physiol.* **21**, 1737–1740.

Ishay, J. (1976). Comb building by the Oriental hornet (*Vespa orientalis*). *Anim. Behav.* **24**, 72–83.

Ishay, J. (1978). Acoustical communication in wasp colonies (Vespinae). *Proc. Int. Congr. Entomol., 15th, 1976* pp. 406–435.

Ishay, J., and Brown, M. B. (1975). Patterns in the sounds produced by *Paravespula germanica* wasps. *J. Acoust. Soc. Am.* **57**, 1521–1525.

Ishay, J., and Ikan, R. (1968). Food exchange between adults and larvae in *Vespa orientalis* F. *Anim. Behav.* **16**, 298–303.

Ishay, J., and Perna, B. (1979). Building pheromones of *Vespa orientalis* and *Polistes foederatus.* *J. Chem. Ecol.* **5**, 259–272.

Ishay, J., Bytinski-Salz, H., and Shulov, A. (1968). Contributions to the bionomics of the Oriental hornet (*Vespa orientalis* F.). *Isr. J. Entomol.* **2**, 45–106.

Ishay, J., Motro, A., Gitter, S., and Brown, M. B. (1974). Rhythms in acoustical communication by the Oriental hornet, *Vespa orientalis.* *Anim. Behav.* **22**, 741–744.

Iwata, K. (1971). Ethological notes on four Japanese species of *Vespa* (Hymenoptera). *In* "Entomological Essays to Commemorate the Retirement of Professor K. Yasumatsu" (S. Asahina *et al.,* eds.), pp. 219–223. Hokuryukan, Tokyo.

Iwata, K. (1976). "Evolution of Instinct: Comparative Ethology of Hymenoptera." Amerind, New Delhi.

Jacobson, R. S., Matthews, R. W., and MacDonald, J. F. (1978). A systematic study of the *Vespula vulgaris* group with a description of a new yellowjacket species in eastern North America (Hymenoptera: Vespidae). *Ann. Entomol. Soc. Am.* **71**, 299–312.

Jeanne, R. L. (1970). Descriptions of the nests of *Pseudochartergus fuscatus* and *Stelopolybia testacea* with a note on a parasite of *S. testacea* (Hymenoptera, Vespidae). *Psyche* **77**, 54–69.

Jeanne, R. L. (1972). Social biology of the Neotropical wasp *Mischocyttarus drewseni.* *Bull. Mus. Comp. Zool.* **144**, 63–150.

Jeanne, R. L. (1973). Aspects of the biology of *Stelopolybia areata (Say) (Hymenoptera: Vespidae).* *Biotropica* **5**, 183–198.

Jeanne, R. L. (1975a). The adaptiveness of social wasp nest architecture. *Q. Rev. Biol.* **50**, 267–287.

Jeanne, R. L. (1975b). Social biology of *Stelopolybia areata* (Say) in Mexico (Hymenoptera: Vespidae). *Insectes Soc.* **22**, 27–34.

Jeanne, R. L. (1975c). Behavior during swarm movement in *Stelopolybia areata* (Hymenoptera: Vespidae). *Psyche* **82**, 259–264.

Jeanne, R. L. (1977a). A specialization in nest petiole construction by queens of *Vespula* spp. (Hymenoptera: Vespidae). *J. N. Y. Entomol. Soc.* **85**, 127–129.

Jeanne, R. L. (1977b). Ultimate factors in social wasp nesting behavior. *Proc. Int. Congr.—Int. Union Study Soc. Insects, 8th, 1977* pp. 164–168.

Jeanne, R. L. (1977c). Behavior of the obligate social parasite *Vespula arctica* (Hymenoptera: Vespidae). *J. Kans. Entomol. Soc.* **50**, 541–557.

Jeanne, R. L. (1978). Intraspecific nesting associations in the Neotropical social wasp *Polybia rejecta* (Hymenoptera: Vespidae). *Biotropica* **10**, 234–235.

Jeanne, R. L. (1979). Construction and utilization of multiple combs in *Polistes canadensis* in relation to the biology of a predaceous moth. *Behav. Ecol. Sociobiol.* **4**, 293–310.

Jeanne, R. (1980). Evolution of social behavior in the Vespidae. *Ann. Rev. Entomol.* **25**, 371–396.

Jeanne, R. L. (1981). Chemical communication during swarm emigration in the social wasp *Polybia sericea* (Oliver). *Anim. Behav.* **29**, 102–113.

Jeanne, R. L., and Fagen, R. (1974). Polymorphism in *Stelopolybia areata* (Hymenoptera: Vespidae). *Psyche* **81**, 155–166.

Jeanne, R. L., and Castellóu Bermudez, E. G. (1980). Reproductive behavior of a male neotropical social wasp, *Mischocyttarus drewseni* (Hymenoptera: Vespidae). *J. Kansas Entomol. Soc.* **53**, 271–276.

Kalmus, H. (1954). Finding and exploitation of dishes of syrup by bees and wasps. *Br. J. Anim. Behav.* **2**, 136–139.

Kayes, B. M. (1978). Digestive proteases in four species of *Polistes* wasps. *Can. J. Zool.* **56**, 1454–1459.

Kemper, H. (1961). Nestunterschiede bei den sozialen Faltenwespen Deutschlands. *Z. Angew. Zool.* **48**, 31–85.

Kemper, H., and Döhring, E. (1967). "Die sozialen Faltenwespen Mitteleuropas." Parey, Berlin.

Kojima, J., and Yamane, Sk. (1980). Biological notes on *Vespa luctuosa luzonensis* from Leyte Island, the Phillipines, with descriptions of adults and larvae (Hymenoptera: Vespidae). *Insecta Matsumurana* [N.S.] **19**, 79–87.

Kirkton, R. M. (1968). Building up desirable wasp populations. *Arkansas Farm Res.* **17**, 8.

Kirkton, R. M. (1970). Habitat management and its effects on populations of *Polistes* and *Iridomyrmex*. *Proc. Tall Timbers Conf.* **2**, 243–246.

Krispyn, J., and Hermann, H. (1977). The social wasps of Georgia: Hornets, yellowjackets, and Polistine paper wasps. *Res. Bull.—Ga. Agric. Exp. Stn.* **207**, 3–39.

Kundu, H. L. (1965). Notes on observations of the behaviour of a social wasp *Polistes hebraeus* (Hymenoptera). *Proc. XII Internat. Congr. Entomol.* p. 304.

Lacey, L. A. (1979). Predacão em gerinos por uma *vespa* e outras associacões de insetos com ninhos de duas espécies de ràs da Amazônia. *Acta Amazônia* **9**, 755–762.

Landolt, P. J., and Akre, R. D. (1978). Biology of the social wasp. *Mischocyttarus flavitarsis* (Hymenoptera: Vespidae; Polybiini). *Proc. Wash. State Entomol. Soc.* **40**, 534–537.

Landolt, P. J., and Akre, R. D., (1979). Occurrence and location of exocrine glands in some social Vespidae (Hymenoptera). *Ann. Entomol. Soc. Am.* **72**, 141–148.

Landolt, P. J., Akre, R. D., and Greene, A. (1977). Effects of colony division on *Vespula atropilosa* (Sladen) (Hymenoptera: Vespidae). *J. Kans. Entomol. Soc.* **50**, 135–147.

Lawson, F. R., Rabb, R. L., Guthrie, F. E., and Bowery, T. G. (1961). Studies of an integrated control system for hornworms on tobacco. *J. Econ. Entomol.* **54**, 93–97.

Lecomte, J., Bourdon, V., Damas, J., Leclercq, M., and Leclercq, J. (1976). Présence de noradrenaline conjuguée dans les parois du nid de *Vespula* germanica Linné. *C. R. Seances Soc. Biol. Ses Fil.* **170**, 212–215.

Lester, L. J., and Selander, R. K. (1981). Genetic relatedness and social organization of *Polistes* colonies. *Am. Natur.* **117**, 147–166.

Lin, N., and Michener, C. D. (1972). Evolution of sociality in insects. *Q. Rev. Biol.* **47**, 131–159.

Litte, M. (1977). Behavioral ecology of the social wasp *Mischocyttarus* mexicanus (Hym., Vespidae). *Behav. Ecol. Sociobiol.* **2**, 229–246.

Litte, M. (1979). Mischocyttarus flavitarsis in Ariżona: Social and nesting biology of a polistine wasp. 2. *Tierpsychology* **50**, 282–312.

Løken, A. (1964). Social wasps in Norway (Hymenoptera, Vespidae). *Nor. Entomol. Tidsskr.* **12**, 195–218.

Long, G. E., Roush, C. F., and Akre, R. D. (1979). A linear model of development for colonies of *Vespula pensylvanica* (Hymenoptera: Vespidae) collected from Pullman, Washington. *Melanderia* **31**, 27–36.

Lord, W. D., Nicolson, D. A., and Roth, R. R. (1977). Foraging behavior and colony drift in *Vespula maculifrons* (Buysson) (Hymenoptera: Vespidae). *J. N. Y. Entomol. Soc.* **85**, 186.

Lövgren, B. (1958). A mathematical treatment of the development of colonies of different kinds of social wasps. *Bull. Math. Biophys.* **20**, 119–148.

MacDonald, J. F., and Matthews, R. W. (1975). *Vespula squamosa:* A yellowjacket wasp evolving toward parasitism. *Science* **190**, 1003–1004.

MacDonald, J. F., and Matthews, R. W. (1981). Nesting biology of the eastern yellowjacket, *Vespula maculifrons* (Hymenoptera: Vespidae). *J. Kansas Entomol. Soc.* **54**, 433–457.

MacDonald, J. F., Akre, R. D., and Hill, W. B. (1974). Comparative biology and behavior of *Vespula atropilosa* and *V. pensylvanica* (Hymenoptera: Vespidae). *Melanderia* **18**, 1–66.

MacDonald, J. F., Akre, R. D., and Hill, W. B. (1975a). Nest associates of *Vespula atropilosa* and *V. pensylvanica* in southeastern Washington State (Hymenoptera: Vespidae). *J. Kans. Entomol. Soc.* **48**, 53–63.

MacDonald, J. F., Akre, R. D., and Hill, W. B. (1975b). Locations and structure of nests of *Vespula atropilosa* and *V. acadica* (Hymenoptera: Vespidae), *J. Kans. Entomol. Soc.* **48**, 114–121.

MacDonald, J. F., Akre, R. D., and Matthews, R. W. (1976). Evaluation of yellowjacket abatement in the United States. *Bull. Entomol. Soc. Am.* **22**, 397–401.

MacDonald, J. F., Akre, R. D., and Keyel, R. (1980). The German yellowjacket (*Vespula germanica*) problem in the United States. *Bull. Entomol. Soc. Am.* **26**, 436–442.

Machado, V. L. L. (1977a). Aspectos da Biologia de *Protopolybia pumila* (Saussure, 1863) (Hym., Vespidae). *Rev. Bras. Biol.* **37**, 771–784.

Machado, V. L. L. (1977b). Estudos biológicos de *Polybia occidentalis occidentalis* (Oliver, 1791) (Hym., Vespidae). *An. Soc. Entomol. Bras.* **6**, 7–24.

MacLean, B. K., Chandler, L., and MacLean, D. B. (1978). Phenotypic expression in the Paper Wasp *Polistes fuscatus* (Hymenoptera: Vespidae). *Great Lakes Entomol.* **11**, 105–116.

Madden, J. L. (1981). Factors influencing the abundance of the European wasp (*P. germanica* [F.]). *J. Aust. Entomol. Soc.* **20**, 59–66.

Maher, G. D. (1976). Some notes on social interactions in *Polistes exclamans* (Hymenoptera: Vespidae). *Entomol. News* **87**, 218–222.

Marino Piccioli, M. T. (1968). The extraction of the larval peritrophic sac by the adults in *Belonogaster. Monit. Zool. Ital.* [N.S.] *Suppl.* **2**, 203–206.

Marino Piccioli, M. T., and Pardi, L. (1970). Studi sulla biologica di *Belonogaster* (Hymenoptera, Vespidae). *Ital. J. Zool.* [N.S.] **9**, Suppl. III, 197–225.

Marino Piccioli, M. T., and Pardi, L. (1978). Studies on the biology of *Belonogaster* (Hymenoptera, Vespidae). *Ital. J. Zool.* **29**, 179–228.

Maschwitz, U. W. (1964). Gefahrenalarmstoffe und Gefahrenalarmierung bei sozialen Hymenopteren. *Z. Vergl. Physiol.* **47**, 596–655.

Maschwitz, U. W. (1966). Alarm substances and alarm behavior in social insects. *Vitam. Horm. (N.Y.)* **24**, 267–290.

Maschwitz, U. W., Beier, W., Dietrich, I., and Keidel, W. (1974). Futterverstandigung bei Wespen der Gattung *Paravespula. Naturwissenschaften* **61**, 506.

Matsuura, M. (1968). Life of hornets. Part II. Royal court around queens of *Vespa* species (in Japanese). *Jpn. Bee J.* **21**, 21–25. (cited by Spradbery, 1973).

Matsuura, M. (1969). Behaviour of post-hibernating female hornets *Vespa*, in the pre-nesting stage, with special reference to intra- and interspecific dominance relationships. *Jpn. J. Ecol.* **19**, 196–203.

Matsuura, M. (1970). Data for the nests of the giant paper wasp. *Polistes gigas* Kirby, in Formosa with special consideration on colony size. *Life Study (Fukui)* **14**, 35–40.

Matsuura, M. (1971a). (Nesting sites of the Japanese *Vespa* species) (In Japanese, English Summary). *Kontyû* **39**, 43–54.

Matsuura, M. (1971b). Nest foundation by the female wasps of the genus *Vespa* (Hymenoptera; Vespidae). *Kontyû* **39**, 99–105.

Matsuura, M. (1973a). Colony development of *Vespa crabro flavofasciata* Cameron in early stage of nesting. *Life Study (Fukui)* **17**, 1–12.

Matsuura, M. (1973b). Intracolonial polyethism in *Vespa*. III. Foraging activities. *Life Study (Fukui)* **17**, 81–99.

Matsuura, M. (1973c). Nesting habits of several species of the genus *Vespa* in Formosa. *Kontyû* **41**, 286–293.

Matsuura, M. (1974). Intracolonial polyethism in *Vespa*. I. Behaviour and its change of the foundress of *Vespa crabro flavofasciata* in early nesting stage in relation to worker emergence. *Kontyû* **42**, 333–350.

Matsuura, M. (1977). Observations on the hornets hunting the cicadas. *Rostria* **27**, 216–218.

Matsuura, M., and Sakagami, S. F. (1973). A bionomic sketch of the giant hornet, *Vespa mandarinia*, a serious pest for Japanese apiculture. *J. Fac. Sci., Hokkaido Univ., Ser. 6*, **19**, 125–162.

Matthews, R. W. (1968a). *Microstigmus comes:* Sociality in a sphecid wasp. *Science* **160**, 787–788.

Matthews, R. W. (1968b). Nesting biology of the social wasp *Microstigmus comes* (Hymenoptera: Sphecidae, Pemphredoninae). *Psyche* **75**, 23–45.

Matthews, R. W. (1970). A new thrips-hunting *Microstigmus* from Costa Rica (Hymenoptera: Sphecidae, Pemphredoninae). *Psyche* **77**, 120–126.

Maxwell-Lefroy, H., and Howlett, F. M. (1909). "Indian Insect Life." W. Thacker, London.

Metcalf, R. A., and Whitt, G. S. (1977a). Intra-nest relatedness in the social wasp *Polistes metricus*. *Behav. Ecol. Sociobiol.* **2**, 339–351.

Metcalf, R. A., and Whitt, G. S. (1977b). Relative inclusive fitness in the social wasp *Polistes metricus*. *Behav. Ecol. Sociobiol.* **2**, 353–360.

Miller, C. D. F. (1961). Taxonomy and distribution of Nearctic *Vespula*. *Can. Entomol. Suppl.* **22**, 1–52.

Miyano, S. (1980). Life tables of colonies and workers in a paper wasp, *Polistes chinensis antennalis*, in central Japan (Hymenoptera: Vespidae). *Res. Pop. Ecol.* **22**, 69–88.

Montagner, H. (1966). Le mécanisme et les consequences des comportements trophoallactiques chez les guêpes du genre *Vespa*. These de l'Université de Nancy, Perigueux.

Morimoto, R. (1960). On the social cooperation in *Polistes chinensis antennalis* Pérez (studies on the social Hymenoptera of Japan. IX). *Kontyû* **28**, 198–207.

Morris, R. F. (1972). Predation by wasps, birds, and mammals on *Hyphantria cunea*. *Can. Entomol.* **104**, 1581–1591.

Myers, J. G. (1929). The nesting-together of birds, wasps and ants. *Proc. Entomol. Soc. London* **4**, 80–90.

Myers, J. G. (1935). The nesting associations of birds with social insects. *Trans. R. Entomol. Soc. London* **83**, 11–22.

Nakasuji, F., Yamanaka, H., and Kiritani, K. (1976). Predation of larvae of the tobacco cutworm *Spodoptera litura* (Lepidoptera, Noctuidae) by *Polistes* wasps. *Kontyû* **44**, 205–213.

Naumann, M. G. (1968). A revision of the genus *Brachygastra* (Hym., Vespidae). *Univ. Kans. Sci. Bull.* **47**, 929–1003.

Naumann, M. G. (1970). The nesting behavior of *Protopolybia pumila* in Panama (Hymenoptera: Vespidae). Ph.D. Dissertation, University of Kansas, Lawrence.

Naumann, M. G. (1975). Swarming behavior: Evidence for communication in social wasps. *Science* **189**, 642–644.

Nelson, J. M. (1966). The biology and distribution of parasites and inquilines in *Polistes* (Paperwasp) nests. M.S. Thesis, Southern Illinois University, Carbondale.

Nelson, J. M. (1968). Parasites and symbionts of nests of *Polistes* wasps. *Ann. Entomol. Soc. Am.* **61**, 1528-1539.

Nelson, J. M. (1969). External morphology of larvae in the genus *Polistes* (Hymenoptera: Vespidae) in the Nearctic region. Ph.D. Dissertation, Southern Illinois University, Carbondale.

Nelson, J. M. (1971). Nesting habits and nest symbionts of *Polistes erythrocephalus* Latreille (Hymenoptera: Vespidae) in Costa Rica. *Rev. Biol. Trop.* **18**, 89-98.

Nelson, J. M., and Crowell, R. (1970). Notes on the nests and nesting habits of *Polistes* (Hymenoptera: Vespidae) in Big Bend National Park, Texas. *W. Va. Acad. Sci., Biol. Sect.* **42**, 57-60.

Noonan, K. M. (1978). Sex ratio of parental investment in colonies of the social wasp *Polistes fuscatus. Science* **199**, 1354-1356.

Pagden, H. T. (1958). Some Malayan social wasps. *Malay. Nat. J.* **12**, 131-148.

Pagden, H. T. (1976). A note on colony founding by *Ropalidia (Icarielia) timida* van der Vecht. *Proc. K. Ned. Akad. Wet., Ser. C* **79**, 508-509.

Pardi, L. (1948). Dominance order in *Polistes* wasps. *Physiol. Zool.* **21**, 1-13.

Pardi, L. (1977). Su alcuni aspetti della biologia di *Belonogaster* (Hymenoptera, Vespidae). *Boll. Ist. Entomol. Univ. Studi Bologna* **33**, 281-299.

Pardi, L., and Marino Piccioli, M. T. (1970). Studi sulla biolologia di *Belonogaster* (Hymenoptera, Vespidae). Z. Differenziamento castale incipiente in *B. griseus* (Fab.). *Monit. Zool. Ital.* [N.S.] *Suppl.* **3**, 235-265.

Pekkarinen, A. (1973). Suomen yhteiskunta-ampiaisista. *Luonnon Tutkija* **77**, 12-19.

Perna, B., Croitoru, N., and Ishay, J. (1977). Dominance and hierarchy in *Polistes* colonies: Correlation with photoelectric properties of the 'yellow strips.' *Proc. Int. Congr.—Int. Union Study Soc. Insects, 8th, 1977* p. 155.

Poinar, G. O., Jr., and Ennik, F. (1972). The use of *Neoplectana carpocapsae* (Steinernematidae: Rhabditoidea) against adult yellowjackets (*Vespula* spp., Vespidae: Hymenoptera). *J. Invertebr. Pathol.* **19**, 331-334.

Post, D. C., and Jeanne, R. L. (1980). Morphology of the sternal glands of *Polistes fuscatus* and *P. canadensis* (Hymenoptera: Vespidae). *Psyche* **87**, 49-58.

Post, D. C., and Jeanne, R. L. (1981). Colony defense against ants by *Polistes fuscatus* (Hymenoptera: Vespidae) in Wisconsin. *J. Kansas Entomol. Soc.* **54**, 599-615.

Potter, N. B. (1964). A study of the biology of the common wasp, *Vespula vulgaris* L., with special reference to the foraging behaviour. Ph.D. Dissertation, Bristol University, England.

Pratte, M., and Gervet, J. (1980). Influence des stimulations sociales sur la Persistance du Comportement de Foundation chez la Guepe Poliste, *Polistes gallicus* L. (Hymen. Vesp.). *Insectes Soc.* **27**, 108-126.

Rabb, R. L. (1960). Biological studies of *Polistes* in North Carolina (Hymenoptera: Vespidae). *Ann. Entomol. Soc. Am.* **53**, 111-121.

Rabb, R. L., and Lawson, F. R. (1957). Some factors influencing the predation of *Polistes* wasps on the tobacco hornworm. *J. Econ. Entomol.* **50**, 778-784.

Rau, P. (1933). "The Jungle Bees and Wasps of Barro Colorado Island." Phil Rau, Kirkwood, Missouri.

Rau, P. (1938). Studies in the ecology and behavior of *Polistes* wasps. Part I. *Bull. Brooklyn Entomol. Soc.* **33**, 224-235.

Rau, P. (1939). Studies in the ecology and behavior of *Polistes* wasps. Part II. *Bull. Brooklyn Entomol. Soc.* **34**, 36-44.

Rau, P. (1941). The swarming of *Polistes* wasps in temperate regions. *Ann. Entomol. Soc. Am.* **34**, 580-584.

Reed, H. C. (1978). The nesting ecology of paper wasps (*Polistes*) in a Texas urban area. M.S. Thesis, Texas A&M University, College Station.

Reed, H. C. (1982). Behavior and biology of *Vespula acadica* (Sladen) and the obligate social parasite, Vespula austriaca (Panzer). Ph.D. Dissertation. Washington State University, Pullman, Washington.

Reed, H. C., and Vinson, S. B. (1979a). Nesting ecology of paper wasps (*Polistes*) in a Texas urban area (Hymenoptera: Vespidae). *J. Kans. Entomol. Soc.* **52**, 673–689.

Reed, H. C., and Vinson, S. B. (1979b). Observations of the life history and behavior of *Elasmus polistis* Burks (Hymenoptera: Chalcidoidea: Eulophidae). *J. Kans. Entomol. Soc.* **52**, 247–257.

Reed, H. C., Akre, R. D., and Garnett, W. B. (1979). A North American host of the yellowjacket social parasite *Vespula austriaca* (Panzer) (Hymenoptera: Vespidae).*Entomol. News* **90**, 110–113.

Richards, O. W. (1962). "A Revisional Study of the Masarid Wasps (Hymenoptera: Vespidae)." British Museum, London.

Richards, O. W. (1969). The biology of some W. African social wasps. *Mem. Soc. Entomol. Ital.* **48**, 79–93.

Richards, O. W. (1971). The biology of the social wasps (Hymenoptera: Vespidae). *Biol. Rev. Cambridge Philos. Soc.* **46**, 483–528.

Richards, O. W. (1973). The subgenera of *Polistes* Latreille (Hymenoptera: Vespidae). *Rev. Bras. Entomol.* **17**, 85–104.

Richards, O. W. (1978a). "The Social Wasps of the Americas Excluding the Vespinae." Br. Mus. (Nat. Hist.), London.

Richards, O. W. (1978b). The Australian social wasps (Hymenoptera: Vespidae). *Aust. J. Zool., Suppl.* **61**, 1–132.

Richards, O. W., and Richards, M. J. (1951). Observations on the social wasps of South America (Hymenoptera: Vespidae). *Trans. R. Entomol. Soc. London* **102**, 1–170.

Rivnay, E., and Bytinski-Salz, H. (1949). The Oriental hornet (*Vespa orientalis* F.) its biology in Israel. *Bull., Agric. Res. Stn., Rehovoth* **52**, 1–32 (in Hebrew).

Robson, C. (1898). *Vespa austriaca,* a cuckoo-wasp. *Sci. Gossip* [N.S.] **5**, 69–73.

Rogers, C. J. (1972). Flight and foraging patterns of ground-nesting yellowjackets affecting toxic baiting control programs. *Proc. Pap. Annu. Conf. Calif. Mosq. Vector Control Assoc.* **40**, 130–132.

Roubaud, E. (1916). Recherches biologiques sur les guêpes solitaires et sociales d'Afrique. *Ann. Sci. Nat., Zool. Biol. Anim.* [10]**1**, 107–157.

Roush, C. F., and Akre, R. D. (1978a). Nesting biologies and seasonal occurrence of yellowjackets in northeastern Oregon forests (Hymenoptera: Vespidae). *Melanderia* **30**, 57–94.

Roush, C. F., and Akre, R. D. (1978b). Impact of chemicals for control of the Douglas-fir tussock moth upon populations of ants and yellowjackets (Hymenoptera: Formicidae, Vespidae). *Melanderia* **30**, 95–110.

Sakagami, S. F., and Fukushima, K. (1957a). *Vespa dybowskii* André as a facultative temporary social parasite. *Insectes Soc.* **4**, 1–12.

Sakagami, S., and Fukushima, K. (1957b). Some biological observations on a hornet, *Vespa tropica* var. *pulchra* (Du Buysson), with special reference to its dependence on *Polistes* wasps (Hymenoptera). *Treubia* **24**, 73–82.

Sakagami, S. F., and Yoshikawa, K. (1968). A new ethospecies of *Stenogaster* wasps from Sarawak, with a comment on the value of ethological characters in animal taxonomy. *Annot. Zool. Jpn.* **41**, 77–84.

Sandeman, R. G. (1938). The swarming of the males of *Vespula sylvestris* (Scop.) around a queen. *Proc. R. Entomol. Soc. London, Ser. A* **13**, 87–88.

Saslavasky, H., Ishay, J., and Ikan, R. (1973). Alarm substances as toxicants of the Oriental hornet, *Vespa orientalis. Life Sci.* **12**, 135–144.

Schaefer, P. W. (1977). Attacking wasps, *Polistes* and *Therion,* penetrate silk nests of fall webworm. *Environ. Entomol.* **6**, 591.

Scheven, J. (1958). Beitrag zur Biologie der Schmarotzerfeldwespen *Sulcopolistes atrimandibularis* Zimm., *S. semenowi* F. Morawitz und *S. sulcifer* Zimm. *Insectes Soc.* **54**, 409–437.

Schremmer, F. (1972). Beobachtungen zur Biologie von *Apoica pallida* (Olivier, 1791), einer neotropischen sozialen Faltenwespe (Hymenoptera: Vespidae). *Insectes Soc.* **19**, 343–357.

Schremmer, F. (1977). Das Baumrinden-Nest der neotropischen Faltenwespe *Nectarinella championi,* umgeben von einem Leimring als Ameisen-Abwehr (Hymenoptera: Vespidae). *Entomol. Ger.* **3**, 344–355.

Schremmer, F. (1978). Eine neotropische Faltenwespen-Art (Hymenoptera: Vespidae), die Buckelzirpen-Larven (Homoptera: Membracidae) bewacht und deren Honigtau sammelt. *Entomol. Ger.* **4**, 183–186.

Smithers, C. N., and Halloway, G. A. (1978). Establishment of *Vespula germinica* (Fabricius) (Hymenoptera: Vespidae) in New South Wales. *Aust. Entomol. Mag.* **5**, 55–60.

Snelling, R. R. (1954). Wasps of the genus *Polistes* in California and Arizona (Hymenoptera: Vespidae). *J. Kans. Entomol. Soc.* **27**, 151–155.

Snelling, R. R. (1970). The social wasps of lower California, Mexico (Hymenoptera: Vespidae). *Contrib. Sci. Los Angeles County Mus. Nat. Hist.* **197**, 1–20.

Snelling, R. R. (1974). Changes in the status of some North American *Polistes* (Hymenoptera: Vespidae). *Proc. Entomol. Soc. Wash.* **76**, 476–479.

Soika, A. G. (1975). Vespidological notes XXXVII, New *Polistes* from the Australian continent (Hymenoptera: Vespidae). *Boll. Soc. Entomol. Ital.* **107**, 20–25 (in Italian).

Spradbery, J. P. (1971). Seasonal changes in the population structure of wasp colonies (Hymenoptera: Vespidae). *J. Anim. Ecol.* **40**, 501–523.

Spradbery, J. P. (1973). "Wasps: An Account of the Biology and Natural History of Solitary and Social Wasps." Univ. of Washington Press, Seattle.

Spradbery, J. P. (1975). The biology of *Stenogaster concinna* van der Vecht with comments on the phylogeny of Stenogastrinae (Hymenoptera: Vespidae). *J. Aust. Entomol. Soc.* **14**, 309–318.

Stage, G. I., and Slobodchikoff, C. N. (1962). New distribution and host record of *Bareogonalos canadensis* (Harrington) (Hymenoptera: Trigonalidae and Vespidae). *Pan-Pac. Entomol.* **38**, 97–98.

Starr, C. K. (1976). Nest reutilization by *Polistes metricus* (Hymenoptera: Vespidae) and possible limitation of multiple foundress associations by parasitoids. *J. Kans. Entomol. Soc.* **49**, 142–144.

Starr, C. K. (1978). Nest reutilization in North American *Polistes* (Hymenoptera: Vespidae): Two possible selective factors. *J. Kans. Entomol. Soc.* **51**, 394–397.

Strassman, J. E. (1979). Honey caches help female paper wasp (*P. annularis*) survive Texas winters. *Science* **204**, 207–209.

Strassman, J. E. (1981). Evolutionary implications of early male and satellite nest production in *P. exclamans* colony cycles. *Behav. Ecol. Sociobiol.* **8**, 55–64.

Suzuki, T. (1980). Flesh intake and production of offspring in colonies of *Polistes chinensis antennalis* (Hymenoptera, Vespidae). *Kontyû* **48**, 149–159.

Suzuki, T. (1981). Flesh intake and production of offspring in colonies of *Polistes chinensis antennalis* (Hymenoptera, Vespidae). II. Flesh intake and production of reproductives. *Kontyû* **49**, 283–301.

Taylor, L. H. (1939). Observations on social parasitism in the genus *Vespula* Thomson. *Ann. Entomol. Soc. Am.* **32**, 304–315.

Thomas, C. R. (1960). The European wasp (*Vespula germanica* Fab.) in New Zealand. *Inf. Ser.—N. Z. Dep. Sci. Ind. Res.* **27**, 1-74.

Tissot, A. N., and Robinson, F. A. (1954). Some unusual insect nests. *Fla. Entomol.* **37**, 73-92.

Turillazzi, S., and Pardi, L. (1977). Body size and hierarchy in polygynic nests of *Polistes gallicus* (L.) (Hymenoptera: Vespidae). *Monit. Zool. Ital.* [N.S.] **11**, 101-112.

Turillazzi, S., and Pardi, L. (1981). Ant guards on nests of *Parischnogaster nigricans serrei* (Buysson) (Stenogastrinae). *Monitore Zool. Ital.* [N.S.] **15**, 1-7.

Turillazzi, S., and Ugolini, A. (1978). Nest defense in European *Polistes* (Hymenoptera: Vespidae). *Monit. Zool. Ital.* [N.S.] **12**, 72.

van der Vecht, J. (1936). Some further notes on *Provespa,* Ashm. (Hym., Vespidae). *J. F. M. S. Mus.* **18**, 159-166.

van der Vecht, J. (1941). The Indo-Australian species of the genus *Ropalidia* (= *Icaria*) (Hymenoptera: Vespidae) (First part). *Treubia* **18**, 103-191.

van der Vecht, J. (1957). The vespinae of the Indo-Malayan and Papuan areas (Hymenoptera: Vespidae). *Zool. Verh. Leiden* **34**, 1-83.

van der Vecht, J. (1959). Notes on oriental Vespinae, including some species from China and Japan (Hymenoptera: Vespidae). *Zool. Meded.* **36**, 205-232.

van der Vecht, J. (1962). The Indo-Australian species of the genus *Ropalidia* (*Icaria*) (Hymenoptera, Vespidae) (Second Part). *Zool. Verh., Leiden* **57**, 1-72.

van der Vecht, J. (1965). The geographical distribution of the social wasps (Hymenoptera: Vespidae). *Proc. Int. Congr. Entomol., 12th, 1964* pp. 440-441.

van der Vecht, J. (1966). The East-Asiatic and Indo-Australian species of *Polybioides* Buysson and *Parapolybia* Saussure (Hym., Vespidae). *Zool. Verhandel.* **82**, 1-45.

van der Vecht, J. (1967). Bouwproblemen van sociale wespen. *Versl. Gewone Vergad. Afd. Natuurkd., K. Ned. Akad. Wet.* **76**, 59-68.

van der Vecht, J. (1971). The subgenera *Megapolistes* and *Stenopolistes* in the Solomon Islands. *In* "Entomological Essays to Commemorate the Retirement of Professor K. Yasumatsu" (S. Asahina *et al.,* eds.), pp. 87-106. Hokuryukan, Tokyo.

van der Vecht, J. (1975). A review of the genus *Stenogaster* Guerin (Hymenoptera: Vespidae). *J. Aust. Entomol. Soc.* **14**, 283-308.

van der Vecht, J. (1977a). Studies of Oriental Stenogastrinae (Hymenoptera: Vespidae). *Tijdschr. Entomol.* **120**, 55-75.

van der Vecht, J. (1977b). Important steps in the evolution of nest construction in social wasps. *Proc. Int. Congr. Int. Union Study Soc. Insects, 8th, 1977* p. 319.

Wade, W. E., and Nelson, J. M. (1978). Further evidence for separation of the cryptic red wasps: *Polistes carolina* (Linné) and *Polistes perplexus* Cresson (Hymenoptera: Vespidae). *Southwest. Entomol.* **3**, 73-75.

Wafa, A. K., and Sharkawi, S. G. (1972). Contribution to the biology of *Vespa orientalis* Fab. *Bull. Soc. Entomol. Egypte* **56**, 219-226.

Wafa, A. K., El-Borolossy, F. M., and Sharkawi, S. G. (1968). Studies on *Vespa orientalis* F. *Bull. Soc. Entomol. Egypte* **52**, 9-27.

Wagner, R. E. (1978). The genus *Dolichovespula* and an addition to its known species of North America. *Pan-Pac. Entomol.* **54**, 131-142.

Wagner, R. E., and Reierson, D. A. (1969). Yellowjacket control by baiting. I. Influence of toxicants and attractants on bait acceptance. *J. Econ. Entomol.* **62**, 1192-1197.

West Eberhard, M. J. (1969). The social biology of polistine wasps. *Misc. Publ. Mus. Zool. Univ. Mich.* No. 140. 101 p.

West Eberhard, M. J. (1977). Morphology and behavior in the taxonomy of *Microstigmus* wasps. *Proc. Int. Cong. Int. Union Study Soc. Insects, 8th, 1977* pp. 123-125.

West Eberhard, M. J. (1978). Polygyny and the evolution of social behavior in wasps. *J. Kans. Entomol. Soc.* **51**, 832-856.

Weyrauch, W. (1937). Zur Systematik und Biologie der Kuckuckswespen *Pseudovespa, Pseudovespula* und *Pseudopolistes*. *Zool. Jahrb., Abt. Syst.* (Oekol.), *Geogr. Tiere* **70**, 243–290.

White, L. D., Hutt, R. B., and Butt, B. A. (1969). Releases of unsexed gamma-irradiated codling moths for population suppression. *J. Econ. Entomol.* **62**, 795–798.

Williams, F. X. (1928). Studies in tropical wasps—their hosts and associates. *Bull. Exp. Stn. Hawaii. Sugar Planters' Assoc. Entomol. Ser.* **19**, 1–179.

Wilson, E. O. (1971). "The Insect Societies." Belknap Press, Cambridge, Massachusetts.

Wilson, M. V. H. (1977). New records of insect families from the freshwater middle Eocene of British Columbia. *Can. J. Earth Sci.* **14**, 1139–1155.

Windsor, D. M. (1972). Nesting association between two neotropical polybiine wasps (Hymenoptera, Vespidae). *Biotropica* **4**, 1–3.

Yamane, S. (1969). Preliminary observations on the life history of two polistine wasps, *Polistes snelleni* and *P. biglumis* in Sapporo, northern Japan. *J. Fac. Sci., Hokkaido Univ., Ser. 6* **17**, 78–105.

Yamane, S. (1971). Daily activities of the founding queens of two *Polistes* wasps, *P. snelleni* and *P. biglumis* in the solitary stage (Hymenoptera, Vespidae). *Konchu* **39**, 203–217.

Yamane, S. (1972). Life cycle and nest architecture of *Polistes* wasps in the Okushiri Island, northern Japan (Hymenoptera: Vespidae). *J. Fac. Sci., Hokkaido Univ., Ser. 6* **18**, 440–459.

Yamane, S., and Kawamichi, T. (1975). Bionomic comparison of *Polistes biglumis* (Hymenoptera, Vespidae) at two different localities in Hokkaido, northern Japan, with reference to its probable adaptation to cold climate. *Kontyû* **43**, 214–232.

Yamane, S., and Okazawa, T. (1977). Some biological observations on a paper wasp, *Polistes* (*Megapolistes*) *tepidus malayanos* Cameron (Hymenoptera: Vespidae) in New Guinea. *Kontyû* **45**, 283–299.

Yamane, S., and Yamane, Sk. (1979). Polistine wasps from Nepal (Hymenoptera: Vespidae). *Insect Matsumurana* **15**, 1–37.

Yamane, Sk. (1973). Discovery of a pleometrotic association in *Polistes chinensis antennalis* Pérez (Hymenoptera: Vespidae). *Life Study* **17**, 79–80.

Yamane, Sk. (1974). On the genus *Vespa* (Hymenoptera: Vespidae) from Nepal. *Kontyû* **42**, 29–39.

Yamane, Sk. (1975). Taxonomic notes on the subgenus *Boreovespula* Blüthgen (Hymenoptera: Vespidae) of Japan, with notes on specimens from Sakhalin. *Kontyû* **43**, 343–355.

Yamane, Sk. (1976). Morphological and taxonomic studies on vespine larvae, with reference to the phylogeny of the subfamily Vespinae (Hymenoptera: Vespidae). *Insecta Matsumurana* [N.S.] **8**, 1–45.

Yamane, Sk. (1977). A young nest of *Vespa xanthoptera* Cameron built within a bamboo tube in Honshu (Hymenoptera: Vespidae). *New Entomol.* **26**, 19 (in Japanese).

Yamane, Sk., and Kanda, E. (1979). Notes on the hibernation of some vespine wasps in northern Japan (Hymenoptera: Vespidae). *Kontyû* **47**, 44–47.

Yamane, Sk., and Kamijo, K. (1976). Social wasps visiting conifer plantations in Hokkaido, northern Japan (Hymenoptera: Vespidae). *Insecta Matsumurana* [N.S.] **8**, 59–71.

Yamane, Sk., and Makino, S. (1977). Bionomics of *Vespa analis insularis* and *V. mandarinia latilineata* in Hokkaido, northern Japan, with notes on vespine embryo nests (Hymenoptera: Vespidae). *Insecta Matsumurana* [N.S.] **12**, 1–33.

Yamane, Sk., and Yamane, S. (1975). A new trigonalid parasite (Hymenoptera: Trigonalidae) obtained from *Vespula* nests in Taiwan. *Kontyû* **43**, 456–462.

Yamane, Sk., Wagner, R. E., and Yamane, S. (1980). A tentative revision of the subgenus *Paravespula* of eastern Asia (Hymenoptera: Vespidae). *Insecta Matsumurana* [N.S.] **19**, 1–46.

Yamasaki, M., Hirose, Y., and Takagi, M. (1978). Repeated visits of *Polistes jadwige* Dalla Torre (Hymenoptera: Vespidae) to its hunting site. *Jpn. J. Appl. Entomol. Zool.* **22**, 51–55.

Yoshikawa, K. (1962a). Introductory studies on the life economy of polistine wasps. VI. Geographical distribution and its ecological significances. *J. Biol. Osaka City Univ.* **13**, 19-43.

Yoshikawa, K. (1962b). Introductory studies on the life economy of polistine wasps. VII. Comparative consideration and phylogeny. *J. Biol. Osaka City Univ.* **13**, 45-64.

Yoshikawa, K. (1963a). Introductory Studies on the life economy of polistine wasps. III. Social stage. *J. Biol. Osaka City Univ.* **14**, 63-66.

Yoshikawa, K. (1963b). Introductory studies on the life economy of polistine wasps. IV. Analysis of social organization. *J. Biol. Osaka City Univ.* **14**, 67-85.

Yoshikawa, K. (1964). Predatory hunting wasps as the natural enemies of insect pests in Thailand. *Nat. Life Southeast Asia, Tokyo* **3**, 391-398.

Yoshikawa, K., Ohgushi, R., and Sakagami, S. F. (1969). Preliminary report on entomology of the Osaka City University 5th Scientific Expedition to Southeast Asia 1966—with descriptions of two new genera of stenogasterine wasps by J. van der Vecht. *Nat. Life Southeast Asia* **6**, 153-200.

Young, A. M. (1979). Attacks by the army ant *Eciton burchelli* on nests of the social paper wasp *Polistes erythrocephalus* in northeastern Costa Rica. *J. Kans. Entomol. Soc.* **52**, 759-768.

Zikán, J. F. (1949). O genero *Mischocyttarus* Saussure (Hymenoptera, Vespidae), con a descrição de 82 espécies novas. *Bol. Parq. Nac. Itatiaia,* Rio de Janeiro **1**, 1-251.

2

Ants: Foraging, Nesting, Brood Behavior, and Polyethism

JOHN H. SUDD

I. INTRODUCTION

Ants differ from other social insects in a number of ways, some of which have far-reaching effects on their social behavior, and on the interpretation of it. First, there are no nonsocial or presocial ants, although a number of such

SOCIAL INSECTS, VOL. IV
Copyright © 1982 by Academic Press, Inc.
All rights of reproduction in any form reserved.
ISBN 0-12-342204-3

forms exist in wasps and bees. Second, ants are taxonomically a much more compact group; the superfamily Formicoidea contains only the single family Formicidae while Apoidea has been divided into 10 families, three of which contain eusocial forms. Eusociality in bees indeed appears to have evolved independently at least eight times (Wilson, 1971). Termites, like ants, are all eusocial, but make up an entire order of insects, divided into five or six families. In spite of the taxonomic compactness of the ants there are more species of ants than of any of the other groups of social insects, they are probably more numerous as individuals, and as a group more widespread geographically.

The behavior of ants differs in three important features from the behavior of one or more of the other eusocial groups. Ants have a far greater variety of diet than bees, which are specialized to collect pollen as a food source for their young, or termites, which base their nutrition on cellulose. Ants complement, or in some cases supersede, a basically predaceous way of life with many other food sources. In some cases (collection of honeydew from Homoptera, seed collection, and fungus culture) their behavior is specialized. Although some species are dependent on a single type of food source, the majority of species, and indeed of individual colonies, maintain a mixed diet. Therefore a large part of the following discussion concerns foraging and foraging systems, although a detailed account of the fungus-culturing Attine ants and the predaceous Dorylinae may be found in other chapters.

The nests of ants and the manner in which they are constructed are also peculiar to ants. Again there is a wide variation, with the majority of nests excavated in soil or rotten wood. However, there are many specialized cases, particularly among arboreal ants, of nests built in other ways or from other materials. Although solitary bees and wasps build earthen nests all the eusocial forms use compounded substances (e.g., paper, wax, propolis) in the construction of their nests. The nests of termites, although also hypogeal or arboreal, involve more compounded materials than the nests of ants, and tend to have a more elaborate structure. In detail, the nests of ants contrast with those of wasps and many bees in which each larvae is raised in an individual cell. Ant brood is reared in chambers or galleries (the use of the word cell for these chambers is confusing) as a group, often segregated with respect to age, size, or caste. Bumble bees use communal cells for small groups of larvae but these are raised in their natal cell and are not moved about the nest as the brood of ants is. The situation in termites is of course entirely different from any of the Hymenoptera because immature termites are active and mobile and play a part in the social life of the colony. Among ants the larvae are dependent on nurse workers for food and movement and have far more contact with them than the cell-bound larvae of other eusocial Hymenoptera. Conversely the workers have far more contact with the larvae and more channels of communication with them.

Malyshev (1968) made an important point of this method of brood care, and believed that the habit of laying eggs in loose heaps separated the ants behaviorally from all other eusocial Hymenoptera and indeed from all Aculeata. "This difference is undoubtedly very deep-rooted, and we have no grounds for turning to the wasps or bees or, from the stand-point of the evolution of the ants, for looking for solitary forms among the ants or their ancestors: no such forms ever existed." That is, Malyshev believed that the ancestral habits of ants involved a female which laid a batch of eggs in group contact with a prey or preys. He attempted to find this situation in Chalcidoidea parasitizing the egg clumps of phytophagous insects or mantid oothecae, or in Bethyloids parasitizing insects much larger than themselves. Malyshev's further development of this theme led him to suppose that "proformicoid" colonies developed around a large prey in its natural habitat (for instance rotting wood) and that, after exhausting the prey, but before hunting started, the incipient colony fed on fungus growing in the burrow. His evidence that present day queen ants feed on fungus in this way is somewhat limited, though the idea has the attraction of giving ants a mixed diet from the very start of their evolution.

Whatever the importance of their peculiar method of brood care may have been in the origin of ants, there is no doubt that it has given them the possibility, through increased and more varied contact between brood and adults, to control development and thus morphological and functional polymorphism. It has also increased the range of services which workers offer to the brood and may therefore have increased their tendency toward worker polyethism.

This chapter highlights some of the important features of these three categories of ant behavior: foraging, nesting, and brood care. This is done in three stages. The first is the description of the behavioral acts of individuals and the clues and releasers which guide their temporal and spatial occurrence. Second, consideration is given to the distribution of behavioral acts or sets of acts among the different members of the colony. This problem of behavioral variation (or polyethism) involves some discussion of the sources of intracolony variation. Finally, discussion turns to the functioning of these behaviors. However, it is only in foraging that enough information is available to follow this scheme in any detail. Its value, lies in illustrating the type of information which is still lacking.

II. FOOD COLLECTION

A. Diet

Since nearly all aculeate Hymenoptera are insectivorous, it is natural to think that ants were also primitively insectivorous. This notion is confirmed by the fact that the vast majority of ant species take insect prey as their main source of proteins. Yet given the primitive association of ants with soil and rotting wood it is worth considering the possibility that their original diet was more mixed. Malyshev (1968) believed that he had shown that queens of *Lasius niger* feed on fungi which grow in their claustral chambers during the foundation of new colonies. Fungus eating is certainly not confined to the specialized attine fungus-growing ants; it is also a minor feature of the diet of *Formica rufa* group ants (Wellenstein, 1952). The decomposition of plant litter and wood involves a complex of fungal and arthropod species and it would be in keeping with the mixed feeding habits of many modern ants to exploit both. However, evidence that fungi are important in the diet of most ants is unconvincing.

Among predatory ants, anatomical or behavioral specialization to particular prey species appears to be rather uncommon. Nevertheless, some specialist species do exist, mostly as one might expect where the ant fauna is diverse and competition presumably great. For example, *Heteroponera monticola* and *Gnamptogenys perspicans* prey on Diplopoda in Colombian forests and are reported to have developed resistance to the defensive cyanide secretions of this prey (Kempf and Brown, 1970). The dacetine ants are another notable exception. Some members of this tribe have a specialized jaw mechanism which allows the mandibles to be placed in an open position at right angles to the body axis. In this position they are strained by the mandibular adductor muscles, perhaps through a specialized elastic mechanism, and prevented from closing by a portion of the labrum. Sensory hairs which project from the medial surface of the mandibles control the release of the jaws when they touch prey (Brown and Wilson, 1959). Carroll and Janzen (1973) believe that the Dacetini are specialized feeders on the plentiful soil fauna and that they achieve short prey-handling and location times by this method. This does not seem to be entirely born out by the studies of Brown and Wilson (1959) since the specialized long-jawed forms such as *Daceton* and *Strumigenys* are on the whole epigaeic or even arboreal foragers while the short-jawed genera *Smithistruma* and *Trichoscapa* are hypogaeic. Furthermore, although handling time and location time may be short, the actual attack is stealthy, especially in the hypogaeic short-jawed types. This behavior contrasts with that of some other species which also hunt Collembola but make sudden dashes at their prey and often lose them. The long-jawed Dacetini seem in fact to be adapted to a type of prey defence, the sudden jump, rather than to a type of prey. The rarity of this degree of mandibular specialization may be the result of the multiple function of the mandibles of ants in nest excavation and brood handling as well as prey capture. In *Odontomachus,* which has similar mandibles to *Daceton* but is much larger, Sudd (1969) reported that digging was not to any extent abnormal. Studies of competition in ant communities confirm the idea that specialization more often takes the form of reduction in diversity of foraging microhabitats or periods of activity than in reduced prey diversity, though prey size diversity may fall (e.g., Bernstein, 1979; Whitford *et al.*, 1980; Byron *et al.*, 1980). These competitive relations are not simple however (Davidson, 1980), and it is not clear that they lead to genetic change.

B. Prey Recognition

Remarkably little is known about the recognition of prey by ants with a less restricted diet. In many cases the importance of a prey species in the diet of an ant may be determined by its availability in the area searched by that ant, as Gotwald (1974) demonstrated by his comparison of the prey of different African driver ants. The prey of *Anomma gerstaeckeri* and *A. nigricans* in a West African forest habitat contained a high proportion of earthworms, while in a grassland habitat earthworms were rarer as prey and insects and arachnids predominated. Ants which prey on other ants may have specific responses to them or to their pheromones. *Myrmecia gulosa* is reported to react strongly to formic acid-impregnated dummies both in the field and in the laboratory and formic acid and undecane (to which their response was slight) may be alarm pheromones of the species of *Camponotus* on which a colony of *M. gulosa* were known to feed (Haskins *et al.*, 1973). However, the evidence that formic acid is used in prey location by this species is slight and the effects observed could be attributable to general excitement or even to an irritant effect. Other accounts of attacks on ant colonies by *Camponotus rufoglaucus* (Maschwitz, 1974) and *Solenopsis fugax* (Hölldobler, 1973) suggest that this type of chemical recognition may not be uncommon.

Clearer evidence of the existence of specific attractant or phagostimulant substances might be expected from studies of the collection of leaves by attine ants, since in this case the behavior of the prey and the effect of the habitat might be less important. Rockwood (1976) found that though many plants were sampled by *Atta colombica* and *A. cephalotes,* leaves were taken mainly from 31 and 22% of the available plant species by these ants, respectively. However, the formulation of a null hypotheses based on the availability and risk of encounter of different plant species is difficult and complicated by the existence of foraging trails in these species of ant. This means that individual selections of leaves are not independent since one ant leads others to particular plants. As a result there are, for example, differences in the preference of different nests of the related *A. sexdens,* a species which Prado (1973) believes has no innate preferences. Laboratory experiments, designed to discover suitable baits for the control of attine ants, have confirmed this to some extent. Although certain substances, especially grapefruit pith (albedo) are preferentially carried into laboratory nests, the only fractions of extracts of albedo which have been shown to encourage pick-up were carbohydrates and amino acids and not secondary substances. Cherrett and Seaforth (1968) concluded that the attractiveness of grapefruit albedo may lie in the variety of substances it contains rather than in any particular group. They also found that a great heterogeneity of preference existed for the same nest on different days, or where individual nests were homogeneous from day to day differences occurred between different nests, or where different nests were homogeneous preferences varied between the two species studied (*Atta cephalotes* and *Acromyrmex octospinosus*). In a later study Littledyke and Cherrett (1975) found that various sugars promoted both the pick-up of paper discs and drinking of fluids. An "inhibitor" (quinine) did not affect pick-up, but did affect drinking. While these experiments may explain why over 90% of loads taken by *Atta cephalotes* are flowers or leaves they cannot account for the selection of any particular plant species, and the effect of quinine did not suggest that inhibitor substances in plants might protect them, though its effect on leaf-cutting behavior was not tested. Lewis *et al.* (1974) have suggested that diel patterns of leaf collection by *A. cephalotes* may reflect changing concentrations of carbohydrate in leaves.

The means by which prey is recognized is even more obscure in the case of more general predators. Büttner (1974a,b) investigated the effects of physical and structural characteristics of the environment and of the prey on the predatory behavior of *Formica polyctena*. He divided prey capture into three phases: orientation, capture, and transport. The orientation phase, which includes the search and the discovery of the prey object was prolonged in thick grass, so that the distance traveled by ants searching for prey averaged 198 cm in grass compared with 86 cm on bare sand. The chance that prey would not be found at all was 42% in grass and only 2% in sand. The search time was also much reduced when the prey contrasted in color with its background. Moving artificial baits were much more rapidly located than stationary ones. There is, therefore, little doubt that vision is important in the location of prey by this species. The chemical nature of the prey (defensive substances aside) seems of relatively little importance, although this may affect the amount of the prey eaten and the eventual fate of objects carried to the nest (Büttner, 1973). This species carries a large variety of edible and inedible substances to its nest, some of which is used as nest material. Since the worker ant which captures the food does not itself eat it, in many cases the situation is very different from that of food plant recognition by a caterpillar.

Movement can hardly be important in food recognition by ants which feed on dead animals or on seeds. Some ants, such as *Lasius emarginatus* (Müller-Schneider, 1971) which combine scavenging and hunting, seem to be attracted by the oily elaiosomes of seeds such as *Scilla bifolia* and *Veronica*. However the ricinoleic acid that these seeds contain is not essential because *Colchicum* seed, which has an elaiosome but does not contain ricinoleic acid, is also collected. *Allium ursinum* and *Crocus biflorus* seeds have no elaisome and are not carried into the nest. The harvesting ants of semiarid and arid lands collect a wider range of seeds but seem to have fairly definite preferences. In times of scarcity the preferences are much reduced and more seeds of nonpreferred species and more material other than seeds is collected (Rissing and Wheeler, 1976). When seeds are abundant the loads of *Pogonomyrmex occidentalis* are composed of 39% seeds and 24% apparently useless plant litter as well as insects (Rogers, 1974). *Pogonomyrmex badius* workers, which have acquired a preference for wheat seed, will collect nonpreferred seeds if these have been rubbed with wheat grains (Nickle and Neal, 1972). It is not known whether such a specific preference is learned or not. In some species the workers' heads are too large for them to feed directly on seeds, but the larvae have elongate heads which are inserted into the cracked seeds (Went *et al.*, 1971). The acquisition of a learned preference by the workers might not be a simple matter in such cases.

C. Searching for Food

Bernstein (1975) has divided foraging methods into three types: individual search, recruitment systems, and group foraging. In a sense, true individual hunting may never occur and it is certainly a difficult task to demonstrate that the search paths of foraging workers in the same area are strictly independent of one another. A moderate degree of dependence could arise from responses to the same prey or to the same concentrations of prey, or to the movements of fellow workers attacking prey, or to pheromones emitted in combatting prey (Robertson, 1971) or deposited around food sources (Hölldobler, 1971b). However, in the harvesting ants of the Californian desert, Bernstein (1975) was able to show a variation in the degree of independence of foragers. *Pogonomyrmex californicus* and *P. rugosa* live at higher altitudes where rainfall and consequently seed production are somewhat greater. They forage singly each traveling a certain distance from the nest in a more or less straight line, and then searching in a circuitous path. When they find seeds they return to the nest, again more or less in a straight line. *Veromessor pergandei* occurs at high altitudes and also at lower ones where conditions are more arid. At high altitudes *V. pergandei* workers forage singly, but at low altitudes in conditions of seed shortage foragers are confined to columns of ants concentrated in areas where seeds are plentiful. The experimental addition of a circle of seeds round the nest was followed by the break-up of the column pattern and the enlargement of the angular sector around the nest, in which foragers were found. Hölldobler (1976) has reported in detail his investigations of *Pogonomyrmex* spp. both in the laboratory and in the Arizona desert, and reviewed much of the earlier work. Foragers of *P. maricopa* move out in all directions from the nest, evidently using the sun as a compass as well as topographical orientations. However, *Pogonomyrex badius* and *P. rugosus* are confined to trunk routes which lead them in fixed directions from the nest. The trunk routes extend up to 40 m from the nest and are remarkably permanent. Each nest has up to five trunks which sometimes branch further, but trunk routes never cross routes belonging to other nests, and only rarely cross those of other species (Fig. 1). Individual workers of *P. barbatus* and *P. rugosus* are, on the whole, faithful to particular routes and foragers marked on one route are only rarely recaptured on other routes. Experiments with artificial landmarks that were moved and alterations to natural landmarks

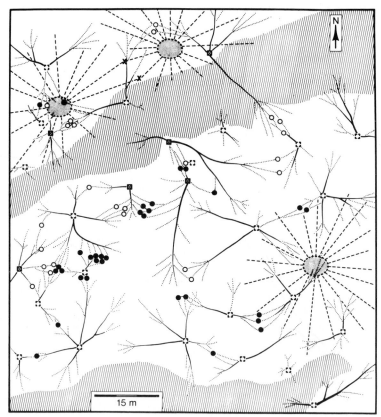

Fig. 1.   The route systems of nests of several species of *Pogonomyrmex* in Arizona, U.S.A. The routes of *P. rugosus* and *P. barbatus* nests never cross those of conspecific nests and rarely those of the other species. *Pogonomyrmex maricopa* nests have no routes. Where routes of *P. rugosus* and *P. barbatus* converge intraspecific (●) or interspecific (○) fights occur: intraspecific fights (×) occur where *P. maricopa* foraging areas touch. (From Hölldobler, 1974.)

showed that the foragers on the route were orientating to the sun, to landmarks, and also to nearer clues, possibly scent marks. Workers of *P.
maricopa,* which never had trunk routes, and of those colonies of *P. barbatus*
which had no routes, oriented on their independent courses by the same
methods. According to Hölldobler the route system channels foraging
workers of *P. rugosus* and *P. barbatus* in definite directions as they leave the
nest and reduces the number of hostile encounters during food collection.
Nests of these two species are considerably closer to nests of their own species
than the mean distance between the nests of *P. maricopa,* the species without
trunk routes.

The same pattern of foragers confined to certain sectors around their nest is
found again in the north temperate *Formica rufa* group. These ants, which are
characteristically found in open woodland or woodland edges, build large
nests from which foragers move out along routes. The routes are more evident
and more precisely followed in some species and some habitats than in others.
In many cases the routes lead to particular trees on which some foragers collect honeydew from Homoptera, and also hunt insects (Fig. 2). However,

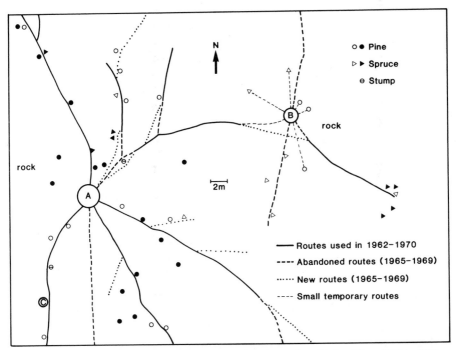

Fig. 2. The route system of a *Formica rufa* colony in southern Finland from 1962 to 1970.
The system includes two nests, A and B, and a deserted nest, C. Nest B was abandoned from 1967
onward, but some of its routes were exploited by nest A. Not all the trees are shown; open sym-
bols represent pines or spruces where aphids were exploited by the ants. (From Rosengren, 1971.)

other foragers leave the foraging routes and hunt on the ground or on low vegetation, as well as collecting plant litter which is incorporated in the thatch of
the nest. Rosengren (1971) conducted very extensive studies of individual
marked workers, and showed that they tended to use and reuse the same route.
Between 65 and 100% of recaptured foragers were found on the same route
where they had been marked (Fig. 3). The percentage fidelity (ratio of marked
ants recaptured where marked to all marked ants recaptured) varied through
the year and was lowest early in the ant season, around May. A group of foragers marked simultaneously in spring retained its fidelity through the following summer, and a fair proportion of workers were recaptured where they had
been marked after as long as 4 months. Rosengren even found evidence of the
fidelity of workers when foraging restarted in spring to the routes where they
had been marked the preceding summer. The routes are thus to some extent
preserved from year to year, since it is the oldest, experienced foragers that are
the earliest to leave the nest in spring. Rosengren's further experiments in
laboratory nests and by altering landmarks around natural nests showed that
the orientation of ants on the routes was primarily controlled by remembered
landmarks. However, further research has shown that scent trails are used by
F. rufa group workers in at least some circumstances (Henquel, 1976;
Horstmann, 1975, 1976a). It must also be added that natural routes in some
species (e.g., *F. polyctena*) and in many seed-collecting ants may form cleared

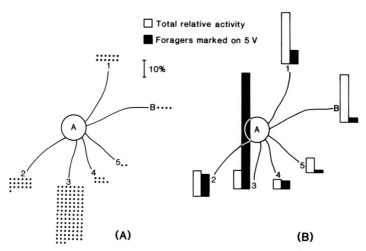

Fig. 3.   Route fidelity in *Formica rufa* foragers from nest A (see Fig. 2). On May 5, 1964, 824
workers were marked on route 3. (A) The distribution of recaptures by routes between May 6 and
May 17, each dot represents the recapture of a marked ant. (B) The percentage distribution of
recaptures on each route compared with the percentage of total forager traffic (sum of 5 min
counts in each direction).

zones on the ground and that this may play a large part in the orientation of workers. This is discussed in detail in Section II,E. It should be noted here that in both cases there is a markedly nonrandom distribution of foragers around the nest (in other words foraging is not independent) but that continuous communication between individual foragers plays a smaller part in establishing this than might be expected.

D. Recruitment

When a number of predatory insects hunt outward from a common nest, they are likely to influence each other even in the absence of specialized communication systems. In laboratory nests of some species, an increase in the number of foragers leaving the nest as a response to an increased number returning appears to occur (Ayre, 1969) without any directional or other guidance; however, it is not clear that this happens in the field. Temporal variation in the rate at which workers pass along a trail (Rosengren, 1971) suggests that workers tend to leave the nest, or to travel on trails in groups or pulses. However, at high rates of flow this sort of nonuniform flow could arise as a result of traffic problems on the routes themselves and other interpretations are possible (Plowright, 1979). In many cases, foragers leave the nest and may be guided to places where food has been found by specialized signals (pheromones, movements, or sounds). The effect of the signal may depend not only on the signal itself (which pheromone, movement, or sound is emitted), but on the place where it is emitted and also on the size, age, or mood (caste in the general sense of Wilson, 1976c) of the responding worker. Thus the mandibular gland secretion of *Pogonomyrmex badius* can attract and excite other workers to prey outside the nest over distances up to 100 mm, as well as mediating alarm and defense behavior in the nest itself (Wilson, 1971). The same species also has specialized chemical signals using pheromones produced in the venom gland and in Dufour's gland (Hölldobler, 1971b) which it uses to communicate directional information. Very often the functions of increasing the number of foragers (recruitment) and of promoting their arrival at a particular place (guidance) cannot be separated clearly.

Directional information is imparted in two ways by ants: by direct means when one ant leads (tandem following Wilson, 1971) or carries another (Duelli, 1973), and by indirect means usually involving chemical trails. Tandem following has been described in a number of species in several subfamilies. In *Bothroponera tesserinodis* (Maschwitz *et al.*, 1974) one ant induces another to follow it by a light pull with its mandibles. The second ant then follows its leader, keeping close antennal contact with it. This tactile contact is necessary to keep the leader running, but both tactile contact and a surface pheromone are needed by the follower. Secretions of the sting or hindgut (which produce chemical trails in other species) are in no way implicated. *Bothroponera tesserinodis* lives in small colonies and tandem running seems to be used to bring workers (possibly inexperienced ones with no established preference for a particular area) to places where prey may be found. Once there they hunt separately from the leader. In *Camponotus sericeus* (Hölldobler *et al.*, 1974) tandem running is brought about in a similar way but the hindgut is the source of the pheromone in this case. In this species and in *Leptothorax acervorum* and *L. nylanderi* (Möglich, 1975) tandem recruitment seems to function as a means of leading recruits into new feeding grounds, or even to new nest sites, and not as a means of guiding ants to a specific find of food. Carrol and Janzen (1973) warned against viewing tandem running as related in an evolutionary sense to trail laying. This warning is emphasized by the fact that, in *Bothroponera tesserinodis* at least, the behavior depends on cuticular secretions which are poorly situated to leave trails. Sudd (1960a) showed that if *Monomorium pharaonis* workers are following a nestmate closely on a journey from the nest, then they are influenced in their choice at a branching point in a trail by the choice made by a nestmate. Since it is not known what part is played by trail pheromone concentrations in this, its relation to tandem following is perhaps not as close as Sudd originally suggested. Duelli (1973) was able to show that carriage of one worker of *Cataglyphis bicolor* over a distance of 10 m by another could act as a means of communicating a direction of travel. He was able to show that the horizontal positions of the head of carrier and carried ant were substantially the same so that the carried ant could acquire the sun-compass orientation.

In true trail laying the secretions of special glands are placed on the substratum and are followed by other ants to a source of food, or in some cases (exploratory trails of Wilson, 1962) to new areas. The simplest instances are found in the recruitment and guidance of members of a hungry laboratory colony over a short distance by a scout which has found food. Under these conditions, Wilson (1962) showed that extracts of Dufour's gland alone were sufficient both to excite workers of *Solenopsis invicta* to leave the nest and to guide them to a piece of food. In his experiments, the number of workers recruited depended on the amount of pheromone blown into the nest. Wilson postulated that in natural conditions the amount of pheromone released in the nest depends on the number of successful scouts or recruits which have fed and returned to the nest. The number of recruits reaching the food rose at first according to a positively accelerating curve and leveled off at a value that was dependent on the size of the piece of food. Wilson concluded that the pheromone was "the paramount and probably sole, mode of communication in the mass response" leading to the build up of numbers at the food. In the laboratory, evaporation of the pheromone limited the length of the primary trail (laid by the first discoverer of a food bait) to about 50 cm, but longer trails could be laid by groups of workers.

The development of methods for demonstrating trail pheromones has distracted attention to some extent from the complexity of the behavior involved, much of which was clearly described in Wilson's original report (Wilson, 1962). *Solenopsis invicta* uses its sting to deposit the pheromone, and forms intermittent long streaks on the substratum (Hangartner, 1969a). *Myrmica rubra* also uses the secretion of Dufour's gland to attract other workers to prey, but closely followed trails are less usual in this ant. The movements of its gaster as it deposits pheromone can be recorded by the traces it leaves on smoked paper. Near to the prey there are short traces (type II) which may represent the deposition of Dufour's gland secretion. On the track of the ant's return from the prey to the nest longer marks (type I) are found (Cammaerts-Tricot, 1974b). Since elongated artificial streaks of venom gland extract are followed more closely than similar streaks of Dufour's gland extract, Cammaerts-Tricot (1974a) concluded that they represent deposition of the secretion of this gland. On returning from the nest to the prey, scouts make short (type II) marks which are taken to indicate the deposition of the attractant secretion of Dufour's gland. The secretion of Dufour's gland is complex; however, the attractant substance is thought to be acetaldehyde in synergy with ethanol. Other constituents seemed to release the deposition of further quantities of Dufour's gland secretion by ants attracted to type I traces and thus reinforce their effect (Cammaerts-Tricot *et al.,* 1976). Multiple pheromones seem to occur also in *Monomorium pharaonis* (Ritter *et al.,* 1975) and in *Lasius fuliginosus* (Huwyler *et al.,* 1975) though it is not clear whether the ants are able to control the composition of the mixture. Löfquist (1977) discussed this problem in relation to the alarm pheromones of *Formica* spp. Cammaerts-Tricot's conclusions depends on her ability to make separate extracts from venom and Dufour's glands in *Myrmica rubra,* a species which is larger than many trail layers. Her assignment of functions to the component of the secretions are necessarily based on laboratory tests which may not always reflect the function of the substance in natural conditions. The behavior of an ant following a trail is particularly hard to analyse since it contains elements both of orientation and of sustained movement. Bossert and Wilson (1963) gave a more satisfying account of the orientation of male moths; in this case the pheromone induces movement which is normally orientated by wind direction. Ant trails are more complicated since, whatever other means of orientation may be used simultaneously, there is no doubt that the trail itself is responsible for much of the orientation. Hangartner (1967) has given a convincing account of biantennal orientation to the scent gradient across the trail of *Lasius fuliginosus.*

Apart from varying the chemical composition of the trail, ants which lay trails can control the amount of substance discharged. In Wilson's original scheme each trail-laying ant deposited a fixed amount of substance and quantitative variation occurred as the number of recruits involved increased or decreased. Hangartner (1969b, 1970) showed that the amount of material deposited by *Acanthomyops interjectus* from its hindgut is sometimes so great that a spreading circular mark is left on a sooted surface, as well as the scrapelike marks shown by some other species. By offering sucrose solutions to individual workers from starved laboratory colonies and recording their traces, Hangartner was able to show three ways in which the ants could vary the amount of pheromone deposited according to the concentration of sugar given them. When he used a 1 *M* solution, the percentage of ants which subsequently left marks on a sooty surface, the number of marks left on a 35-mm length of trail, and the percentage of marks which were of the spreading type (strong marks) was higher than when 0.01 *M* sucrose was used. Unfortunately the response of *A. interjectus* workers to the marks in the laboratory was not very clear and Hangartner was unable to show that strong marks elicited a stronger response. A similar mechanism is described in *Atta cephalotes* by Jaffe and Howse (1979) and Jaffe (1980). Many ants use special structures to deposit pheromone (e.g., the sting in many myrmicines, and perianal brushes in formicines) and these may allow greater control of deposition. *Crematogaster ashmeadi* (Leuthold, 1968) deposits pheromone from its tibial gland through a hollow tendon reaching the foot (Pasteels *et al.,* 1971).

A further source of variation in pheromone amount is the fading of the trail. In his original work, Wilson (1962) showed that the trails of *Solenopsis invicta* faded rather rapidly having a life of only about 100 sec on a glass surface in the laboratory, a little longer on more absorbent surfaces. He emphasized that the ephemeral nature of the trail was important in controlling the number of recruits. The attractiveness of the trail should reflect fairly rapidly changes in the number of ants returning from feeding on the bait, in order to reduce a large overshoot effect. In some subterranean ants (*Acanthomyops interjectus,* Hangartner, 1969b; *Lasius flavus,* Hangartner, 1967) the response of an individual ant to its own trails was low when tested in the laboratory. Possibly lower concentrations and a higher rate of fading are necessary in tunnels compared with ants which hunt above ground.

The behavior of a scout ant has interesting features, quite apart from those which are directly concerned with producing a trail. When a scout ant has identified an object as food, its behavior often varies with the size of the object. Very small objects may be carried while the scout continues to search for food, and somewhat larger ones may be carried intact to the nest without much recruitment taking place. At larger sources of food the scout may or may not feed or drink to various degrees of repletion before returning to the nest and recruiting nestmates (Sudd, 1960b, 1965). The details of the ethology of scouts requires further study.

Wilson (1962) was correct in stating that the pheromone of Dufour's gland could account for all the recruitment and guidance shown by *Solenopsis in-*

victa, but he was unable to show that it was solely responsible in natural communication. It is not, for instance, certain that the amounts of pheromone that Wilson injected into laboratory nests were similar in amount to those produced by one forager or a few foragers. Later Wilson continued to discount the importance of tactile contacts, motivated perhaps by the principle of parsimony, maintaining that ants were not very responsive to tactile stimulation in the nest. Subsequently Hölldobler (1971a) showed that *Camponotus socius* uses a pheromone to place sign posts around food and also lays a trail of hindgut material to the nest. However, extracts of hindgut material do not cause any nestmates to leave the nest and the trail seems to play no part in recruitment, as opposed to guidance. Foragers which return to the nest after feeding perform an excited display within the nest. Ants with a sealed anus can successfully recruit other workers, so if any pheromones are involved in recruitment they are not the ones responsible for the trail. After displaying in the nest the scout returns to the food, reinforcing her trail as she goes. The reinforcement is necessary for recruits to follow the trail, but they need not follow the scout since artificial reinforcement of the trail with hindgut extract is effective. A similar display occurs in *Formica fusca* (Möglich and Hölldobler, 1975): the existence of recruitment can be demonstrated by the difference in the number of foragers leaving the nest when scouts are or are not allowed to return to the nest after feeding. In the nest they perform a moderately elaborate diplay. Szlep-Fessel (1970), while studying a *Pheidole* sp., described similar dances which were particularly effective for recruitment of major workers (which are perhaps less responsive to pheromones). However, after observing another species of *Pheidole,* Wilson (1976a) ascribed the recruitment of major workers for nest defense to a special pheromone.

In *Camponotus socius* a single scout can recruit up to 30 workers by its dances (Hölldobler, 1971a). It is not known whether the dancer monitors the number of potential recruits it encounters or the number which respond to its dances. It would be interesting to study the duration of dances to discover what terminates a dancing episode, and leads the scout to return to its trail. For instance, does a dancer which is prevented from contacting potential recruits attempt a continued display or return to the food? Is the same true of ants with partial dependence on pheromones? These are questions that can only be answered after further investigation.

E. Collection of Honeydew and Nectar

The collection of honeydew, the sugary liquid feces of aphids, is widespread among ants. The closeness and complexity of the relationship between ant species and the aphids they attend varies a good deal. In its simplest form the ant simply collects honeydew as it would any sweet fluid or nectar.

However, many aphids have well developed defenses which range from flight to exudation of wax. In many cases the specialized behavior of ants toward aphids can be seen as a result of a conflict between attraction to honeydew and repulsion by siphuncular wax [as in *Tuberolachnus salignus* attended by *Formica aquilonia* (Paul, 1974)]. Workers of the *Formica rufa* group attending aphids often include aggressive and predaceous individuals (Horstmann, 1973). However, among the Japanese *F. yessensis,* Higashi (1974) has shown that the amount of wear on the mandibles is lower in aphid tenders than in any of the other occupational groups he examined, and this may suggest a lower degree of aggressiveness. In this ant, as in *F. polyctena* (Otto, 1958), aphid tenders are also on average smaller than hunters. Kleinjan and Mittler (1975) have shown that the mandibular gland secretion of *Formica fusca* is capable of suppressing the production of alate forms in *Aphis fabae*. The effect of dendrolasin (a component of the mandibular gland secretion of *Lasius fuliginosus* and a chemical related to juvenile hormone and to aphid defensive secretions) is similar. Kleinjan and Mittler proposed that mandibular gland secretion is applied to aphids by the ant's antenna, i.e., during the antennal stroking which characterizes the behavior of ants toward aphids. The antenna might acquire mandibular gland secretion when it is cleaned by the foreleg strigil which in turn is cleaned by the mouth or directly from the mouthparts. Dendrolasin was originally considered a defensive chemical secreted as a response to irritation (Pavan, 1962; Maschwitz, 1964). Other terpenoids have been found in abdominal secretions of *Lasius* spp. (Bergström and Lofqvist, 1970), but their relation to aphid culture is obscure. The suppression of alate production is thus indirectly associated with attacking behavior on the part of the attendant ants. Predatory attacks on aphids are common in some species. Pontin (1961) showed that aphids compose about one-third of the prey items he was able to collect from among ant brood in *Lasius flavus* and about one-quarter in *Lasius niger*. Pontin (1978) argued that *Lasius flavus* consumes a large proportion of first instars of its protected aphids, and Skinner and Whitaker (1981) showed that *Formica rufa* treats two species of aphid on *Acer* differently, preying on *Drepanosiphum platanoidis* and protecting *Periphyllus testudinaceus*. However, Wisniewski (1967b) found no remains of aphids in his extensive searches of the mounds of *Formica polyctena*. It is interesting that Brian (1970, 1973a) has found that attacks by *Myrmica rubra* workers on their own larvae can alter the course of their development causing them to pupate without entering diapause and to produce workers instead of sexual forms. In this case mandibular gland secretion has been shown to have no effect (Brian, 1975a). The alarm responses of attended aphids are often reduced. Four species of Macrosiphini, which are not myrmecophilous, did not respond to antennal stroking from ants by secreting a drop of honeydew, but rather fled. They were attacked by the ants and killed. These aphid species dispersed when they were exposed to the secretion aphid siphuncular glands, or to synthetic *trans-β*-farnesene, a component of the secretion. In contrast, the myrmecophilous *Aphis fabae, Chaetophorus viminalis,* and *C. populicola* only responded by walking away slowly or by small movements of their body. Aphids of this species which had shortly before been in groups attended by *Formica subsericea* responded only by moving their bodies. Attendant ants themselves responded to the pheromone by aggressive behavior, but were less sensitive than the aphids to it (Nault *et al.,* 1976).

In general, antennal contacts from an ant promote the production of a droplet of honeydew by a myrmecophilous aphid. Thus, although *Formica lugubris* attendants spent only 43% of their time actually in contact with aphids, they were able to collect 84% of the droplets produced. In this general sense at least the antennal and palpal contacts constitute a communication channel between ant and aphid (Douglas and Sudd, 1980). Further investigation of this behavior, like investigations of antennal movements of food exchange between ants, encounters difficulties of recording and measuring the rapid antennal movements involved (Torossian, 1973). It is by no means clear that antennal movements of ants attending aphids are distinct from those of ants feeding from nonliving sources of food, and there is little evidence of incorporation of behavior elements foreign to the feeding situation, even in a ritualized form. There is indeed some evidence that the ability of individual ants to respond to aphids varies (Banks, 1958; Douglas and Sudd, 1980), and an element of individual learning may be involved. In some cases where the relationship of ant and aphis is obligate, one would expect a coevolved system of behavioral interactions to be present.

Aphids produce small quantities of honeydew (probably less than is necessary to induce an ant to return to the nest) at intervals over a period of time. They usually occur in groups, and are attended by more than one ant at a time. Attendance on aphids thus involves relations of each aphid with several ants and of each ant with several aphids and with its sister ants. Douglas and Sudd (1980) found that attendant ants spend 4% of their time in exchanging food with other ants; this is probably comparable with the time spent in actually drinking honeydew from aphids.

The discovery of aphid groups by ants which attend aphids away from their nest has not been studied. Herzig (1938) reported that *Lasius niger* searched for aphids and laid trails to them as it did to any other source of food. Carroll and Janzen (1973) believed that ants which orientate by individual memory and visual orientations would be better suited to attending aphids because of the permanence of location which they ascribe to aphids. Tweed (1980) confirmed Otto's (1957) finding that individually marked ants return to the same twig in search of aphids. Tweed also showed that this fidelity was retained over winter and was important in the discovery of aphids in spring. However, in the case of the aphid *Symmodobius oblongus* attended by *Formica lugubris,* Douglas and Sudd (1980) found that not only did aphid groups move about on small *Betula* saplings but that 7 of 11 saplings studied lost their infestation of aphids at some time during the summer. Therefore, in this particular instance, it was necessary for ants to maintain a more or less constant search for new groups of aphids during summer. The ants appeared to have little control over the movements of their aphids. In other, perhaps more highly evolved associations, Pontin (1960) was able to show that *Lasius flavus* workers kept aphid winter eggs in their nests, but believed the ants did not help aphids to return to their host plant in summer.

The relationship between ants and aphids is certainly wide open to behavioral investigation. There is clearly a system of communication channels coordinating the behavior of ants and aphids, in which tactile, chemical, and, in many cases, probably visual signals are involved. In addition the relationship seems to vary in closeness, and presumably in its degree of coordination according to different examples. There is also some evidence of short-term individual learning and habituation both on the part of ants and of aphids.

F. Functioning of Foraging Systems in the Field

The studies of individual behavior of foraging ants, which have been summarized in the foregoing section, have shown a number of features. Fidelity to location (Ortstreue) or to particular routes may restrict the areas where individual ants forage within the colony foraging ground. Guidance may be supplied by scent trails with varying degrees of accuracy or by other mechanisms such as tandem running. The number of foragers can be affected by chemical or other signals (as well as by innate or learned rhythms and the direct effects of temperature, light, or availability of food). In this section the question of how these features function in foraging systems in nature is discussed. It must be admitted at the outset that relevant studies are not common; as Carroll and Janzen (1973) noted many studies of foraging avoid the relationship between the distribution of food and the distribution of foraging ants. In addition some field studies can be biased by preconceived ideas of mass or group hunting, when in fact these are arbitrary points in a continuum of behavior. Indeed, for most ant species the study of foraging in the field is a difficult matter and the results often disappointing. A full description of a system entails estimating the potential forager population, the number of actual foragers from time to time, and the spatial distribution and movements of both actual and potential foragers. It is then necessary to study the way in which spatial distribution and movement respond to the discovery of prey. In practice, it is usually possible to study the number of workers arriving at only one sample point (for example a food source) and the other points are neglected. More rarely the spatial distribution of foragers is studied at one sample time (Huber, 1965) or integrated over a period of time (Brian, 1964). It is very difficult to build a picture of foraging activity as it changes in space and time from such data.

The shortage of information regarding foraging is serious, even for those ants which appear to forage singly. According to Hermann (1975) *Paraponera clavata* workers leave the nest to forage in the early evening, first forming a "preforaging group" near the nest. Hermann observed 99 foragers leave one nest, which he later excavated during the day to discover 131 workers. Thus from that nest, on that night, 75% of the forager population was foraging. Although foragers did not always leave the nest singly, Hermann believed that they separated later to hunt as individuals in the tree tops. Hermann also reported that foragers which caught large prey occasionally dismembered it and carried only part of it to the nest. In this case, the forager did not appear to return to the place where it had left part of its prey. Much of the material brought into the nest was tree sap, which may possibly supply energy to an inefficient hunting system.

More coordinated systems may show some or all of the following features: (1) the location of finds is remembered by successful foragers as a learned mnemotaxis or a pheromone trace or both (little is known of the use foragers make of their own pheromone trails) (Carroll and Janzen, 1973); (2) this memory or trace is transferred in some way to other foragers; (3) the departure of foragers is coordinated with prey finds so that the number of active foragers is increased when food is easy to find. A corollary of the last is that not all of the potential foragers actually forage spontaneously, so that there is a recruitable reserve. Variations in the importance of these three factors lie behind the arbitrary division of foraging systems into mass recruitment, group recruitment, and partition of foraging grounds (Carroll and Janzen, 1973; Möglich and Hölldobler, 1975).

Comprehensive field information is for the most part lacking for the more extreme forms of mass recruitment. According to Carroll and Janzen (1973), in this type of system "the workers at the food increase gradually to an asymptote." The system, they believe, is advantageous when many workers provide a reserve of recruits, or when a single worker's contribution of trail pheromone is small, or when there is considerable competition for food which must be retrieved quickly when found, or when a spread of small workers at the end of an inaccurate trail is effective in capturing moving prey, or when large items of food occur erratically near the nest. However, most studies have concentrated on the arrival of recruits at a bait, usually a single artificial one. There seem to be few estimates of the number of density of scouts in any species, presumably because of their small size in many species. Sudd (1960c) reported that *Monomorium pharaonis* scouts left the nest of a large colony at a rate of about 5 each min but gave no estimates of their total number or proportion to the rest of the colony. The ability of this ant to recruit 200 or more workers to a bait within 10 min suggests that a large reserve exists in the nest. The second characteristic of multiple trail laying is again derived from laboratory studies. In field conditions, *M. pharaonis* operates trails up to 40 ft long (Sudd, 1960c). In *Solenopsis invicta* Horton *et al.* (1975) found that recruitment time is shorter as the distance from the nest increases. This seems to suggest that reserves are not always located in the nest. This is not the case in natural infestations of houses by *M. pharaonis* where scouts can be followed back to the nest above ground over long distances. As far as the ability to monopolize a food source or food item is concerned, Brian *et al.* (1966) have shown that this can be achieved by a variety of methods. The ability to mass workers by recruitment is indeed important in the relations of *Tetramorium caespitum* and *Lasius alienus,* but other factors (e.g., the ability to approach underground and the possession of chemical means of offense and defense) can be more effective. In fact the capture of moving prey may be more efficiently conducted using more rapid systems, especially when it is not necessary for the scout to return to the nest because many foragers are already in the area. For instance, in *Leptogenys occellifera* (Maschwitz and Mühlenberg, 1975) where colonies are large, ants that find food lay trails only to the nearest permanent route and the pheromone mixture has both attractive and guiding properties with different fading times. However, where large items of food occur erratically near the nest, extreme mass recruitment may well be the best possible strategy. This is almost the same as saying that many extreme mass-recruiting species live by scavenging in buildings.

According to Carroll and Janzen (1973), in group recruitment there is simply a single pulse of recruits and not a progressive increase in the numbers of ants at the food. The number of recruits is not necessarily less than in mass recruitment; for example, in *Solenopsis invicta* the number recruited by a single scout is often about ten (Wilson, 1962) and in *Camponotus socius* as many as 30 can be recruited by a single display, and these in turn may recruit more ants (Hölldobler, 1971a). The distinction between mass and group recruitment is reduced to a question of whether or not successive pulses coalesce. This might depend on whether scouts feed on the prey before recruiting other ants, on their rate of feeding, and on the nature of the prey. There is a danger of losing sight of the very wide variation which exists in recruitment behavior and its relation to group raiding in doryline and ponerine ants. *Leptogenys occellifera* recruits nestmates from the nearest frequented permanent route, but in *L. diminuata* (Maschwitz and Mühlenberg, 1975), *L. (Lobopelta) attenuata,* and *L. nitida* (Fletcher, 1971) scouts apparently return to the nest and recruit a group there. This raises the question of whether this difference depends on the behavior of the scout or on the density of other workers outside the nest. Chabab and Rettenmeyer (1975) describe the ability of *Eciton* to attract and guide scores of fellow workers to prey by a chemical trail and momentary contact. They noted that this recruitment mechanism is similar to that involved in mass (or group) recruitment, but that the speed and amplification factor is higher. It is important to record the parameters of behavior in each case and to avoid becoming imprisoned in a system of typical examples.

The distinction between recruitment systems and "partition of foraging grounds" (Dobrzanska, 1958) is somewhat restricting as well, particularly if the two are seen as mutually exclusive. Rosengren (1971) has emphasized that there is no reason why both recruitment and route fidelity should not occur in the same species. In fact, it is now clear that not only some form of recruitment (Rosengren, 1971) but also scent trail guidance (Horstmann, 1975, 1976a; Henquel, 1976) occurs in *Formica polyctena*, a classic example of a species that uses partition. Carroll and Janzen (1973) stated that partition of foraging grounds seems to be typical of "large and generalized colonies of temperate zone forest ants" which seems to be a circumlocution for the *Formica rufa* group. Exploitation of only part of the available foraging ground is in fact known in other species, for example the American harvester ants (Hölldobler, 1976). African *Camponotus* species exploit only about 10% of the available area at one time (Levieux, 1971), and only 20–30% of available plant species are exploited by *Atta* spp. (Rockwood, 1976). Knowledge of the occurrence and nature of spatial partition is limited by techniques for marking ants individually. Route fidelity can be demonstrated most easily in ants which can be seen hunting for food in moderate numbers and whose workers are large enough to be marked quickly. New techniques for marking small ants individually and for recapturing them might show that the phenomenon is more wide-spread than currently believed. Attine ants in particular are promising subjects for study of the movements of the first ants to discover unexploited areas as well as the established trails to exploited plants. Their system has some of the features of mass recruitment based on the model of *Solenopsis invicta,* in addition to other novel features (Jaffe and House, 1979; Jaffe, 1980). Nothing is yet known about the early establishment of trails, however.

The foraging systems of American harvesting ants are now thought to be much more complex than earlier laboratory studies of their trail laying behavior suggested (Hölldobler, 1971b). It is clear from field studies that individual responses vary according to species, location, and food availability (see Section I,C). In times of scarcity in the Californian desert, *Veromessor pergandei* foragers are confined to a limited sector around the nest, but spread more widely when seeds are plentiful. When the area is most restricted they may be said to forage in columns. The columns or sectors are not constant in position but rotate around the nest. (No continuous observations over a long time period have been made in this desert habitat.) Thus, a circular area is searched sector by sector (Bernstein, 1975). In the Mohave Desert the same species foraged "in columns" with few changes of location when seeds were plentiful, and in columns with frequent changes in location and also with a great deal of individual foraging when seeds were scarce (Rissing and Wheeler, 1976). Some of the apparent contradictions in these two accounts are no doubt attributable to different interpretations of plenty and scarcity. In *Pogonomyrmex barbatus* and *P. rugosus* (Hölldobler, 1976), finders of food lay trails that attract and guide recruits. These trails later become trunk routes and guide unrecruited searching scout ants. Foraging in columns and individually has been described for ants of New Mexico by Whitford (1976). If trail laying and following by individual ants are the behavioral elements of the foraging system the probability of new finds and their location must be of prime importance. In extreme scarcity, with virtually no finds, both recruitment trails and routes would fade through lack of renewal. Scout ants would presumably scatter over long distances in search of food. Their distribution would be sparse and they might be hard to discover. In moderate scarcity successful searches would still be rare and nests with only one unfaded route would be common. The sector served by this route would be almost totally depleted of seeds, and scouts guided by the route at first would leave it and move some distance circumferentially before making a new find. Finds would be rare and unlikely to occur more than once in each period of observation. If, as is the case of *Pogonomyrmex* (Hölldobler, 1976), scouts are able to return to the nest by a sun-compass mechanism the trail would be corrected to a more or less radial line centered on the nest and to one side of the previous trail. After the exploitation of the find more scouts would leave the nest guided by the trunk route and leave it to search for food on either side at random. They would be unlikely to find food in the heavily exploited area of the previous route and the next trail and its subsequent route would continue the rotation. Unfortunately available studies do not give the information needed to test this suggested scenario. The regularity of the rotation (its departure from equiprobable deviation to either side) has not been statistically established. The distribution of solitary (independent) foragers has been too little studied. The hypothesis outlined predicts equal distribution of foragers on either side of a column: collection of solitary foragers by pitfall trapping without baits could falsify this hypothesis.

The foraging activity of wood ants (the *Formica rufa* group of north and central Europe) has received more attention as a system than that of most ants. This is partly because of the interest of foresters in the role of these ants in controlling or reducing damage to trees by caterpillars, and partly because in some ways the system is easier to study than the behavior of individuals. In general the wood ant system is characterized by large numbers of foragers (180,000–360,000 for a large nest of *F. polyctena*) (Horstmann, 1974a) and the maintained high flow of foragers from the nest (up to 10 per sec) cannot be immediately related to finds of food (Horstmann, 1974b). Foragers commonly travel up to 30 m from the nest and can be found at greater distances. They leave the nests along definite routes and show route fidelity (Section II,C). In a large-scale study, Horstmann (1973) showed that 75% of returning foragers came from trees and that 60–90% of them were carrying honeydew in their crops. Ants carrying prey might have captured it on trees or on the ground and prey could be carried as juices in the crop as well as visibly in the jaws. About 12% of ants from the ground and 6% of ants from trees carried prey, but this proportion could be much higher when caterpillars were particularly plentiful (e.g., in an outbreak of *Tortrix viridiana*). Since the majority of ants foraged on trees, more than one-half of the prey was captured there. Food intake (measured as nitrogen) consisted of 49% whole prey, 25% prey juices in crop, and 26% honeydew. The calorific content and wet weight of honeydew brought in is, of course, greater than that of prey (Horstmann, 1974b). Route fidelity is probably responsible for much of the assignment of ants to different tasks, although this is difficult to test because of the relatively large proportion of unladen ants. When a sample of ants, caught on the ground with prey, was marked and released a higher proportion of them was recaptured with prey than was the case for all ants. The same was true for ants marked while carrying nest material. Specialized hunters and nest-material collectors therefore exist. However, this may simply be because they visit more promising areas. On trees the situation is different. Only 2.6% of the ants marked while carrying prey on trees were recaptured with prey and the proportion was close to that for all ants. However, at a time when prey was particularly abundant on trees a much higher proportion were recaptured with prey. In general, there is a high degree of fidelity to routes and to foraging on the ground or on trees. Ants on the ground tend to be faithful to hunting or to nest material collection. In contrast, ants on trees do not appear to be permanently attached either to prey capture or to honeydew collection (Horstmann, 1974b). Ants tending aphids normally attack and capture prey insects that they encounter or that are offered to them by an experimenter. The exact degree of spatial fidelity is not known; some observations by Otto (1958) suggest that individually marked ants return to the same twig or group of twigs to tend aphids. Tweed (1980) confirmed this and recaptured ants marked the previous summer at the same twig in spring.

The method by which individual wood ants become faithful to a particular route or area is not known. Rosengren (1971) showed that fidelity is the result of learned orientations to landmarks rather than a response to pheromone trails, and he suggested that inexperienced ants adopt particular routes and become faithful to them as a result of social factors. He was able to show in laboratory colonies of *Formica rufa* and in *F. polyctena* in the field that placement of artificial baits on a route increases the traffic flow on that route. The increase in traffic is not attributable to ants already marked on other routes, but rather to the arrival of larger numbers of unmarked ants. It is unlikely that these were attracted from areas between routes because there was a consistent excess on the route of ants running outward over ants running homeward. The average survival in an isolation nest of ants captured at the artificial bait, or between the nest and the bait, was higher than that of ants collected beyond the bait or on other routes; this suggests that they contained younger ants. Therefore, it appears that young inexperienced ants can be recruited to particular routes by older ones. Horstmann (1975, 1976a) and Henquel (1976) have shown in laboratory conditions that *Formica polyctena* uses some form of scent trail to mark tunnels in which food has been found in artificial nests. The significance of this in natural conditions is not clear, and scent trails could be used at any point in the system, e.g., on tree branches rather than at the origin or branching of routes. Their experiments were, significantly, only successful in spring and scent may be used particularly in the extension and reconstitution of the route system by surviving workers at this time.

Once a worker has found food on a route it may learn the landmarks or bearings by an associative process. The nature of the reinforcement is a little unusual because foraging ants do not eat their prey themselves and need not even ingest juices from it. Alternatively, inexperienced workers may have a sensitive period at which they acquire an affiliation to a particular route. Zakharov (1973) suggests that workers emerge into a particular column or team in the nest and adopt its route affiliation. None of these suggestions allow very much flexibility within the foraging system. Old ants, it appears cannot learn new "tricks," and flexibility only comes, if it comes at all, with their death and replacement by younger ants. However, flexibility exists where it is most often needed if Carroll and Janzen (1973) are correct in assuming that *F. rufa* group systems are specifically adapted to local and seasonal outbreaks of exploitable caterpillar prey. The presence in trees of large numbers of foragers which are supported by honeydew collection but able to switch to predation when more prey is available, allows this to happen.

The study of ant foraging as a system demands the collection of immense amounts of data on the numbers and distribution of foragers. The behavior of a system, in particular its response to changed food availability, cannot be predicted from laboratory studies of communication channels and other behavior, although these are essential preliminary investigations. Further advancement of knowledge will be assisted by the development of equipment for automatic data capture (Lewis *et al.,* 1974; Kruk de Bruyn and Tissing, 1975).

III. NEST CONSTRUCTION

The nests of ants, like those of other social insects, function to protect the brood and queens from enemies and other hazards, to control the distribution of the brood and food sources such as aphids, and may also allow some amelioration or even control of microclimate. Nest building activities can be classified as follows:

1. The excavation of material to form cavities (shafts, tunnels and chambers);
2. The deposition of materials to fill or enclose cavities (mounds, chambers, craters);
3. The import of materials from outside the nest for special purposes;
4. The compounding of materials to form special substances, with particular physical properties.

The first two activities are found in all ants that build nests, and can to some extent at least be studied as the behavior of individual ants. They are also both found in the behavior of the tiphiid wasp *Methocha* and no doubt in the presocial ancestors of ants. Wilson and Farish (1973) have described the excavation techniques of *Methocha californica*. The adult female wasp invades the burrow of a cicindelid larva, kills the larva and lays a single egg on it. Then it closes the burrow by depositing soil in it with its mandibles. The wasp, given a larva outside the burrow, will kill it, dig a burrow, and seal the prey in it. If one follows Wilson's relation of *Methocha* and the Mesozoic *Sphecomyrma* (Wilson, 1971), one may conclude that early ants probably already had the ability to dig burrows and to manipulate soil. However, wasplike mandibles of *Sphecomyrma* are poorly adapted for extensive digging.

It is reasonable to suppose that early ants dug in soil, humus, or rotten wood. Many species of ants still confine themselves to nesting in some of these materials. For instance the successful ponerine *Odontomachus haematodes* is able to nest in friable material whether this is dead wood, the crowns of palm trees, or soil (Colombel, 1970), provided the material is reasonably moist. Unlike social wasps and bees, few ants make use of compounded materials such as paper or wax, which have an appreciable energy content. Their use among ants is typical in arboreal species.

A. Excavation Behavior

The general methods used by ants in digging have been described by Sudd (1969) for 15 species of ants. When single ants are placed in sand they usually dig into by removing particles of soil with their mandibles. The particles are placed below the ant's thorax or may be carried away to a distance. The ant then inserts its head into the space where it has removed soil and removes more soil. Spangler (1973) has suggested that some ants make vibratory movements with their bodies to break up compacted soil. When the ant deposits soil under its thorax (apparently this is when the particles it is able to remove are relatively small) a heap of soil accumulates and the ant may make raking movements with its forelegs to control the heap. Eventually it picks up the heap in its mandibles and carries it out of the excavation. While Sudd (1969) described some differences in the relative frequencies and sequences of the component acts of digging in different species, recent data (J. H. Sudd, unpublished), suggest that these are influenced by the particle size of the sand provided. Thus, it seems less likely that specific differences exist, except perhaps in those arboreal species whose digging behavior is defective.

In some desert ants special anatomical arrangements help in the transport of sand. In *Messor arenarius* long hairs on the gula, mandibles, and clypeus allow workers to carry large amounts of sand provided it is moist. Dry sand, such as the blown sand which fills the nest entrance each day, does not adhere to the ventral hairs of the gular psammophore, and only the dorsal, clypeal hairs can be used (Delye, 1971). It is not clear from this what advantage the ant obtains from the possession of the ventral hairs. Over short distances, *Messor* and other Myrmicinae as well as *Cataglyphus* species disperse dry sand by raking movements like those used by many ants in excavation.

Single ants in laboratory experiments usually continue digging in a single place until a tunnel has been formed. The ant then enters the tunnel, travels all or part of the way down it, removes soil, carries it back up the tunnel, and deposits it. The shape of the excavation depends on where soil has been removed; although even in such simplified situations some soil is deposited in branches of tunnels. Rather surprisingly such tunnels, although subject to variation in shape, show consistent differences in shape in different species. *Formica lemani* workers dig straight vertical unbranched tunnels, whereas *Lasius niger* dig wavering branched ones (Sudd, 1970a). *Lasius niger* is apparently less strongly attracted to the ends of tunnels and digs part-way down the side more often than *F. lemani;* this may be the way in which branches arise (Sudd, 1970b). However, the straightness of the tunnels of *Formica lemani* may be attributable to its possession of a marked orientation to gravity, because it corrects its tunnels rather accurately for experimentally imposed deviations from the vertical (Sudd, 1972). Sudd (1975) used a simple computer model to simulate the production of straight and wavy tunnels, but was not able to produce branching tunnels.

In natural conditions, the only excavations made by single ants are those made by newly fertilized queens. Isolated queens of *Lasius niger* taken after a mating flight dig fairly readily in moist sand. Their method of starting a tunnel is similar to that of isolated worker ants; the queen removes soil with her mandibles and carries it out. When the excavation is about as deep as the length of the queen's body, differences from worker behavior appear. The queen is more reluctant than workers to enter the tunnel and dig at its end. Instead she works from the tunnel entrance so that the tunnel becomes a flask-shaped chamber with a narrow neck. When the queen does enter the chamber she is likely to turn around and carry soil out by walking forward instead of carrying it out walking backward as workers often do. She then deposits the soil in the tunnel entrance or just outside it, sometimes back-filling the tunnel to form a closed chamber. In the laboratory situation this does not always happen: chambers are dug and abandoned and sometimes the queen closes the chamber from the outside instead of the inside. Two out of ten queens successfully sealed themselves in, and one of these laid eggs which eventually produced a few workers. These workers dug typical wavering tunnels from the queen's chamber. These limited observations suggest that the construction of a claustral chamber, in which the queen seals herself and raises her first brood, can result from fairly simple differences between the method of excavation of workers and queens (J. H. Sudd, unpublished work).

Ants which nest in wood may use substantially the same methods of excavation that other ants use in soil, as *Cataulacus parallelus* does (Sudd, 1969). Some which nest in trees are unable to excavate sound wood. *Crematogaster scutellaris* can only tunnel in decayed wood or cork-oak bark and the queen never digs tunnels but establishes nests in pre-existing tunnels made by other insects (Casewitz-Weulersse 1972, 1973). However, *Camponotus vagus* (Benois, 1972) and other species (Sanders, 1970) can dig in sound wood. *Camponotus herculeanus* often removes the soft, spring wood from the heart wood of pines so that a system of concentric chambers separated by wooden lamellae is formed. The sapwood is left so that the tree survives (Fuchs, 1976a).

It is characteristic of ant excavation that the greater part of the excavated soil is carried out of the burrow in grains or crumbs, and not thrown out in jets with the legs as it is in some wasps. The deposit of soil on the surface may be indetectable (perhaps because it is dispersed by further transport), or it may be organized to form an irregular, or circular, or crescentic crater surrounding the nest entrance, or even as a mound containing tunnels or chambers. Simple experiments with single ants have thrown little light on this except that *Myrmica ruginodis* workers deposited soil preferentially on horizontal surfaces rather than on sloping ones. This occurred whether the slope was toward or away from the tunnel mouth (J. M. Stockdale and J. H. Sudd, unpublished).

B. Construction and Functions of Nests

Nests of ant colonies in nature are the result of the digging activity of numbers of ants. Sudd (1971) extended his work with single ants to pairs, and found that they dug significantly wider tunnels. Chambers, that is a considerable dilation of a tunnel, have never been produced by small groups of worker ants without brood. Experimental study of the formation of nests requires the development of larger observation nests containing soil. It is difficult to observe the behavior of individual ants in these, even though complex and apparently normal nests are often built. A fairly well established social response in several species is a preference for digging in soil previously occupied by their own colony (Hangartner *et al.,* 1970; Hubbard, 1974). Presumably this acts to keep the nest compact in the absence of a structural envelope; many excavations are later filled with excavated soil, and a nest often occupies a core of worked soil.

The form of complete natural nests is often adaptive. Dlussky (1974) described several types of nest architecture used by desert ants in Uzbekistan and Turkmenistan; some are superficial and confined to wet areas, others are deep and tap underground supplies of water. However, Delye (1968) emphasized that among Saharan ants deep nests may tap only aquifers with a high salt content. In Britain, nests of *Lasius niger* in the moister parts of heaths are more superficial than those of *Formica fusca* found in drier areas (Brian and Downing, 1958; Nielsen and Jensen, 1975). Other species dig deep winter nests and abandon their shallow summer ones (Ito, 1973; Imamura, 1974). The interaction of individual orientations, such as those studied by Sudd (1970a,b), with these environmental factors has not been studied.

Above-ground features of nests are also known to be adaptive traits. Deposits of soil in the form of craters around nest entrances are, in general, typical in warmer climates, and mounds are typical in cool, temperate climates. The desert ant, *Cataglyphis bicolor,* builds crater nests in warmer deserts but builds mounds in Afghanistan in places were the contrast in day and night temperatures is extreme (Schneider, 1971). The functions of the crater are not clear. In many drier climates they may merely be evidence of the amount of excavation needed to make a nest with a microclimate suitable for ant brood. It is possible that a ring of soil at its maximum angle of rest disposes of this soil at minimum energy cost. If one accepts this view, then one must explain why some ants use more than this minimum amount of energy in dispersing soil, and why some build craters which consist of an incomplete (cresentic) ring of soil. Wehner (1970) showed that *Cataglyphis bicolor* workers each have an individual preference, based on visual orientation, for depositing soil by day, but dump soil in random directions at night. In *Trachymyrmex septentrionalis,* Tschinkel and Bhatkar (1974) have shown that the orientation of the crescentic crater depends on the slope of the ground and is adjusted if this is altered by experimental interference. However, the response of the ants depositing soil in the crescent is not directly to the slope (which is presumably masked by the soil already deposited) but rather to landmarks.

Although the preferred orientation of soil depositers may have been carried over from some other activity, it seems likely that the crater is orientated to prevent soil or water from entering the nest. Bhatkar (1974) also showed that the angle of the inner slope of the crater was species specific, although this may result from specific differences in the size of particle deposited. Sudd (1970c) showed that different species of ants removed soil of a characteristic range of grain sizes, with a mean size approximately equal to 0.6 times the headwidth of the ant.

There are no clear descriptions of the mound construction behavior of north European ants such as *Lasius flavus* and *L. niger* beyond a general belief that they extend the mound by laying soil in dew. This seems very likely in the case of the looser structures built around grass stems by *L. niger* and *Myrmica rubra* but does little to explain the apparent differences in the strengths of the mounds in these species, which makes the mounds of *L. flavus* more resistant to trampling by cattle (Brian, 1964). It is strange that the detailed description given by Huber (1810) of mound construction in "la fourmi jaune" and "la fourmi noir-cendrée" (evidently *L. flavus* and *Formica fusca* group, respectively) has not been confirmed.

C. The Import of Material for Nesting

As with human settlements, ant colonies are net importers of materials so that a localized increase in some minerals and organic matter results (Gentry and Stiritz, 1972; Czerwinski *et al.,* 1972; Rogers and Lavigne, 1974). Most of these imports are made up by food, especially insect prey. However, many ants bring in other materials from the foraging range. *Iridomyrmex purpureus* carries pebbles to its nest and lays them on the excavated soil (Greenslade, 1974) and Ettershank (1971) has discussed the functions of this "decoration." The decorated areas may mark a defended area on which ants attack vertebrate intruders. The stones may be used to close nest entrances before rainstorms and to protect the nest surface from erosion. The stones present an aerodynamically rough surface which may increase heat loss to the air by preventing laminar flow. The surface of the nest is in fact a few degrees cooler than the surrounding soil surface, although at levels below the ground the nest is always warmer than the surrounding soil. Finally, the rough, cool surface may collect dew from the surrounding air, which can then be absorbed by the underlying soil in the nest.

The characteristic mounds of the *Formica rufa* group are made of various materials, mostly vegetable litter but also stones, resin, and dead insects, imported from the foraging area. In *Formica polyctena,* about 1% of workers returning to the nest carry nest material. Since 45% of workers are unladen, nest material makes up nearly 2% of all loads. The proportion varies considerably with the time of year and is greater in summer than in spring (Horstmann, 1973). In ten nests from a pinewood, Wisniewski (1967a) found differences in the percentage composition of the inner and outer layers of the nest material. The inner layers contained 60% twigs, mostly less than 50 mm in length and the outer layers only about 30% twigs. The outer layers contained more of the other parts of plants, especially pine needles, resin, and buds. The outer layers also contained 19% inorganic material, mostly less than 2 mm in diameter, while the inner layers contained only 12%. The outer layers were therefore composed of more closely packed fine material with good resistance to weather and the inner layers were more open. However, the specific gravity of the layers hardly differed, weighing about 0.21 g/ml. The segregation of material in the nest must be the result of the continuous turning over of material by ants on the nest mound described by Chauvin (1960). The litter mound has a higher thermal diffusivity than the surrounding soil and conducts solar heat to the interior of the nest (Brandt, 1980). This is due to its low specific gravity, its fibrous nature, and the continuous reworking and defense of the mound against damage by birds and other predators which the workers provide, at least in summer. It can also be built into steep mounds, especially when partly supported by low shrubs. Litter mounds are therefore able to intercept more radiation than earth mounds.

A rather special case of the import of materials is provided by the construction of tents over Homoptera tended by some ants. When these tents are built of imported soil (e.g., those built by *Crematogaster striatula, Camponotus acvapimensis* and *Pheidole megacephala* on Cocoa trees) they may contribute to the spread of plant diseases by returning spores from the soil to the trees (Evans, 1973).

D. Compounding of Nest Materials

Tents constructed over Homoptera and some types of arboreal nests often consist of soil or plant fibers reinforced in some manner. A possible origin of tent building behavior may be found in the habit common to many ants of depositing soil in liquids inside or outside their nest. In spite of the collection of honeydew by attendant ants, Homoptera and the surrounding plant are often contaminated with honeydew and this may release covering behavior. This type of behavior can be interpreted in other ways: Morrill (1972) has described *Pogonomyrmex badius* placing sand on honey and carrying the saturated sand to its nest. This species has a much reduced proventricular valve and is not well adapted to carry liquids.

Maschwitz and Hölldobler (1970) have shown that *Lasius fuliginosus* builds its carton nest by impregnating pieces of wood with liquid from its crop and not, as has been suggested (Pavan, 1962), with mandibular gland secretion. When a laboratory colony was supplied with fragments of cellulose, one group of ants gnawed it into 1-mm pieces and carried it into the nest. A second group of Innendienst workers took it, regurgitated onto it syrup they had received from Aussendienst workers, and incorporated old carton with it. The cellulose, impregnated with syrup, was invaded from the old carton by the fungus *Cladosporium myrmecophilum* which held it together. The compounded carton has good thermal properties and allows *L. fuliginosus* to raise, and to a certain extent to regulate, its nest temperature (Kravchenko, 1973). The use of a food substance in the construction of the nest would seem to be wasteful, but in fact it is no more wasteful than the similar use of silk by *Polyrachis simplex* (Ofer, 1969) or *Oecophylla,* or indeed the use of silk to protect the pupae in ants, other Hymenoptera, and Lepidoptera. It is possible that the saving in energy costs of transport in an arboreal ant outweighs the expenditure of energy-rich materials. Indeed the energetics of nest construction may be worth investigation. Single workers of *Myrmica* in the laboratory are able to excavate about 1 g of sand per day, raising it an average of about 10 mm, with a useful output therefore of almost 1 erg or 2.3×10^{-8} cal (Sudd, 1969). The respiratory cost of this is not known but presumably efficiency is rather low. In ants which, like *Lasius fuliginosus,* have access to large quantities of honeydew, this may be better spent as a material than as energy. The payoff on the expenditure would presumably be the more rapid and more certain production of sexual forms, to which a secure and thermally efficient nest contributes.

IV. BROOD CARE BEHAVIOR

The brood care behavior of ants contrasts with that of social bees and wasps because the brood is kept in the nest in common areas and not in individual cells. The queen lays eggs in groups, not individually into empty cells, and nurse workers are able to recognize brood and move it around the nest according to its stage of development and the variation in nest climate.

A. Recognition of Brood

Although ant larvae have a characteristic shape and, in many species, a characteristic cover of bristles (see Vol. I, Chapter 7 this series), these are not of importance in the recognition of larvae in the species where this has been studied. Workers of *Myrmica rubra* recognize larvae by a chemical signal spread fairly generally over the larval cuticle. Larval movements, the presence of a head and exudates from the mouth and anus are unimportant, and although shaved larvae are adopted more slowly the hairs are not essential for adoption (Brian, 1975b). Objects recognized as brood are carried into brood chambers of the nest, retrieved if they are placed outside the nest, and carried away if the nest is disturbed. However, care is needed when these criteria are applied because food is placed with larvae in many species (Glancey *et al.*, 1970; Walsh and Tschinkel, 1974). In *Solenopsis invicta* a nonvolatile substance widely distributed on prepupae and pupae (unfortunately larvae seem not to have been tested in this species) releases such behavior (Walsh and Tschinkel, 1974). Bigley and Vinson (1975) have reported that triolein is the substance responsible, since filter paper smeared with synthetic triolein or with triolein extracted from sexual prepupae is carried into the nest, and kept there for 24 hr or more. Paper impregnated with soy–oil was eaten. Paper treated with triolein was placed near sexual pupae and kept separate from worker brood. This substance cannot therefore be analogous with the larval chemical of *Myrmica rubra* (which is not recognized by *M. scabrinodis*) (Brian, 1975b). Brian points out that all instars must have recognition signals since they are normally sorted into age groups which are often eggs and microlarvae, larger larvae and workers with callow adults. Brood in general might be distinguished from adults by its immobility. Large queen larvae of *Myrmica rubra* produce a special pheromone on the ventral side of the thorax, which in queen-right colonies elicits attacks by workers, especially older workers which are no longer fully committed to nursing (Brian, 1970, 1973a, 1974). According to Jaisson (1972), *Formica polyctena* workers recognize callows by a specific chemical signal.

B. Transport of Brood

The division of the brood into age classes is without a doubt a method of separating the feeding stages (e.g., larvae) from the nonfeeding stages (e.g., eggs, microlarvae, prepupae, and pupae). However, larvae are moved to different parts of the nest according to temperature responses of the transporting workers. In *Myrmica rubra* workers placed with brood in a temperature gradient aggregate themselves and the larvae in regions of about 19°–22°C. Larvae were moved first from the regions above 22°C and then the workers themselves stopped visiting these parts. The collection of larvae from the cooler regions was slower and workers were reluctant to enter the parts below 5°C (Brian, 1973b). The behavior of *Formica polyctena* in moving brood has been extensively studied and has been reviewed by Schmidt (1974). In winter, with the exception of a few older workers, workers and queen remain below ground level. In spring, queens and workers move into the parts of the nest above ground since these become warmer and it is here that spring eggs and sexual brood are produced. By early summer the queens and the younger nurse workers withdraw to deeper cooler parts of the nest where summer eggs produce worker larvae. Horstmann (1976b) has shown this movement to be a response to gravity, not to temperature. All brood is carried into the warmer outside parts of the nest as pupae. Therefore, it appears that workers with different temperature preferences are differentially attracted to brood of different ages, or else different temperature responses are released in the same worker by different stages of brood. Less far-reaching transport occurs in other species. *Aphaenogaster subterranea* larvae are carried by workers to prey brought into the nest by foragers and are placed on the prey. In experiments, only larvae actually placed with their heads on the prey were able to feed. Those placed with their trunk in contact with prey or 5 mm away from it were unable to reach the prey in the 2 hr for which they were observed. Workers normally leave larvae on prey for about 30 min and then replace them in the brood pile (Buschinger, 1973).

The varied contacts between nurse and other workers and larvae are of importance in the control of caste. Because larvae are in a group and not in individual cells the distribution of food to larvae may be less equitable than in bees and wasps. Differentiation of the brood into workers and sexuals is dependent at least in part on larval growth rates and thus on larval nutrition (see Vol. I, Chapt. 5, this series). It is tempting to relate the wide range of worker size in ants to their general method of feeding larvae. However, as Wilson (1971) points out there is considerable evidence that the composition of the worker force is regulated in the colony to maintain more or less constant portions of each size class. Studying the dimorphic *Pheidole pallidula,* Passera (1974) has shown that the development of major workers (soldiers) is dependent on a high protein diet, which must be unevenly distributed to produce some third stage larvae 1.5 mm long and others 3 mm long. However, the production of soldier pupae is influenced by the number of soldiers in the worker force. As a result small colonies set up with a variety of proportions of workers and soldiers tend to an equilibrium of about 5% soldiers and 95% workers.

Mature larvae of *Formica polyctena* attract nurse workers, apparently by searching movements of their heads. In laboratory nests these workers collect sand grains and then move the mature larvae onto them. The larvae make a groove in the sand and begin to spin their cocoons. Sand grains or nest materials are essential for anchoring the first threads of the cocoon (Schmidt and Gürsch, 1971).

V. POLYETHISM

A. Behavioral Repertoires

The use of the term polyethism to replace the older "division of labor" (with which it is synonymous) (Wilson, 1963) was introduced by Weir (1958). In Weir's experiments, *Myrmica* workers were shown to differ in their response to disturbance of their laboratory colony: some workers approached or attacked the disturbance at once, others approached less readily or actually fled. Since the classes of workers separated in this way showed other differences in behavior, but no morphological differentiation (apart from differences in pigmentation related to their age) Weir described them as polyethic, by analogy with polymorphic. Another approach to the problem entails monitoring the behavior of undisturbed ants and examining the results for a consistent bias of particular ants, or categories of ants, toward particular behaviors. These two approaches in fact display polyethism in two complementary ways. Polyethism results in members of the colony responding in more than one way to stimuli: as knowledge of the social control of behavior by pheromones increases there is a need to understand the range of response if "present tendencies to reductionism" (Carroll and Janzen, 1973) are to be contained. Further, polyethism ensures the performance of all the colony tasks by distributing them among the worker force. However, there is some difficulty in preparing a list of all the colony tasks. Wilson and Fagen (1974) therefore proposed the production of an ethogram by the following method. They observed a laboratory colony of *Leptothorax curvispinosus* over a total period of 51 hr. This led them to define a total of 27 behavioral acts. Fagen and Goldman (1977) argued that the frequency of behavioral acts is distributed according to a negative binomial or log-normal Poisson function and that the probability that more behavioral acts would be discovered by further observation can be estimated by fitting such a distribution function to the empirical results. Thus Wilson and Fagen (1974) estimated that they had discovered 77% of the repertoire of *L. curvispinosus* and that its total repertoire consisted of 27–35 acts (95% confidence limits) with a most probable size of 29 acts. Wilson (1976b) extended this method to *Zacryptocerus varians*. He found workers of this ant performing 40 acts and estimated that its total repertoire was 42 (95% confidence, limits 38–46). In both cases the repertoire was relatively small and only a small proportion of acts occurred rarely. In Rhesus monkeys, by contrast, the total number is much larger and a larger proportion of behaviors are rare. Wilson and Fagen conjectured that these characteristics result from the small size of the insect nervous system which has insufficient space to store rarely used patterns. The method has so far only been applied to studies of the behavior of ants in artificial nests. Although a behavioral inventory could be made for ants foraging in the field, it would be impossible to put inventories from the field and from artificial nests together to get a total inventory without the use of special weighting factors which would be difficult to estimate. Therefore, an obvious criticism is that the repertory of ants studied in such nests will underrepresent, for instance, foraging or digging behavior. The relative frequency of food collection behavior in *Zacryptocerus* for example was 0.1428 and in *Lepthothorax* 0.054. Wilson and Fagen (1974) countered this criticism by saying that most of an ant's life is spent in the nest anyway, that is they estimate the weighting to be given to out-of-nest activities as effectively zero. This assumes that the density of behavioral acts is the same in any context; their analysis is not based on times but on acts. It could be argued that more frequent and more varied acts will be required outside the relatively even conditions of a nest. In natural conditions a forager of *Formica polyctena* can spend 9–10 hr on a single excursion from the nest, so that one-third of its behavior might be carried on outside the nest. However, the method is applicable to any situation where a definable body of ants can be observed in an artificial nest (Wilson, 1976b; Otto, 1958), struggling with prey (Büttner, 1973), or attending aphids (Otto, 1958). In a more recent paper, Wilson (1976c) tried to extend the range of the repertory by subjecting a nest of *Pheidole dentata* to a variety of stresses (e.g., disturbance, attack by other ants, burial), but he did not make the acquisition of food more difficult. Possibly a *Zacryptocerus* worker spends more time, and performs more feeding acts, when it has to get food from aphids than in Wilson's experimental situation.

Another criticism is that the categorizations used by Wilson and his co-workers are functional, not to say teleological, in nature, rather than merely descriptive. Wilson and Fagen (1974) argued that this is mainly a verbal problem, and that one investigator's act B and C is simply included in another investigator's act A. However, this depends on the level of the analysis, since two acts at one level may both incorporate a lower level act so that act Z is sometimes a component of act Y and sometimes of act X. Where this is true, for example, ants feeding from other workers, from aphids or from inert food, or licking brood all use similar antennal movements, the frequency of distribution of acts might have a very different form at different levels.

Generally speaking the total behavioral repertoire is not evenly distributed over the workers of the colony. Wilson (1968) has argued that tasks would be most efficiently performed (or in his terms contingencies would be most effectively met) by workers specialized so that each task would be performed only by a particular subset of workers specially qualified in some way for that task. As a result the number of castes should equal the number of tasks. The colony's investment in each caste would depend on the cost of leaving the task unperformed (the contingency unmet), and the probability of the contingency arising. Wilson extended his idea of caste to include any subset of workers specially qualified in this way; that is, castes could be distinguished by their behavior and not by morphology or size only. Carroll and Janzen (1973) noted that Wilson's argument assumes that the contingencies which the colony has to meet are uncorrelated, while, in fact, a failure to make nest repairs is likely to increase the chance of predatory attack.

The intracolony differences in behavior that polyethism implies can hardly be due to genetic differences within the colony. Ant colonies seem to depend primarily on the aging process and on environmentally controlled differences in growth and development and perhaps on learning to introduce behavioral variability into their colony populations. However, the ability to develop different castes in response to these extrinsic factors may well be genetically controlled.

B. Age Polyethism

The known occurrence of age polyethism was reviewed by Wilson (1971); it probably is common in some form to all species of ant. The pattern is best known from the studies of Otto (1958) on a laboratory colony of *Formica polyctena*. Workers which have newly emerged from the pupal stage are at first inactive, then engage in tasks within the nest (care of eggs and queens). After a variable time these workers switch to activities outside the nest (mainly food collection and nest defense). It is convenient and appropriate to use Otto's terms Innendienst and Aussendienst for these two groups, respectively. The position of nest construction behavior in this sequence is a little uncertain, for the situation is complicated by the involvement of older Aussendienst workers in the collection of nest material in *Formica polyctena*. However, in *Formica yessensis* Higashi (1974) found the degree of mandibular wear in workers manipulating material on a natural nest to be similar to that of nurse ants. The change in behavior from Innendienst to Aussendienst is fairly well correlated with changes in the vestigial ovaries of worker ants. Innendienst workers normally have developing or ripe ova while the ovaries of Aussendienst ants contain degenerating ova. The state of the ovaries can therefore be used to assign workers, whose true age is unknown, to approximate age classes (e.g., Möglich and Hölldobler, 1974). Other anatomical characters can also be used, for example, the darkening of the cuticle (Weir, 1958; Wilson, 1976c) or the amount of mandibular wear (Higashi, 1974).

It is not clear whether the changes in the ovary (and the possible corresponding changes in the corpus allatum, Otto, 1958) are directly related to the change from Innendienst to Aussendienst or whether both, like the changes in pigmentation, are merely correlates of an underlying aging process. The change is in any case not rigidly fixed in time but can be accelerated or delayed according to the conditions and requirements of the colony. However, newly emerged workers kept without insect food or without older workers will take up Aussendienst duties although their ovaries fail to develop. The idea that the ovarian and behavioral changes are vestiges of temporal changes in solitary ancestors does not seem to have been formulated. The relation between ovarian cycles and nesting and provisioning behavior has been briefly discussed by Evans and West Eberhard (1973). Since the corpus allatum may control the beginning of adult behavior in female insects (e.g., Riddiford, 1973) as well as the maturation of oocytes, it may also be responsible for the age related development of behavior in worker ants. In general, the idea that the changes in behavior with time result from learning processes of individual ants (e.g., Heyde, 1924) may be discounted; however, it is possible that learning plays an important part in polyethism (see Section V,D).

Otto (1958) showed that some head glands also showed changes in size and complexity in the course of adult life, and it may be argued that age polyethism results from the physical inability of some ages of ant to perform certain tasks. However, as Wilson (1971) emphasized, the changes in the glands are by no means closely correlated with the tasks performed or with the time of change from Innendienst to Aussendienst. However, in some cases the ability to produce pheromones or to respond to them is apparently lacking in young ant workers. In *Myrmica rubra* the size of the poison gland, Durfour's gland, and the mandibular gland increase with age. The poison gland reservoir is empty in callow workers. However, they are able to respond to the secretions of the poison gland and of Defour's gland, although they are less sensitive to it than older workers are (Cammaerts-Tricot, 1974c; Cammaerts-Tricot and Verhaege, 1974). *Neivamyrmex* (Topoff *et al.,* 1972a) and *Eciton* (Topoff *et al.,* 1972b) callow workers are able to follow scent trails but do not do so readily. Callow workers of *Formica* spp. and *Myrmica rubra* are not usually attacked by other species or by members of stranger colonies of their own species (Jaisson, 1972). Callows of *Lasius flavus* are never attacked in strange nests of their own species but callows reared to a stage of dark pigmentation in isolation are attacked both by stranger colonies and by their parent colony (Anderson, 1970). This may be because callows are not able to produce a colony odor, or because they produce a special masking odor. Callows are possibly detected by their behavior rather than by odor. In general, therefore, callows differ in their readiness to perform certain pieces of behavior rather than in their physical ability, or equipment. Digging is shown from the earliest stages in *Myrmica rubra* (G. B. Frith, unpublished) and *Manica rubida* (Heyde, 1924). Brood care is also shown from the earliest stages in the latter species and although active attack is lacking even the youngest workers are able to bite (Heyde, 1924). The differences between older age groups are similar in nature to these between callows and older ants, since both Weir (1958) and Otto (1958) found differences in readiness to perform certain behavior rather than in ability. The simplest hypothesis explaining these differences would be that young workers are less aggressive than older ones, and that the older ones are more likely to respond photopositively, or to become photopositive when they are hungry, or when the colony is "hungry" (Otto, 1958). Readiness to leave a laboratory nest and to attack a pair of forceps is, of course, the easiest to observe of the behavioral characteristics of worker ants, just as pigmentation is the easiest morphological character in most species.

The 87 individually marked Innendienst ants studied by Otto (1958) included 18 which were most often engaged in doing nothing, and eight of these were never seen tending eggs or larvae, although these were always available to them. The most obvious common feature between these workers and the other Innendienst workers that did tend brood, therefore, is that they did not leave the nest, and not that they were all attracted to brood care. Wilson (1976c) found that the behavior of *Pheidole dentata* workers in a laboratory nest could be arranged in three categories; those carried out by workers of the three lightest pigmentation classes (A tasks), those performed predominantly by the third to fifth pigmentation classes (B tasks), and those carried out almost entirely by the sixth darkest and oldest class (C tasks). The A tasks involved attendance on queen and brood, B tasks care of mature larvae, and C tasks foraging, excavation and colony defence. One task (grooming pupae) fell in both A and B tasks, and five (food exchanges and self- and allogrooming) were performed by all ages of minor workers without bias. Thus the minor workers fall into three "temporal castes." Wilson suggested that this increases spatial efficiency since "it is more efficient for a particular ant grooming a larva to regurgitate to it as well." Unfortunately Wilson's results are not concerned with "particular ants" but only with members of the same pigmentation group, since he has not yet been able to extend these interesting studies to individually marked ants. Otto's (1958) data for analogous observations on individually marked *Formica polyctena* workers suggested that an appreciable proportion of individuals are specialists (see Section IV,D). Wilson's results are consistent with the idea that age polyethism is based on a spatial orientation, with Innendienst ants less likely to leave the nest; however, other work, summarized previously, shows that callow workers at least are less likely to respond to trail pheromones or to be aggressive, when tested in the same conditions as older workers. Both spatial and motivational factors are therefore involved.

The time at which workers changed from Innendienst to Aussendienst in *Formica polyctena* varied in 12 workers from day 46 to day 71. The change might involve an interval of several days while the ant engages in "guard duty" at the nest entrance or could occur without this interval. Some ants spontaneously returned to Innendienst after being seen in Aussendienst activities. A reversal could be induced by taking a group of ants collected at aphids in the field and putting them in a nest with brood but no Innendienst ants (Otto, 1958). There is abundant evidence (summarized by Otto, 1958) that the change can be accelerated, delayed, or reversed, according to conditions in the colony in several other species.

C. Caste Polyethism

In ants such as *Messor, Pogonomyrmex, Pheidole,* and *Camponotus,* which have a marked differentiation into two or more morphological castes, it is possible to list the tasks done by major workers (soldiers) and minor workers in the field. In other species, which have a considerable but continuous range of worker size, workers captured in different activities can be measured. However, this is open to criticism where the age of the workers is not known, as must be the case in general in field observations. If age polyethism exists in either caste, unless the age distribution is the same in each caste, age and caste polyethism will be confounded. If the production of soldiers is dependent on the nutritional status of the colony the age of major workers could vary with the development of the colony, with the season of the year or as a result of chance events. By applying his behavioral repertoire methods to *Pheidole dentata* in a laboratory nest Wilson (1976c) has shown that the behavioral repertoire was different in the two worker castes. While minor workers had an estimated total of 27 tasks (95% confidence range 26–28), the estimated size of the repertoire of major workers was only 9 (95% confidence range 8–10). Specifically majors engaged only in self-grooming, licking larvae, exchanging food with minor workers, feeding inside and outside the nest, foraging, carrying dead nestmates, and eating dead nestmates. In addition they responded strongly to an attack by *Solenopsis geminata* workers but less strongly to mechanical disturbance of the nest. Wilson (1976a) has shown that the recruitment of major workers to nest defense when *S. geminata* attacks is rather specific and does not occur when some other species attack. Minor workers use a poison-gland pheromone to recruit soldiers who respond aggressively to a combination of either *S. geminata* body odor or venom and movement. Wilson believed that major workers showed no age polyethism; since he did not determine the age of the major workers he must have based his conclusion on the small size and restricted nature of their behavioral repertoire. Higashi (1974) found that workers of *Formica yessensis* visiting aphids never included the largest sizes of worker, while hunters were larger. This agrees with results for *F. polyctena* (Otto, 1958), although Otto found that outgoing hunters included more small workers than returning hunters with prey, while Higashi found no differences in size between these two groups. Since there is no segregation of tree-visiting *Formica polyctena* in hunters and aphid tenders (Horstmann, 1973), this may only reflect the relative success of larger and smaller workers in capturing prey or in the distance they travel to trees.

D. The Existence of Specialists within Age and Caste Groups

When the effects of caste (or of worker size in species with a continuous range of sizes) and of age have been accounted for, any remaining consistent differences in behavior between individuals can be called specialization. The main evidence for the existence of such specialists comes from Otto's extensive study of *Formica polyctena,* in which he marked hundreds of ants individually in a laboratory nest. His results have been imperfectly understood by English speaking workers and are briefly reviewed here. For every ant that he was able to observe for a reasonably long time and for every task, Otto calculated a ratio M, where M = (number of times seen doing that task)/(number of times seen doing other tasks). Ants whose value of M for a particular task was greater than 0.5 were said to show a weak preference for the task, over 1.0 a strong preference, over 2.0 to be specialized and over 4.0 markedly specialized. These M values correspond to probabilities of performance of the task of 33, 50, 67, and 80% respectively. For this purpose doing nothing was treated as a task. Otto did not give a detailed summary of the occurrence of specialists, but his protocols suggest that as many as 40% of Innendienst ants showed a preference or specialization toward brood care. A further 10% preferred to do nothing. About 15% showed no preference and frequently changed tasks; Otto refers to these as "unstet" (inconstant). Some tasks such as attendance on the queen were not performed by specialists but by all ants. Among Aussendienst workers some appeared to specialize in tending aphids, in hunting, or in collecting nest material. However, Otto suggests that this is merely the expression of a fidelity to place (a specialization in locality rather than in occupation). This has subsequently been confirmed by Rosengren (1971) and Horstmann (1973) (see Section II,C). Otto also noted the heterogeneity of response of ants in natural conditions as evidence of specialization. For instance only about one ant in seven would retrieve pupae in his experiments. Since these ants were not individually marked it is not possible to say whether some ants consistently ignored pupae. The probability of a response is also commonly affected by the animal's previous activity, as the sequence analyses of Fuchs (1976b) have recently shown.

There is no direct evidence to show how this degree of specialization comes about, either for Innendienst or Aussendienst ants. Otto favored a genetic origin on the grounds that the most extreme specialization in Innendienst ants was apparent even in their earliest days after emergence, and that ants of similar age and thus similar experience showed diverse preferences or specializations. There are many difficulties in accepting this hypothesis of genetic variability. Queen and workers are cytologically similar and a genetic situation which produced a wide range of worker genotypes would also produce highly variable queens. Male ants are haploid and contribute nothing to the genetic variability of their daughters. Workers from the extremes of this range could become specialists within a larger nest population, but single queens from the extremes would produce an imbalanced (heavily skewed) distribution of worker types. Even queens with a genotype near to the mean worker genotype might produce imbalanced ranges in the initial stages of colony growth, when worker populations are still small. It is true that Crozier (1973) showed from allozymic studies of *Apaenogaster rudis* that the queens from 47 colonies were all heterozygotes for 2 alleles, while workers were approximately 50% homozygotic and 50% herozygotic. Thus selection appeared to have acted differentially on queens and workers. It is also true that a large proportion of queens fail to establish colonies, and that groups of queens are more successful than single queens in some species (Waloff, 1957; Wilson, 1966), and this could represent the disadvantage of the single, specialist endowed queen in colony foundation. However, in general, the haploid–diploid mechanism of sex determination in ants as in other Hymenoptera acts to reduce genetic variability within colonies. Hamilton (1964) and others (Trivers and Hare, 1976) have argued that this increased relationship (and therefore reduced variability) between sisters has been important in the evolution of social life. Possibly the necessity to incorporate nongenetic variation into the worker population can be related to the widespread age related polyethism in Hymenopterous societies, since aging is common to all animals, or at least to those that survive.

An alternative to the hypothesis that specializations arise from genetic differences is that they are the result of learning. Wilson (1975) suggests that the role of learning in the rapid course of a honeybee's life must be "narrowly directed and stereotyped at best," and he refers only to the role of learning in orientation and foraging and, indirectly, in the dialects of communicatory dances. Rapid localized learning, akin to imprinting, could, however, play a role in producing specialization. Jaisson (1975) investigated the ontogeny of the response of *Formica polyctena* workers to cocoons. He took callow workers and kept them in groups of 250–300 for the first 15 days after eclosion. Groups supplied with cocoons of their own species or of other species preferentially retrieved cocoons of that species from a collection of mixed cocoons. Where cocoons of another species were given during the training period, that species was preferred subsequently to cocoons of *Formica polyctena*. The preference lasted for 5–6 months. Moreover, callow workers which had not been supplied with any pupae during the first 15 days of adult life never retrieved cocoons when they were tested subsequently. Early experience with cocoons is thus essential for the development of one part of cocoon tending behavior. Jaisson relates his findings to imprinting. In the conditions of an ant brood chamber it is possible that some callows are more likely to experience cocoons than others, and a fortuitous difference of this kind could have far-reaching effects through the rapid and localized process of imprinting. However, Jaisson only investigated the retrieval of cocoons and not the whole range of behavior included in pupal care, and there is no evidence that similar learning might be involved in the care of larvae and eggs, to which most specialists were attached (Otto, 1958).

The acquisition of specialization allows a form of response to colony need, since it might be that the chance of experience with pupae, and thus of becoming a specialist, increases when many pupae are present. However, the adjustment would be inefficient, because of the time lags involved. Imprinting appears to be a method of increasing behavioral variability by magnifying chance effects. To that extent it resembles caste determination where small differences in growth rate can divert ants into various developmental paths.

Wilson (1968) predicted that as many castes (in his sense this includes temporal castes) should evolve as there are tasks. However, in his study of *Pheidole dentata* this was not so, and there were only three temporal castes (or two and a transitional stage) in minor workers and one in major workers. Specialization within these could provide more "castes." While this would achieve one function of polyethism (variability of response to a situation), it could only provide the increased economy on which Wilson's argument (Wilson, 1968) is based if the imprinted worker was more efficient than others in performing the imprinted task. While it is probably true that Aussendienst ants learn good foraging grounds (Rosengren, 1971), and perhaps learn techniques for the manipulation of loads (Chauvin, 1974; Dobrzanski, 1968) evidence of this in Innendienst workers is lacking.

REFERENCES

Anderson, R. E. (1970). An investigation into the colony odour of the ants *Lasius flavus* Fab. Ph.D. Thesis, University of Cambridge, England.

Ayre, G. L. (1969). Comparative studies on the behaviour of three species of ant. II. Trail formation and laying. *Can. Entomol.* **101** (2), 118–127.

Banks, C. J. (1958). The effects of the ant *Lasius niger* (L.) on the behaviour and reproduction of the Black Bean Aphid, *Aphis fabae* Scop. *Bull. Entomol. Res.* **49**, 701–714.

Benois, A. (1972). Etude ecologique de *Camponotus vagus* Scop (= *pubescens* Fab.) dans la région d'Antibes. *Insectes Soc.* **19**, 111–129.

Bergstrom, G., and Löfqvist, J. (1970). Chemical basis for odour communication in four species of *Lasius* ants. *J. Insect Physiol.* **16**, 2353–2375.

Bernstein, R. (1975). Foraging strategies of ants in response to variable food density. *Ecology* **56**, (1), 213–219.

Bernstein, R. A. (1979). Evolution of niche breadth in populations of ants. *Am. Nat.* **114** (4), 533–544.

Bhatkar, A. P. (1974). Oriented mounds of terrestrial ants. *Fla. Entomol.* **57** (2), 153.

Bigley, W. S., and Vinson, S. B. (1975). Characterisation of a brood pheromone isolated from the sexual brood of the imported fire ant *Solenopsis invicta. Ann. Entomol. Soc. Am.* **68** (2), 301–304.

Bossert, W. H., and Wilson, E. O. (1963). The analysis of olfactory communication among animals. *J. Theor. Biol.* **5**, 443–469.

Brandt, D. C. (1980). The thermal diffusivity of the organic matter of a mound of *Formica polyctena* Foerst in relation to the thermoregulation of the brood. *Netherl. J. Zool.* **30** (2), 326–344.

Brian, M. V. (1964). Ant distribution in a Southern English heath. *J. Anim. Ecol.* **33**, 451–461.

Brian, M. V. (1970). Communication between queens and larvae in the ant *Myrmica. Anim. Behav.* **18**, 467–472.

Brian, M. V. (1973a). Caste control through worker attack in the ant *Myrmica. Insectes Soc.* **20**, 87–102.

Brian, M. V. (1973b). Temperature choice and its relevance to brood survival and caste determination in the ant *Myrmica rubra. Physiol. Zool.* **46** (4), 245–252.

Brian, M. V. (1974). Brood-rearing in small cultures of the ant *Myrmica rubra. Anim. Behav.* **22**, 879–889.

Brian, M. V. (1975a). Caste determination through a queen influence on diapause in *Myrmica rubra. Entomol. Exp. Appl.* **18** (4), 429–442.

Brian, M. V. (1975b). Larval recognition by workers of the ant *Myrmica rubra. Anim. Behav.* **23** (4), 745–756.

Brian, M. V., and Downing, B. M. (1958). The nests of some British ants. *Proc. Int. Congr. Entomol., 10th, 1956* Vol. 2, pp. 539–540.

Brian, M. V., Hibble, J., and Kelly, A. F. (1966). The dispersion of ant species on a southern English heath. *J. Anim. Ecol.* **35** (2), 281–290.

Brown, W. L., and Wilson, E. O. (1959). The evolution of the Dacetine ants. *Rev. Biol.* **34** (4), 278–294.

Buschinger, A. (1973). Transport und Ansetzen von Larven an Beutestücke bei der Ameise *Aphaenogaster subterranea. Zool. Anz.* **190** (1), 63–66.

Büttner, K. (1973). Untersuchungen über den Einfluss des Beutetiers auf dem Erbeutungsvorgang bei der Waldameise *Formica polyctena. Z. Angew. Entomol.* **74** (2), 177–196.

Büttner, K. (1974a). Die Bedeutung abiotischer Faktoren für die Erbeutung von Insekten durch Waldameisen (*Formica polyctena*). Teil I. Der Einfluss von Temperatur, Luftfeuchtigkeit und Beleuchtungs-stärke. *Waldhygiene* **10** (5), 129–140.

Büttner, K. (1974b). Teil II. Der Einfluss des Erbeutungsortes. *Waldhygiene* **10** (5), 141–155.

Byron, P. A., Byron, E. R., and Bernstein, R. A. (1980). Evidence of competition between two species of desert ants. *Insectes Soc.* **27** (4), 351–360.

Cammaerts-Tricot, M. C. (1974a). Recrutement d'ouvrières chez *Myrmica rubra,* par les phéromones de l'appareil à venim. *Behaviour* **50** (1 and 2), 111–122.

Cammaerts-Tricot, M. C. (1974b). Piste et phéromone attractive chez la *Myrmica rubra. J. Comp. Physiol.* **88**, 373–382.

Cammaerts-Tricot, M. C. (1974c). Production and perception of attractive pheromones by differently aged workers of *Myrmica rubra. Insectes Soc.* **21** (3), 235–248.

Cammaerts-Tricot, M. C., and Verhaeghe, J. C. (1974). Ontogenesis of trail pheromone production in the workers of *Myrmica rubra. Insectes Soc.* **21**, 275–282.

Cammaerts-Tricot, M. C., Morgan, E. D., Tyler, R. C., and Brackman, J. C. (1976). Dufour's gland secretion of *Myrmica rubra:* Chemical, electrophysiological and ethological studies. *J. Insect Physiol.* **22** (7), 927–932.

Carroll, C. R., and Janzen, D. H. (1973). The ecology of foraging ants. *Ann. Rev. Ecol. Syst.* **4**, 231–257.

Casewitz-Weulersse, J. (1972). Habitat et comportement nidificateur du *Crematogaster scutellaris* Olivier. *Bull. Soc. Entomol. Fr.* **77**, 12–19.

Casewitz-Weulersse, J. (1973). Sur la présence de *Crematogaster scutellaris* Olivier dans les suberaies de Sardigne. *Bull. Mus. Natl. Hist. Nat., Zool.* **119**, 45–63.

Chabab, R., and Rettenmeyer, C. W. (1975). Mass recruitment by Army ants. *Science* **188**, 1124–1125.

Chauvin, R. (1960). Facteurs d'asymmetry et facteurs de régulation dans la construction du dôme chez *Formica rufa. Insectes Soc.* **8**, 201–205.

Chauvin, R. (1974). La motivation de traction chez *Formica rufa. Insectes Soc.* **21** (2), 157–162.

Cherrett, J. M., and Seaforth, L. E. (1968). Phytochemical arrestants for the leaf-cutting ants *Atta cephalotes* (L.) and *Acromyrmex octospinosus. Bull. Entomol. Res.* **59**, 615–621.

Colombel, P. (1970). Recherches sur la biologie et l'éthologie de *Odontomachus haematodes*. Etudes des populations dans leur milieu naturel. *Insectes Soc.* **17**, 183–198.

Crozier, R. H. (1973). Apparent differential selection at an iso-enzyme locus between queens and workers of the ant *Aphaenogaster rudis*. *Genetics* **73** (2), 313–318.

Czerwinski, J., Jakubczyk, H., and Petal, J. (1972). Influence of ant-hills on meadow soils. *Pedobiologica* **11**, 227–280.

Davidson, D. W. (1980). Some consequences of diffuse competition in a desert ant community. *Am. Nat.* **116** (1), 92–105.

Delye, G. (1968). Recherches sur l'écologie, la physiologie et l'éthologie des fourmis du Sahara. Thèse, Université d'Aix-Marseille.

Delye, G. (1971). Observations sur le nid et le comportement constructeur de *Messor arenarius*. *Insectes Soc.* **18**, 15–20.

Dlussky, G. M. (1974). Nest structure in desert ants. *Zool. Zh.* **53** (2), 224–236; *Entomol. Abstr.* (1975) **6**, 445.

Dobrzanska, J. (1958). Partition of foraging grounds and modes of conveying information among ants. *Acta Biol. Exp.* **18**, 55–67.

Dobrzanski, J. (1968). Über das Lernvermögen von Ameisen. *Naturwissenschaften* **55** (2), 89.

Douglas, J. M., and Sudd, J. H. (1980). Behavioural coordination between an aphis (*Symydobius oblongus*) and the ant that attends it (*Formica lugubris*): an ethological analysis. *Anim. Behav.* **28** (4), 1127–1139.

Duelli, P. (1973). Astrotaktisches Heimfindevermögen tragender und getragener Ameisen. *Rev. Suisse Zool.* **80** (3), 712–719.

Ettershank, G. (1971). Some aspects of the ecology and microclimatology of the meat-ant *Iridomyrmex purpureus* (Sm.). *Proc. R. Soc. Victoria* **84**, 137–152.

Evan, H. C. (1973). Invertebrate vectors of *Phytophthora palmivora* causing Black-Pod Disease of Cocoa in Ghana. *Ann. Appl. Biol.* **75** (3), 331–345.

Evans, H. E., and West-Eberhard, M. J. (1973). "The Wasps." Univ. of Michigan Press, Ann Arbor.

Fagen, R. M., and Goldman, R. N. (1977). Behavioural catalogue methods. *Anim. Behav.* **25** (2), 261–274.

Fletcher, D. J. C. (1971). The glandular structure and social functions of trail-pheromones in two species of ant (*Leptogenys* H.-F.). *J. Entomol., Ser. A: Gen. Entomol.* **46**, 27–37.

Fuchs, S. (1976a). The response to vibrations of the substrate and reactions to the specific drumming in colonies of Carpenter Ants (*Camponotus,* Formicidae, Hymenoptera). *Behav. Ecol. Sociobiol.* **1**, 155–184.

Fuchs, S. (1976b). An informational analysis of the alarm communication by drumming in nests of Carpenter Ants (*Camponotus,* Formicidae, Hymenoptera). *Behav. Ecol. Sociobiol.* **1**, 315–336.

Gentry, G. B., and Stiritz, K. L. (1972). The role of the Florida Harvester Ant, *Pogonomyrmex badius* in old field mineral nutrient relationships. *Environ. Entomol.* **1**, 39–41.

Glancey, R. M., Stringer, C. E., Craig, C. H., Bishop, P. M., and Martin, B. B. (1970). Pheromone may induce brood-tending in the fire-ant *Solenopsis saevissima*. *Nature (London)* **226**, 803–804.

Gotwald, W. H. (1974). Predatory behaviour and food preferences of Driver Ants in selected African habitats. *Ann. Entomol. Soc. Am.* **67** (6), 877–886.

Greenslade, P. J. M. (1974). Some relations of the meat-ant, *Iridomyrmex purpureus,* with soil in S. Australia. *Soil Biol. Biochem.* **6** (1), 7–14.

Hamilton, W. D. (1964). The genetic evolution of social behaviour. *J. Theor. Biol.* **7**, 1–52.

Hangartner, W. (1967). Spezifitat und Inaktivierung des Spurphenomenons von *Lasius fuliginosus* und Orientierung der Arbeiterinnen im Duftfeld. *Z. Vergl. Physiol.* **57**, 103–136.

Hangartner, W. (1969a). Structure and variability of the individual odour trail in *Solenopsis geminata* Fab. *Z. Vergl. Physiol.* **62** (1), 111–120.

Hangartner, W. (1969b). Trail-laying in the subterranean ant *Acanthomyops interjectus*. *J. Insect Physiol.* **15**, 1–4.

Hangartner, W. (1970). Control of pheromone quantity in odour trails of *Acanthomyops interjectus*. *Experientia* **26**, 664–665.

Hangartner, W., Teechson, J. M., and Wilson, E. O. (1970). Orientation to nest material by the ant *Pogonomyrmex badius*. *Anim. Behav.* **18**, 331–334.

Haskins, C. P., Hewitt, R. E., and Haskins, E. F. (1973). Release of aggressive capture behaviour in the ant *Myrmecia gulosa* F. by exocrine products of the ant *Camponotus*. *J. Entomol., Ser. A: Gen. Entomol.* **47** (2), 125–139.

Henquel, D. (1976). Sur l'existence d'une piste chimique chez *Formica polyctena* dans des conditions de vie semi-naturelle. *Insectes Soc.* **23** (4), 577–583.

Hermann, H. R. (1975). Crepuscular and nocturnal activities of *Paraponera clavata*. *Entomol. News* **86** (5–6), 94–98.

Herzig, J. (1938). Ameisen und Blattläuse (Ein Beitrag zur Ökologie aphidophiler Ameisen). *Z. Angew. Entomol.* **24**, 367–435.

Heyde, K. (1924). Die Entwicklung der psychischen Fähigkeiten bei Ameisen und ihr Verhalten bei abgeänderten biologischen Bedingungen. *Biol. Zentralbl.* **44**, 623–654.

Higashi, S. (1974). Worker polyethism related with body size in a polydomous red wood ant (*Formica yessensis* Forel). *J. Fac. Sci., Hokkaido Univ., Ser. 6* **19** (3), 695–705.

Hölldobler, B. (1971a). Recruitment behaviour in *Camponotus socius*. *Z. Vergl. Physiol.* **75**, 123–142.

Hölldobler, B. (1971b). Homing in the Harvester Ant *Pogonomyrmex badius*. *Science* **171**, 1149–1151.

Hölldobler, B. (1973). Chemische Strategie beim Nahrungserwerb der Diebsameise (*Solenopsis fugax* Latr.) und der Pharoameise (*Monomorium pharaonis* (L.)). *Oecologia* **11** (4), 371–380.

Hölldobler, B. (1974). Home range orientation and territoriality in harvesting ants. *Proc. Natl. Acad. Sci. U.S.A.* **71**, 3274–3277.

Hölldobler, B. (1976). Recruitment behaviour, home range and territoriality in Harvester Ants. *Behav. Ecol. Sociobiol.* **1** (1), 3–44.

Hölldobler, B., Möglich, M., and Maschwitz, U. (1974). Communication by tandem running in the ant *Camponotus sericeus*. *J. Comp. Physiol.* **90**, 105–127.

Horstmann, K. (1973). Untersuchungen zur Arbeitsteilung unter den Aussendienstarbeiterinnen der Waldameise *Formica polyctena*. *Z. Tierpsychol.* **32**, 532–543.

Horstmann, K. (1974a). Die Umlaufzeit bei den Aussendienstarbeiterinnen der Waldameise. *Waldhygiene* **10** (8), 241–246.

Horstmann, K. (1974b). Untersuchungen über dem Nahrungserwerb der Waldameise. III. Jahresbilanz. *Oecologia* **15** (2), 187–204.

Horstmann, K. (1975). Frielanduntersuchungen zum Rekrutierungverhalten bei der Waldameise, *Formica polyctena* Förster. *Waldhygiene* **11** (2), 33–40.

Horstmann, K. (1976a). Uber die Duftspurorientierung bei der Waldameise (*Formica polyctena* Forst.). *Insectes Soc.* **23**, 227–242.

Horstmann, K. (1976b). Über die Struktur des Waldameisennestes und ihre Bedeutung für den Nahrungstransport. *Mitt. Dtsch. Entomol. Ges.* **35**, 91–98.

Horton, P. M., Hayes, S. B., and Holman, J. R. (1975). Food carrying ability and recruitment in the red imported fire ant. *J. Ga. Entomol. Soc.* **10** (3), 207–213.

Hubbard, M. D. (1974). Influence of nest material and colony odour on digging in the ant *Solenopsis invicta*. *J. Ga. Entomol. Soc.* **9** (2), 127–132.

Huber, P. (1810). "Vie et moeurs des fourmis indigènes." Paris & Geneva.

Huber, R. (1965). Untersuchungen zur Representanz und Einstellung der Waldameisen. *Collana Verde* **16**, 175–186.

Huwyler, S., Grob, K., and Viscontini, M. (1975). The trail pheromone of the ant *Lasius fuliginosus. J. Insect Physiol.* **21** (2), 299–304.

Imamura, S. (1974). Observations on the hibernation of the polydomous ant, *Formica yessensis* For. *J. Fac. Sci., Hokkaido Univ., Ser. 6* **19** (2), 438–444.

Ito, M. (1973). Population trends and nest structure of *Formica yessensis. J. Fac. Sci., Hokkaido Univ., Ser. 6* **19** (1), 270–275.

Jaffe, K. (1980). Theoretical analysis of the communication system for chemical mass recruitment in ants. *J. Theor. Biol.* **84**, 589–609.

Jaffe, K., and Howse, P. E. (1979). The mass recruitment system of the leaf-cutting ant, *Atta cephalotes* L. *Anim. Behav.* **27** (3), 930–939.

Jaisson, P. (1972). Mise en évidence d'une phéromone d'attractivitée produite par la jeune ouvrière *Formica polyctena. C.R. Hebd. Seances Acad. Sci., Ser. D* **274**, 429–437.

Jaisson, P. (1975). L'imprégnation dans l'ontogénèse des comportements de soins aux cocons chez la jeune fourmi rousse (*Formica polyctena* Forst.). *Behaviour* **52**, 1–37.

Kempf, W. W., and Brown, W. L. (1970). Two new ants of the tribe Ecatommini from Colombia. *Stud. Entomol.* **13**, 311–320.

Kleinjan, J. E., and Mittler, T. E. (1975). A chemical influence of ants on wing development in aphids. *Entomol. Exp. Appl.* **18** (3), 384–388.

Kravchenko, M. A. (1973). (Thermoregulation in nests of *Lasius fuliginosus.*) *Zool. Zh.* **52** (3), 454–457: *Entomol. Abstr.* **5**, 258.

Kruk de Bruin, M., and Tissing, O. (1975). A T.V. scanning line comparator for recording locomotor activity of animals. *Oecologia* **20** (2), 189–191.

Leuthold, R. H. (1968). A tibial scent gland, scent trail and trail-laying behaviour in the ant *Crematogaster ashmeadi* Mayr. *Psyche* **75** (3), 233–248.

Levieux, J. (1971). Mise en évidence de la structure des nids et l'implantation des zones de chasse de deux espèces de *Camponotus* à l'aide de radio-isotopes. *Insectes Soc.* **18**, 29–48.

Lewis, T., Pollard, G. V., and Sibley, G.C. (1974). Microenvironmental factors affecting diel patterns of foraging in the leaf-cutting ant *Atta cephalotes. J. Anim. Ecol.* **43**, 143–153.

Littledyke, M., and Cherrett, J. M. (1975). Variability in the selection of substrate by the leaf-cutting ants *Atta cephalotes* and *Acromyrmex octospinosus. Bull. Entomol. Res.* **65** (1), 33–47.

Löfqvist, J. (1977). Toxic properties of the chemical defence system in the competitive ants *Formica rufa* and *F. sanguinea. Oikos* **28**, 137–151.

Malyshev, S. I. (1968). "Genesis of the Hymenoptera and the Phases of their Evolution." Methuen, London.

Maschwitz, U. (1964). Gefahralarmstoffe und Gefahrenalarmierung bei sozialen Hymenopteren. *Z. Vergl. Physiol.* **47** (6), 596–655.

Maschwitz, U. (1974). *Camponotus rufoglaucus,* eine weglagernde Ameise. *Zool. Anz.* **191** (5–6), 364–368.

Maschwitz, U., and Hölldobler, B. (1970). Der Kartonnestbau bei *Lasius fuliginosus* Latr. *Z. Vergl. Physiol.* **66** (1–2), 176–189.

Maschwitz, U., and Mühlenberg, M. (1975). Zur Jagdstrategie einiger orientalischen *Leptogenys*-Arten. *Oecologia* **20** (1), 65–83.

Maschwitz, U., Hölldobler, B., and Möglich, M. (1974). Tandemlaufen als Rekrutierungsverhalten bei *Bothroponera tesserinoda* Forel. *Z. Tierpsychol* **35** (2), 113–123.

Möglich, M. (1975). Recruitment of *Leptothorax. Proc. Symp. Pheromones Def. Secretion Social Insects, 1974* pp. 235–242.

Möglich, M., and Hölldobler, B. (1974). Social carrying behavior and division of labour during nest moving in ants. *Psyche* **81** (2), 219–236.

Möglich, M., and Hölldobler, B. (1975). Communication and orientation during foraging in the ant *Formica fusca* L. *J. Comp. Physiol.* **101**, 275–288.

Morrill, W. L. (1972). Tool-using behaviour of *Pogonomyrmex badius*. *Fla. Entomol.* **55**, 59–60.

Müller-Schneider, P. (1971). Beiträge zur Kenntnis der Samenverbreitung durch Ameisen. *Ber. Schweiz. Bot. Ges.* **80**, 289–297.

Nault, L. R., Montgomery, M. E., and Bowers, W.S. (1976). Ant-aphid association: Role of Aphid alarm pheromone. *Science* **192**, 1349–1351.

Nickle, D. A., and Neal, T. M. (1972). Observations of foraging behaviour of Southern Harvester Ant, *Pogonomyrmex badius*. *Fla. Entomol.* **55**, 65–66.

Nielsen, M. G., and Jensen, T. F. (1975). Økologische studier over *Lasius alienus*. *Entomol. Medd.* **43** (1), 5–16.

Ofer, Y. (1969). Biology of a Weaver Ant, *Polyrachis simplex*. *Proc. Int. Congr. Int. Union Study Soc. Insects, 6th, 1969* pp. 201–206.

Otto, D. (1958). Über die Arbeitsteilung im Staate von *Formica rufa rufo-pratensis minor* Gössw. *Wiss. Handl. Dtsch. Akad. Landwirtschaftswiss. Berlin* **30**, 1–169.

Passera, L. (1974). Différenciation des soldats chez la fourmi *Pheidole pallidula* Nyl. *Insectes Soc.* **21** (1), 71–86.

Pasteels, J. M., Crewe, R. M., and Blum, M. S. (1971). Etude histologique et examen au microscope electronique à balayage de la glande secretant la phéromone de piste chez deux *Crematogaster* nord-americains. *C.R. Hebd. Seances Acad. Sci., Ser. D* **271**, 835–838.

Paul, R. (1974). Observations of ant aggression towards aphids. *Entomol. Mon. Mag.* **110**, 53.

Pavan, M. (1962). Données chimiques et biologiques sur les secretions des Formicidae et Aphidae. *Symp. Genet. Biol. Ital.* **12**, 3–21.

Plowright, R. C. (1979). Social facilitation at nest entrances of bumble bees and wasps. *Insectes Soc.* **26** (3), 223–231.

Pontin, A. J. (1960). Observations on the keeping of aphid eggs by ants of the genus *Lasius* (Hym., Formicidae). *Entomol. Mon. Mag.* **96**, 189–199.

Pontin, A. J. (1961). The prey of Lasius niger (L.) and L. flavus (L.) (Hym., Formicidae). *Entomol. Mon. Mag.* **97**, 135–137.

Pontin, A. J. (1978). The number and distribution of subterranean aphids and their exploitation by the ant *Lasius flavus*. *Ecol. Entomol.* **3**, 203–207.

Prado, L. (1973). Le comportement de decoupage des feuilles chez *Atta sexdens*. *Insectes Soc.* **20** (2), 133–144.

Riddiford, L. M. (1973). *In* "Experimental Analysis of Insect Behaviour" (L. B. Browne, ed.), 286–296. Springer-Verlag, Berlin and New York.

Rissing, S. W., and Wheeler, J. (1976). Foraging responses of *Veromessor pergandei* to changes in seed production. *Pan-Pac. Entomol.* **52** (1), 63–72.

Ritter, F. J., Bruggeman-Rotgans, I. E. M., Verkuil, E., and Persoons, W. (1975). The trail pheromone of *Monomorium pharaonis*, components of the odour trail and their origin. *Proc. Symp. Pheromones Def. Secretion Social Insects, 1974* pp. 99–103.

Robertson, P. (1971). Pheromones involved in aggressive behaviour in the *Myrmecia gulosa*. *J. Insect Physiol.* **17**, 691–715.

Rockwood, L. L. (1976). Plant selection and foraging patterns in two species of leaf-cutting ants. *Ecology* **57** (1), 48–61.

Rogers, L. E. (1974). Foraging activity of the Western Harvester Ant (*Pogonomyrmex occidentalis*) in the short-grass plains ecosystem. *Environ. Entomol.* **3** (3), 420–424.

Rogers, L. E., and Lavigne, R. J. (1974). Environmental effects of Western Harvester Ants on the short-grass plains ecosystem. *Environ. Entomol.* **3** (6), 994-997.

Rosengren, R. (1971). Route fidelity, visual memory and recruitment behaviour in foraging wood-ants of the genus *Formica. Acta Zool. Fenn.* **133**, 1-106.

Sanders, C. J. (1970). The distribution of Carpenter Ant (*Camponotus* spp.) colonies in the spruce-fir forests of Northwestern Ontario. *Ecology* **51**, 865-873.

Schmidt, G. H. (1974). "Sozialpolymorphismus bei Insekten." Wiss. Verlagsges., Stuttgart.

Schmidt, G. H., and Gürsch, E. (1971). Analyse der Spinnenbewegungen der Larve von *Formica pratensis* Retz. *Z. Tierpsychol.* **28**, 19-32.

Schneider, P. (1971). Vorkommen und Bau von Erdhügelnestern bei der Afghanischen Wüstenameise *Cataglyphis bicolor* Fab. *Zool. Anz.* **187**, 202-213.

Skinner, G. J., and Whitaker, J. B. (1981). An experimental study of the inter-relationship between the wood-ant, (*Formica rufa*) and some tree-canopy herbivores. *J. Anim. Ecol.* **50** (1), 313-326.

Spangler, H. G. (1973). Vibration aids soil manipulation in Hymenoptera. *J. Kans. Entomol. Soc.* **46** (2), 157-160.

Sudd, J. H. (1960a). Interaction between ants on a scent trail. *Nature (London)* **183**, 1588.

Sudd, J. H. (1960b). The transport of prey by an ant, *Pheidole crassinoda* Em. *Behaviour* **16**, 295-308.

Sudd, J. H. (1960c). The foraging method of *Mononorium pharaonis* (L.). *Anim. Behav.* **8**, 67-75.

Sudd, J. H. (1965). The transport of prey by ants. *Behaviour* **25**, 234-271.

Sudd, J. H. (1969). The excavation of soil by ants. *Z. Tierpsychol.* **26**, 257-276.

Sudd, J. H. (1970a). Specific patterns of excavation in isolated ants. *Insectes Soc.* **17**, 253-260.

Sudd, J. H. (1970b). The response of isolated digging ants to tunnels. *Insectes Soc.* **17**, 261-282.

Sudd, J. H. (1970c). Selective removal of soil particles by ants. *J. Niger. Entomol. Soc.* **1**, 122-125.

Sudd, J. H. (1971). The effect of tunnel depth and of working in pairs on the speed of excavation in ants. *Anim. Behav.* **19**, 677-686.

Sudd, J. H. (1972). The response of digging ants to gravity. *Insectes Soc.* **19** (3), 243-250.

Sudd, J. H. (1975). A model of digging behaviour and tunnel production in ants. *Insectes Soc.* **22** (3), 225-236.

Szlep-Fessel, R. (1970). The regulatory mechanism in mass foraging and the recruitment of soldiers in *Pheidole. Insectes Soc.* **17**, 233-244.

Topoff, H., Boshes, M., and Trakimas, W. (1972a). A comparison of trail following between callows and adult workers of the army ant *Neivamyrmex nigrescens. Anim. Behav.* **20**, 361-366.

Topoff, H., Lawson, K., and Richards, P. (1972b). Trail following and its development in the Neotropical Army Ant genus *Eciton. Psyche* **79** (4), 357-364.

Torossian, C. (1973). Etudes des communications antennaires chez les Formicoidea. Analyse du comportement trophallactic des ouvrières de *Dolichoderus quadripunctatus* lors d'échanges alimentaires practiques au benefice des sexués de la colonie. *C.R. Hebd. Seances Acad. Sci., Ser. D* **277**, 2073-2075.

Trivers, R. L., and Hare, H. (1976). Haploidy and the evolution of the social insects. *Science* **191**, 249-263.

Tschinkel, W. R., and Bhatkar, A. (1974). Orientated mound building in the ant *Trachymyrmex septentrionalis. environ. Entomol.* **3** (4), 667-673.

Tweed, R. L. (1980). The collection of honeydew and the foraging system of the wood-ant (*Formica lugubris* Zetterstedt; Hymenoptera, Formicidae). Ph.D. Thesis, University of Hull, England.

Waloff, N. (1957). The effect of the number of queens of the ant *Lasius flavus* on their survival and on the rate of development of the first brood. *Insectes Soc.* **4** (4), 391–405.

Walsh, J. P., and Tschinkel, W. R. (1974). Brood recognition by contact pheromone in the red imported fire ant *Solenopsis invicta. Anim. Behav.* **22** (3), 695–704.

Wehner, R. (1970). Etudes sur la construction des cratères au-dessus des nids de la fourmi *Cataglyphis bicolor. Insectes Soc.* **17**, 83–94.

Weir, J. (1958). Polyethism in the ant *Myrmica. Insectes Soc.* **5** (1), 97–128.

Wellenstein, G. (1952). Zur Ernahrungsbiologie der Roten Waldameise. *Z. Pflanzenkr. (Pflanzenpathol.) Pflanzenschutz* **59**, 430–451.

Went, F. W., Wheeler, J., and Wheeler, G. C. (1971). Feeding and competition in some ants. *BioScience* **22**, 82–88.

Whitford, W. G. (1976). Foraging behaviour of Chihuahuan desert harvester ants. *Am. Midl. Nat.* **95** (2), 455–458.

Whitford, W. G., Depree, E., and Johnson, P. (1980). Foraging ecology of two Chihuahuan desert ant species. *Insectes Soc.* **27** (2), 148–156.

Wilson, E. O. (1962). Chemical communication among workers of the fire ant *Solenopsis saevissima* (F. Smith). I. The organisation of mass foraging. *Anim. Behav.* **10** (1–2), 134–147.

Wilson, E. O. (1963). The social biology of ants. *Annu. Rev. Entomol.* **8**, 345–368.

Wilson, E. O. (1966). Behaviour of social insects. *Symp. R. Entomol. Soc. London* **3**, 81–96.

Wilson, E. O. (1968). The ergonomics of caste in social insects. *Am. Nat.* **102**, 41–66.

Wilson, E. O. (1971). "The Social Insects." Belknap Press, Cambridge, Massachusetts.

Wilson, E. O. (1975). "Sociobiology, the New Synthesis." Belknap Press, Cambridge, Massachusetts.

Wilson, E. O. (1976a). The organisation of colony defence in the ant *Pheidole dentata* Mayr. *Behav. Ecol. Sociobiol.* **1** (1), 63–81.

Wilson, E. O. (1976b). A social ethogram of the neotropical arboreal ant *Zacrytocerus varians* (F. Smith). *Anim. Behav.* **24** (2), 354–363.

Wilson, E. O. (1976c). Behavioural discretization and the number of castes in an ant species. *Behav. Ecol. Sociobiol.* **1**, 141–154.

Wilson, E. O., and Fagen, R. M. (1974). On the estimation of total behavioural repertoires in ants. *J. N.Y. Entomol. Soc.* **82**, 106–112.

Wilson, E. O., and Farish, D. J. (1973). Predatory behaviour in the ant-like wasp *Methocha stygia* (Say). *Anim. Behav.* **21** (2), 292–298.

Wisniewski, J. (1967a). Die Zusammensetzung des Baumaterials der Hesthügel von *Formica polyctena* in Kiefernwaldern. *Waldhygiene* **7**, 117–121.

Wisniewski, J. (1967b). An analysis of animal remnants occurring in anthills of Formica polyctena Forst. *Pol. Pismo Entomol.* **37** (2), 385–390.

Zakharov, A. A. (1973). (The sector-wise distribution of columns of ants within the nest in *Formica rufa.) Zool. Zh.* **52** (4), 519–528; *Entomol. Abstr.* (1974) **5**, 386.

3

Army Ants

WILLIAM H. GOTWALD, JR.

I. INTRODUCTION

"I was glad to go to sleep early, but was scarce soundly asleep when I was turned out of the house by a furious attack of the bashikouay ants. They were

SOCIAL INSECTS, VOL. IV
Copyright © 1982 by Academic Press, Inc.
All rights of reproduction in any form reserved.
ISBN 0-12-342204-3

already all over me when I jumped up, and I was terribly bitten." Thus, DuChaillu (1861), an intrepid explorer of equatorial Africa, described an unpleasant encounter with army ants. He apparently was no less impressed with these diminutive but fearsome creatures than he was with the gorilla, which, with great bravado, he described as a "monstrous and ferocious ape." Few nineteenth century explorers and naturalists in the Old and New World tropics could ignore the cohesive, regimented behavior, let alone the often painful attacks, of these ants; ants that Wheeler (1910) referred to as the "Huns and Tartars of the insect world." Their journals and adventurous accounts are replete with references to the predatory exploits of these ants. Bates (1863), a naturalist who roamed Amazonas collecting biological specimens, noted that when a "pedestrian falls in with a train of these ants," they "swarm up his legs with incredible rapidity, each one driving his pincerlike jaws into his skin, and with the purchase thus obtained, doubling in its tail, and stinging with all its might." Another naturalist, Thomas Belt (1874), recorded how one small species of army ant in Nicaragua would "visit our house, swarm over the floors and walls, searching every cranny, and driving out the cockroaches and spiders, many of which were caught, pulled or bitten to pieces, and carried off." The Rev. Thomas S. Savage (1847), a medical missionary to west Africa, noted in his observations on army ants of the subgenus *Anomma,* that this kind of ant "drives every thing before it capable of muscular motion, so formidable is it from its numbers and bite." Even Charles Darwin (1859) was fascinated with army ants, although his interest was not so much in their spectacular foraging behavior as it was with the striking allometry and polymorphism exhibited by the worker caste. He examined a series of *Anomma* workers and described their morphology with the following analogy:

> . . . the difference [between worker ant sizes] was the same as if we were to see a set of workmen building a house, of whom many were five feet four inches high, and many sixteen feet high; but we must in addition suppose that the larger workmen had heads four instead of three times as big as those of the smaller men, and jaws nearly five times as big.

Although ant taxonomists, particularly Emery (1895), were actively considering the army ants during the latter half of the nineteenth century, especially at the alpha level, little was recorded in any systematic fashion about the biology of army ants. Sumichrast (1868) and Müller (1886) described their observations of New World species as did Savage (1847, 1849), F. Smith (1863), Perkins (1869), and Wroughton (1892) for a few Old World forms, but much of the literature remained in the realm of natural history.

Similarly in the twentieth century, naturalists have alternately cursed and marveled at these ants. Carpenter (1920) and Loveridge (1949, 1953) both recorded observations of army ants, in particular of *Anomma* driver ants, in popular accounts of their field research in Africa. In one narrative, Loveridge (1949) described a remarkable invasion of these ants in his house in which the

"the whitewashed walls were a moving mass of Siafu." The nonscientific literature, particularly of colonial Africa, also includes references to the activities of army ants. For instance, Isak Dinesen (a pseudonym taken by the Danish baroness Karen Blixen-Finecke) (1937, p. 35) in her memorable book, "Out of Africa," noted with annoyance how her dogs had been attacked by these "murderous big ants" and how the ants had to be picked from the dogs "one by one." However, the scientific study of army ant behavior had also begun. Beebe (1919), for example, made some organized observations of the New World genus *Eciton* and was one of the first biologists to examine carefully activities other than foraging. Other biological observations of army ants were recorded in the first 4 decades of the twentieth century by Wheeler (1900), von Ihering (1912), Gallardo (1915, 1929), Luederwaldt (1926), M. R. Smith (1927), and Reichensperger (1934) in the New World and by Brauns (1901, 1903), Vosseler (1905), Arnold (1915), Swynnerton (1915), Burgeon (1924a,b), and Cros (1939) in the Old World. However, it was not until T.C. Schneirla, a psychologist by training, began his research on *Eciton* that an understanding of army ant behavior was significantly extended beyond the fragmentary and often anthropomorphized observations of the naturalists.

Army ants have also captured the imaginations of the indigenous peoples of the Old and New World tropics, and tales of these ants have been incorporated into the oral traditions and folklore of numerous cultures. Whether these ants are referred to as Tauoca in Amazonas, Tepeguas in Mexico, or Ensanafu, Siafu, or Kelelalu in Africa, they are generally regarded with fear engendered respect. For instance, the Ashanti of Ghana allegorically communicate this respect in a tale about the hungry python. They say that before a python takes a large meal, one that might immobilize it and thus make it more vulnerable to army ant attack, it circles the immediate area in search of driver ants or Nkran. If it sees no such ants, it will take its meal, but if the ants are present, then the python will forego, in its own prudent self-interest, the much desired meal. Curiously, even the name of the capital city of Ghana, Accra, may have been derived from Nkran, the Akan word for driver ants, through a former spelling Akra. Reputedly common to the "medical" practices of numerous tribes, such as the Kikuyu (Murray-Brown, 1972, p. 40), is the use of the soldier subcaste of some army ant species in suturing wounds. Alex Haley (1976, p. 12) in his much celebrated book, "Roots," gives a fictional account of the process:

> . . . Grandma Yaisa tightly pressed together the skin's split edges, then pressed one struggling driver ant after another against the wound. As each ant angrily clamped its strong pincers into the flesh on each side of the cut, she deftly snapped off its body, leaving the head in place, until the wound was stitched together."

While it is true that army ants, particularly of the soldier subcaste, can inflict a painful bite on human skin (and draw blood as this author can attest) and while some species are excellent general predators, they do not deserve their reputation as the scourge of tropical forests that not only devour all animal life they contact but defoliate, in the process, all vegetation that stands in their way. Although it is easy to unknowingly step in the midst of a raiding swarm of workers or in an emigration or foraging column, it is just as easy, and decidedly judicious, to step out of their midst. Certainly the only vertebrates caught and killed by army ants are those individuals that for some reason or other are incapacitated and immobilized. Caged vertebrates, such as snakes, commonly fall prey to army ant attack, and there has been one unsubstantiated report, of a human infant that, left shaded beneath a tree while its mother cultivated her crops, died of a driver ant attack. Contrary to some prevailing mythologies, army ants do not defoliate plants, though they may take some vegetation as food. Yet to witness an army ant colony foraging or emigrating can be an awe-inspiring experience not soon forgotten. Army ant colonies display an extraordinary cohesiveness unmatched by other ant species, and for this reason alone, they are worthy of special attention.

II. ARMY ANTS DEFINED

A. Behavioral Characteristics

Generally the term "army ant" refers to any species in the ant subfamily Dorylinae [many other common names, such as legionary ants, driver ants (usually *Anomma*), soldier ants, raiding ants, and others, have been applied to members of this subfamily]. While this is a convenient taxonomic definition, it does not include species that qualify behaviorly, but not morphologically, as army ants. There are two features that characterize army ant behavior: (1) group predation, and (2) nomadism (Wilson, 1958a). All species within the Dorylinae exhibit these patterns that are, for all intents and purposes, inextricably joined together in the army ant adaptive syndrome. However, other ant species, notably of the subfamily Ponerinae, have also achieved army ant lifeways and combine, in varying degrees of complexity, group predation and nomadism.

Group predation, as defined by Wilson (1958a), includes both group raiding and group retrieval of living prey. He pointed out that these two processes involve different innate behavior patterns and are "not invariably linked." Although, for instance, many nondoryline ants may group retrieve prey, few also group raid. Those that combine both behavior patterns and thus qualify as group predators are found in such ponerine genera as *Cerapachys, Phyracaces (= Cerapachys), Leptogenys* (e.g., in the *L. processionalis* group), *Termitopone, Megaponera, Paltothyreus,* and *Simopelta* (Wheeler, 1936; Wilson, 1958a,b; Gotwald and Brown, 1966; Brown, 1975). The convergent development of group predation among nondorylines may be associated with concomitant morphological adaptations among the different castes. For example, the two queens thus far described for the genus *Simopelta* are both dicthadiiform, a peculiar habitus found chiefly among doryline queens in which wings are absent and the petiole and gaster are hypertrophied (Borgmeier, 1950; Gotwald and Brown, 1966). The extent to which nomadism is associated with group predation in these ponerine genera is not well known.

Nomadism or emigration of ant colonies is not in itself unusual. Colonies of numerous species commonly shift their nesting sites in response to unfavorable environmental conditions. Some species, such as *Iridomyrmex humilis,* may normally emigrate one or more times in a single season (Wilson, 1958a), but no species move with the regimented precision and frequency of the doryline army ants. For example, in the New World genus *Eciton,* colonies pass through a functional cycle that includes a nomadic phase during which emigrations to new temporary nesting sites occur on a daily basis (Schneirla, 1971). Among nondoryline group predators, nest-changing emigrations have been observed for such species as *Megaponera foetens* (Arnold, 1914) and *Leptogenys purpurea* (Wilson, 1958a); the emigration of the latter closely resembles that of some dorylines. Noteworthy is the fact that the larvae of the Cerapachyini are slender and cylindrical, as in the dorylines that carry their larvae on emigrations slung longitudinally beneath their bodies, and may circumstantially indicate that these ants are also nomadic (Brown, 1975).

B. Classification of Army Ants

1. Subfamily Dorylinae

Because the males of doryline ants are unusually large, wasplike and infrequently found with the worker caste, the early taxonomic history of the subfamily is complex and sometimes enigmatic. Indeed, the first species of doryline described was based on a male of *Dorylus* that Linnaeus (Linné, 1764) initially placed in the wasp genus *Vespa*. The type specimen of this species, now *D. helvolus,* was collected at the Cape of Good Hope. Later in the twelfth edition of "Systema Naturae," Linnaeus (Linné, 1767) transferred the species from *Vespa* to the genus *Mutilla* and in doing so precipitated a taxonomic controversy that was not to be settled for another 8 decades. In 1793, Fabricius removed the Linnaean species from the genus *Mutilla* and created the genus *Dorylus,* which he placed between the ants and mutillids. It was not until 1858 that the worker of *helvolus* was described by F. Smith and then as *Typhlopone punctata*. However, the relationship of the doryline males to the ants, in the absence of an associated worker caste, remained unrecognized. Shuckard (1840), in what constituted the first monographic review of the dorylines, noted "many points of analogy between *Ponera* and the Dorylidae" but still considered the dorylines as being distinct from the ants. In fact, he speculated that "these extraordinary genera [*Dorylus* and *Labidus*] may possibly be parasites upon the Social Ants," and grouped them as the "Parasiticae" within the "Heterogyna." Even so, he described three species of *Labidus* based on workers in the absence of any direct evidence of their association but considered the workers to be "females." Although Lepeletier de Saint-Fargeau (1836) perceptively placed the males of *Dorylus* and *Labidus* close to the ants, it was not until Savage (1849) collected males and workers together, of an *Anomma* species that he described as *A. rubella,* that the true relationship of *Dorylus* and *Labidus* to the ants became clear. When Savage first observed the males moving in a column of workers he initially supposed them to be "capitives" but later concluded that "they seemed to be no unimportant members of the community." After finding a total of ten dealate males he noted that,

> I was soon convinced that they belonged to the drivers, and proceeded to test the truth of the conclusion. I took one or two [males] from the lines to a distance of six and ten feet. They seemed at once to miss their companions, and manifested great trepidation, and made continuous efforts to find a way of return. At last they reached the lines and instantly resumed their places, displaying at the same time decided gratification.

Taking note of Savage's observations, F. Smith (1858) suggested that *Labidus* might prove to be the male of the ant genus *Eciton,* a conjecture that seemed to Sumichrast (1868) "to be sustained by the fact that in Mexico it is in the season when the sorties of the *Eciton* are the more frequent that the *Labidus* also show themselves." This supposition was finally shown to be true by Mayr (1886) and Müller (1886).

The first doryline female described and recognized as such was that of *D. helvolus* (Trimen, 1880), although Emery (1887) later noted that some doryline queens had been described as workers. Still by the year 1900 few doryline queens had been described so that such descriptions were in themselves of special interest. For instance, Wheeler (1900) described in detail the newly discovered queen of *Eciton sumichrast* (= *Neivamyrmex sumichrasti* Norton) and noted that, "One of the most interesting problems confronting the student of ant life in subtropical and tropical America is the determination of the sexual forms of the foraging, or driver, ants . . ." Even today a great majority of doryline species, from both the New and Old Worlds, are known only from the workers and males.

A comprehensive revision of doryline ant taxonomy will only be possible after the association of all three phena (i.e., workers, females, and males) is established for a majority of species. Even so, numerous systematic studies of the dorylines have been attempted. Emery (1895) taxonomically revised the genus *Dorylus,* and later (1910) reviewed the entire subfamily. A rather unusual interpretation of the subfamily was advanced by Ashmead (1906).

Smith (1942) considered the taxonomy of *Neivamyrmex* army ants in the United States, and Borgmeier (1953, 1955), in a monumental effort, revised the New World dorylines. His conclusions will endure for years to come, although Watkins (1976, 1977) recently supplemented Borgmeier's work. In 1964 Wilson revised the Indo–Australian species of *Aenictus* and *Dorylus*. Only the African forms of *Aenictus* and *Dorylus* remain in taxonomic disarray, and these await completion of a revision in which I am currently engaged. Phenetic studies of *Dorylus* major workers and males have yielded only four integral species clusters, and these correspond to the subgenera *Alaopone, Dorylus, Rhogmus,* and *Typhlopone.* Members of the subgenera *Dorylus* and *Anomma* form a single, diverse, but continuous taxon, while the status of the subgenus *Dichthadia* remains unclear (Gotwald and Barr, 1980; Barr and Gotwald, 1982).

The true army ants traditionally placed in the subfamily Dorylinae are now regarded as belonging to two subfamilies, the Dorylinae and the Ecitoninae. However, as a matter of convenience, throughout this chapter these species are referred to in both subfamilies as dorylines. The army ants are classified as follows, although the status of the subgenera of *Dorylus* will change in the impending revision of that genus:

Subfamily Dorylinae (Old World)
   Tribe Aenictini
      Genus *Aenictus*
   Tribe Dorylini
      Genus *Dorylus*
         Subgenus *Alaopone*
            Anomma
            Dichthadia
            Dorylus
            Rhogmus
            Typhlopone
Subfamily Ecitoninae (New World)
   Tribe Cheliomyrmecini
      Genus *Cheliomyrmex*
   Tribe Ecitonini
      Genus *Eciton*
         Labidus
         Neivamyrmex
         Nomamyrmex

2. Nondoryline Army Ants

Ant species not belonging to the Dorylinae or Ecitoninae that qualify behaviorly as army ants (i.e., supposedly manifest both nomadism and group predation) are confined to the subfamily Ponerinae. Most are termitophagous and myrmecophagous species in the genera *Leptogenys, Megaponera, Onychomyrmex, Simopelta,* and *Termitopone* (Wheeler, 1936; Wilson, 1958a; Gotwald and Brown, 1966; Hermann, 1968a). Army ant lifeways in early stages of development in the cerapachyines, a group now relegated to tribal status within the Ponerinae, and the genus *Acanthostichus* are apparently evident (Brown, 1975). Also of interest are the genus *Aenictogiton* and the small biologically cryptic ants of the subfamily Leptanillinae; both groups were once considered a part of the Dorylinae (Emery, 1910). *Aenictogiton,* a genus containing only seven species, is known only from the male caste collected in central Africa. Although Brown (1975) placed this genus within the Ponerinae in the tribe Aenictogitini, he admitted that such placement was provisional until the workers and queens for the genus could be identified with certainty. The habitus of the males is generally like that of some army ants, and they lack metapleural gland openings as do the males of all army ants.

Although the Leptanillinae have undergone extreme morphological reduction that obscures their affinities, they are nevertheless regarded as close to the Dorylinae (Brown, 1954). Queens, workers, and males are known, and while the queens are dicthadiiform, nothing is known of their biology. The best that can be surmised is that they are probably subterranean (Wheeler, 1910; Brown, 1954). Emery (1904), Kutter (1948), and Petersen (1968) have contributed to the knowledge of the leptanillines and Baroni Urbani (1977) has recently revised the entire subfamily.

All of the doryline and nondoryline ants discussed thus far are tropical or subtropical in distribution and constitute such a diverse taxonomic assemblage that it suggests that the adaptive value attached to adopting army ant lifeways in tropical environments is significant.

III. IDENTIFICATION OF DORYLINE ARMY ANTS

A. Subfamily Characteristics

1. Workers

In doryline workers (Fig. 1C,D), the frontal carinae are raised and lack the lateral expansions typical of most other ants outside the Pseudomyrmecinae. Thus, when viewed dorsally, the antennal insertions are exposed. This condition in combination with the fact that the workers either lack eyes or have eyes that are reduced to an ocelluslike structure, makes it relatively simple to separate these forms from the workers of other subfamilies (Wheeler, 1910, 1922; Bolton, 1973). Additionally, the clypeus is so reduced that the antennal insertions are located close to the anterior margin of the head (Wheeler, 1910; Bolton, 1973). In other respects, worker morphology is more ambiguous. For example, the waist has either one or two segments, and although all workers possess an apparently complete sting apparatus (Hermann, 1969), some species do not sting (Gotwald, 1978).

2. Females (Queens)

All female dorylines are dichthadiiform (Fig. 1A,B), that is, they are either blind or possess reduced or vestigial eyes, and they are apterous and have a hypertrophied petiole and gaster (Wilson, 1971). The dichthadiiform condition is found only in the dorylines and in some of the ponerines that exhibit army ant behavior patterns, such as *Simopelta oculata* (Gotwald and Brown, 1966). The females are always much larger than the workers. Other characters include antennae that have 10–12 segments, a waist that always has one segment, and an alitrunk in which suturing is reduced (Emery, 1910).

3. Males

The males (Fig. 1E,F) are alate and much larger than the workers. Their antennae have 13 segments; they possess large compound eyes and three conspicuous ocelli; their thoracic suturing is not reduced; their waist always has one segment; and their genitalia are completely retractile (Emery, 1910; Wheeler, 1910). Curiously, the metapleural glands, structures found only in ants, are not present in doryline males (Brown, 1968).

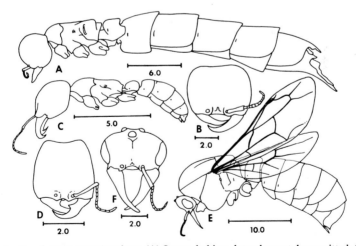

Fig. 1. *Dorylus (Anomma) molesta.* (A) Queen, habitus, lateral aspect, legs omitted; (B) head of queen dorsal aspect; (C) soldier, habitus, lateral aspect, legs omitted; (D) head of soldier dorsal aspect, (E) male, habitus, lateral aspect, legs omitted; (F) head of male dorsal aspect. All scales are in millimeters.

B. Keys to the Tribes, Genera, and Subgenera of Doryline and Ecitonine Army Ants

1. General Considerations

The following keys are adapted from Emery (1910), Wheeler (1910, 1922), Borgmeier (1955, 1958), Raignier and van Boven (1955), Schneirla (1971), Bolton (1973), van Boven (1975), and Watkins (1976); some of the characters used in the keys to the subgenera of *Dorylus* are employed here for the first time. For determinations to the species level, one should refer to Borgmeier (1955) and Watkins (1976) for the New World species and Wilson (1964) for the Indo–Australian species. Complete keys to the African species of *Dorylus* and *Aenictus* are not yet available.

Although keys are provided for the identification of workers, females (or queens), and males, those for the workers of polymorphic species (e.g., *Eciton* and *Dorylus*) refer most often to characteristics that are either exclusive to or best developed in the major workers and/or soldiers.

2. Tribes of Dorylinae and Ecitoninae

Workers

1. Waist, 2 segments (Fig. 2A) 2
   Waist, 1 segment (Fig. 2B) 3
2. Antenna, 10 segments; Old World species . . . . Aenictini (genus *Aenictus*)
   Antenna, 12 segments; New World species . . . . Ecitonini
3. Pygidium impressed, armed with 2 lateral spines, 1 on each side; Old World species (Fig. 2C, D) . . . . Dorylini (genus *Dorylus*)
   Pygidium simple; New World species (Fig. 2E) . . . .
   . . . . Cheliomyrmecini (genus *Cheliomyrmex*)

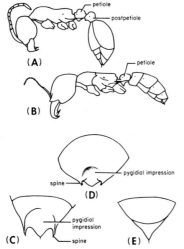

Fig. 2. (A) *Aenictus* worker, habitus, lateral aspect, legs omitted; (B) *Dorylus (Anomma)* worker, habitus, lateral aspect, legs omitted; (C) pygidium of *Anomma* worker, dorsal aspect; (D) pygidium of *Typhlopone* worker, dorsal aspect; (E) pygidium of *Cheliomyrmex* worker, dorsal aspect.

Queens

1. Antenna, 10 segments; Old World Species .... Aenictini (genus *Aenictus*)
   Antenna, 11 or 12 segments; Old or New World species 2
2. Copulatory bursa covered by the pygidium; hypopygium not prominent; New World
   species (Fig. 3A,B) .... Ecitonini
   Copulatory bursa open, not covered by pygidium; hypopygium forked, conspicuously
   extended beyond the pygidium; Old World species (Fig.3C,D,E) ....
   .... Dorylini (genus *Dorylus*)
   (Note: the female of the Cheliomyrmecini is unknown)

Males

1. Wing vein Mfl originating adjacent or distal to crossvein cu-a; Old World species
   (Fig. 3F) 2
   Wing vein Mfl originating considerably proximal to cu-a; New World species (Fig.3G)
   3
2. Stigma of forewing narrow (Fig.4A); total body length (exclusive of mandibles) usually
   greater than 18 mm .... Dorylini
   Stigma of forewing wide (Fig.4B); total body length (exclusive of mandibles) usually less
   than 8.5 mm .... Aenictini
3. Flagellum of antenna only slightly longer than width of the head ...... Cheliomyrmecini
   Flagellum of antenna much longer than width of the head .... Ecitonini

3. Genera of the Ecitonini

Workers

1. Tarsal claws simple, without teeth (Fig.4C) .... *Neivamyrmex*
   Tarsal claws with teeth (Fig.4D) 2
2. Scape and flagellum of antenna wide (apical width of scape greater than one-third its
   length) .... *Nomamyrmex*

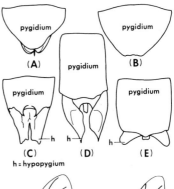

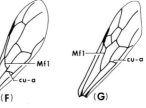

Fig. 3. Pygidia of army ant queens, dorsal aspect: (A) *Eciton*, (B) *Labidus*, (C) *Anomma*, (D) *Rhogmus*, and (E) *Alaopone*. Forewing of army ant males, dorsal aspect: (F) *Typhlopone*, and (G) *Eciton*.

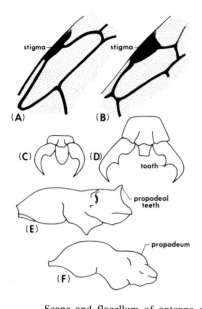

Fig. 4. (A) Wing stigma of *Dorylus* male, dorsal aspect; (B) wing stigma of *Aenictus* male, dorsal aspect; (C) tarsal claws of *Neivamyrmex* worker; (D) tarsal claws of *Eciton* worker; (E) alitrunk of *Eciton* worker, lateral aspect; (F) alitrunk of *Labidus* worker, lateral aspect.

Scape and flagellum of antenna slender (apical width of scape less than one-third its length) 3

3. Propodeum armed posteriorly with teeth or lamellae (Fig.4E); soldiers usually with falcate mandibles . . . . *Eciton*
   Propodeum unarmed (Fig.4F); mandibles of soldiers not falcate . . . . *Labidus*

Queens

1. Tarsal claws simple, without teeth (Fig.4C) . . . . *Neivamyrmex*
   Tarsal claws with teeth (Fig.4D) 2
2. Propodeum armed posteriorly with two horns or blunt teeth (Fig.5A) . . . . *Eciton*
   Propodeum unarmed (Fig.5B) 3
3. Promesonotum strongly convex in lateral view; propodeum sloped obliquely (Fig.5B)
   . . . . *Labidus*
   Alitrunk in lateral view more or less straight . . . . *Nomamyrmex*

Males

1. Legs long, metafemur reaching or surpassing the posterior margin of the second gastral segment 2
   Legs short, metafemur not reaching the posterior margin of the second gastral segment
   . . . . *Neivamyrmex*
2. Apices of lateral aedeagal sclerites (penis valves) without setae (Fig.5C) . . . . *Eciton*
   Apices of lateral aedeagal sclerites with setae (Fig.5D) 3
3. Gastral tergites with clusters of long setae . . . . *Nomamyrmex*
   Gastral tergites without conspicuous clusters of long setae . . . . *Labidus*

4. Subgenera of Dorylus

Workers

1. Antenna, 12 segments . . . . *Dichthadia*
   Antenna, 9-11 segments 2

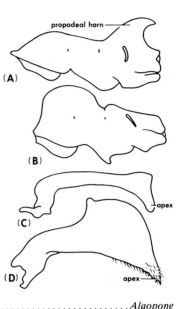

Fig. 5. (A) Alitrunk of *Eciton* queen, lateral aspect; (B) alitrunk of *Labidus* queen, lateral aspect; (C) lateral aedeagal sclerite of *Eciton* male, lateral aspect; (D) lateral aedeagal sclerite of *Labidus* male, lateral aspect.

2. Antenna, 9 segments .... *Alaopone*
    Antenna, 10–11 segments 3
3. Pygidial impression with sharp, well-defined margins (Fig.2C) 4
    Pygidial impression without distinct margins (Fig.2D) 5
4. Antenna short and thick, all segments of the flagellum except the last as wide or wider than they are long; each frontal carina *usually* armed with a caudally projecting spine (Fig.6A) .... *Dorylus*
    Antenna long and slender, at least some segments of flagellum longer than wide; frontal carinae never armed with spines (Fig.6B) .... *Anomma*
5. Subapical tooth of mandible simple (Fig.6C); frontal carina in lateral view drawn to a ventrally directed point (Fig.6F); promesonotal suture only slightly impressed ....
    .... *Typhlopone*
    Subapical tooth either truncate or notched at the middle (Fig. 6D); frontal carina in lateral view rounded, not pointed (Fig.6E); promesonotal suture deeply impressed ... *Rhogmus*

Queens

1. Antenna, 12 segments .... *Dichthadia*
    Antenna, 11 segments 2
2. Posterior margin of pygidium with a deep, median, semicircular notch (Fig. 3C) 3
    Posterior margin of pygidium straight or only slightly concave, without a semicircular notch (Fig. 3D,E)... 4
3. Propodeum in dorsal view wider than the pronotum (Fig.7A); posterior angles of petiole considerably divergent (Fig.7A) .... *Anomma*
    Propodeum not as wide as pronotum (Fig.7B); posterior angles of petiole only slightly divergent (Fig.7B) .... *Dorylus*
4. Posterior margin of pygidium straight; in dorsal view, hypopygium extending only a short distance beyond posterior margin of pygidium and terminating in two rounded, diverging lobes (Fig.3E)... *Alaopone*
    Posterior margin of pygidium slightly concave; hypopygium extending far beyond posterior margin of pygidium and divided by a median cleft so that it terminates in two apically pointed processes (Fig.3D) 5

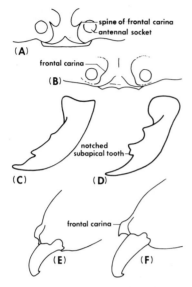

Fig. 6. (A) Frontal carinae of major worker of *Dorylus,* dorsal aspect; (B) frontal carinae of major worker of *Anomma,* dorsal aspect; (C) mandible of major worker of *Typhlopone;* (D) mandible of major worker of *Rhogmus;* (E) anterior half of head, major worker, *Rhogmus,* lateral aspect; (F) anterior half of head, major worker, *Typhlopone,* lateral aspect.

5. Hypopygium forming a dorsoventrally flattened plate, in dorsal view its median cleft ending considerably caudal of the posterior margin of the pygidium . . . . *Typhlopone*
    Hypopygium not forming a flattened plate, in dorsal view its median cleft appears to totally subdivide the hypopygium into two lateral parts (Fig.3D) . . . . *Rhogmus*

Males

1. Forewing with "second recurrent" vein (Fig.7C); aedeagus enlarged distally, in dorsal view apex is knoblike; mandible as in Fig.7D. . . . . *Rhogmus*
    Forewing without second recurrent vein; aedeagus smoothly tapered distally, not terminating in a knoblike enlargement; mandible shaped otherwise .2

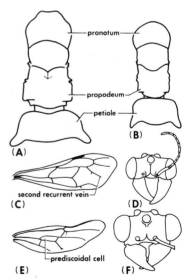

Fig. 7. (A) Alitrunk and petiole of *Anomma* queen, dorsal aspect; (B) alitrunk and petiole of *Dorylus* queen, dorsal aspect; (C) forewing of *Rhogmus* male, dorsal aspect; (D) head of *Rhogmus* male, dorsal aspect; (E) hindwing of *Dichthadia* male, dorsal aspect; (F) head of *Dichthadia* male, dorsal aspect.

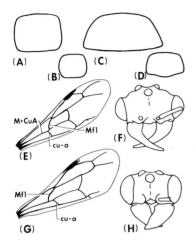

Fig. 8. Petioles of the males of the genus *Dorylus:* (A) *Typhlopone,* (B) *Alaopone,* (C) *Anomma,* and (D) *Dorylus.* (E) Forewing of *Typhlopone* male, dorsal aspect; (F) head of *Typhlopone* male, dorsal aspect; (G) forewing of *Alaopone* male, dorsal aspect; (H) head of *Alaopone* male, dorsal aspect.

2. Hindwing with a prediscoidal cell (Fig.7E); mandible as in Fig. 7F . . . . *Dichthadia*
   Hindwing without prediscoidal cell; mandible shaped otherwise 3
3. Petiole in dorsal view nearly square or round (Fig.8A,B) 4
   Petiole in dorsal view wider than long, its posterior surface concave or at least flattened
   (Fig.8C,D) 5
4. Crossvein cu-a of forewing intersects with M + CuA, i.e., proximal to the intersection of
   Mfl (occasionally the cu-a and Mfl intersections are opposite one another) (Fig.8E);
   anterior margin of labrum medially cleft; mandible as in Fig.8F . . . . *Typhlopone*
   Crossvein cu-a intersects with CuA, i.e., distal to the intersection of Mfl (Fig.8G);
   anterior margin of labrum entire; mandible as in Fig.8H . . . . *Alaopone*
5. Antennal fossa projects beyond the anterior margin of the head forming a prominent
   tooth (Fig.9A); prongs of the subgenital plate are parallel or only slightly divergent at
   their apices (Fig.9C) . . . . *Anomma*

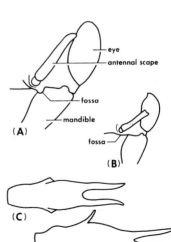

Fig. 9. (A) Head of *Anomma* male, left anterior quadrant, including base of mandible, dorsal aspect; (B) head of *Dorylus* male, left anterior quadrant, including base of mandible, dorsal aspect; (C) subgenital plate of *Anomma* male, dorsal aspect; (D) subgenital plate of *Dorylus* male, dorsal aspect.

Antennal fossa forming a wide but inconspicuous protrusion (Fig.9B); prongs of
subgenital plate are divergent, not parallel (Fig.9D) . . . .*Dorylus*

IV. COLONY COMPOSITION AND CASTE POLYETHISM

A. Army Ant Polymorphism

Even a cursory examination of the genetic females (i.e., workers and
queens) of army ants dramatically demonstrates why the earlier classifica-
tions of dorylines were often incomplete and contradictory. Many of the
species are highly polymorphic and when collected as single, unassociated
specimens they are difficult, more often impossible, to relate to previously
collected conspecific individuals of different castes or subcastes.

For the social insects, polymorphism is defined as the coexistence of two or
more functionally distinct castes within the same sex (Wilson, 1971). Thus in
the dorylines two full castes are present: neuter females or workers and
reproductive females or queens. Soldiers are present in some species and
represent subcastes since they constitute the largest individuals in a continuous polymorphic series of workers (Wilson, 1971). Other worker subcastes among highly polymorphic forms such as *Anomma* and *Eciton* are ar-
bitrarily designated, on the basis of size as either major, or media, or minor
workers. Such distinctions are not made where the size differential between
the smallest and largest workers is slight, such as in *Aenictus.* Although it is
sometimes convenient to refer to the males as yet another caste, such a
designation is not faithful to the definition of caste.

The caste distinctions made between some individuals of a polymorphic
species result not only from obvious size differences but also from the morphological manifestations of allometric growth; allometry being a growth
phenomenon that produces differences in the relative proportions of body
parts that are a function of total body size (Wilson, 1971). Although Darwin
(1859) certainly noted the allometric nature of *Anomma* worker morphology,
it was not until much later, when Huxley (1927) examined what he called
"heterogonic growth" in *Anomma,* that this growth phenomenon was quantitatively analyzed in army ants. While Cohic (1948) claimed that the workers
of *Anomma* could be divided into distinct morphological and functional
types, Hollingsworth (1960) demonstrated that such discontinuities did not
exist. Indeed, he showed that the *Anomma* workers of a single colony could be
placed in a continuous series from the smallest to the largest. On the other
hand, although morphological discontinuities in *Anomma* [in particular in *D.
(A.) wilverthi*] are not evident and only one worker caste exists, van Boven
(1961) determined that four subcaste phases were morphologically distinguishable. These he referred to as minima, minor, media, and major

workers. Hollingsworth (1960) found that allometry in *Anomma* is not simple since "various parts of the body have different allometric constants in different individuals." Because the relative growth curve of the workers changes slope, the *Anomma* species analyzed is said to exhibit diphasic allometry (Hollingsworth, 1960; Wilson, 1971).

All true army ants, with the exception of *Aenictus* and some species of the genus *Neivamyrmex,* are strongly polymorphic (Gotwald and Kupiec, 1975). In terms of total body length, the size differential between the smallest and largest workers for *Eciton burchelli* is 8.1 mm; *D. (A.) wilverthi,* 8.0 mm; *Cheliomyrmex morosus,* 4.12 mm; *Neivamyrmex nigrescens,* 2.8 mm; and for *Aenictus gracilis,* 0.5 mm (Gotwald and Kupiec, 1975). In polymorphic species, the size frequency distribution of workers is asymmetric. For instance, in *E. hamatum,* "small intermediate" workers are most numerous, large intermediates and minors next and major workers least numerous (Schneirla, 1971). In *E. burchelli,* the minor workers predominate, and in both species, the majors constitute less than 2% of the entire colony (Schneirla, 1971). Topoff (1971), who calculated the frequency distributions of total body length for army ant worker pupae, noted that the smaller workers predominate, even in essentially monomorphic species such as *Aenictus laeviceps.*

Among the army ants *Anomma* species have been most thoroughly examined in terms of their polymorphism and allometry. To the previously cited studies of Huxley (1927), Cohic (1948), Hollingsworth (1960), and van Boven (1961) must be added those of Raignier and van Boven (1955), van Boven (1958) and Raignier *et al.* (1974). Polymorphism and allometry have also been studied in the brood of New World species by Schneirla and Brown (1952), Tafuri (1955), Lappano (1958), and Schneirla *et al.* (1968).

B. Army Ant Castes and Subcastes

1. Workers

a. Morphology. Details of doryline worker ant morphology are included in comparative studies of the mouthparts (Bugnion, 1930; Gotwald, 1969), thorax (Tulloch, 1935; Reid, 1941), proventriculus (Eisner, 1957), and poison apparatus (Hermann and Blum, 1967; Hermann, 1969). The morphology of Old World species has been studied by Mukerjee (1933), Cohic (1948), and Hollingsworth (1960) and of New World species by Whelden (1963), Gotwald (1971), and Gotwald and Kupiec (1975). A review of structures and morphological features that are important in either interpreting doryline phylogeny or understanding army ant behavior follows.

The workers of *Dorylus, Aenictus,* and *Cheliomyrmex* are eyeless, while, with few exceptions, those of *Eciton, Labidus, Neivamyrmex,* and

Nomamyrmex possess reduced compound eyes. Although the Old World species and some New World forms are blind, Werringloer (1932) attributed to them a subdermal or integumental light sense. While Schneirla (1971) suggested that eyes are better developed in surface-adapted species (i.e., in species that commonly forage and even nest on or above the soil or substrate surface), Gotwald (1978) pointed out that several species of *Dorylus* (particularly *Anomma*) and *Aenictus* frequently and even habitually forage on the soil surface and climb vegetation. Furthermore, Gotwald (1978) constructed a scenario to account for eyelessness in surface foraging species. In this scenario, eyeless species arose from predatory ancestors that were surface-active and possessed well developed eyes; eyelessness was a consequence of subterranean living that became selectively advantageous for these predorylines. Because numerous extant Old World species of surface foragers are blind, their readaptation to surface lifeways may be a recent event. Gotwald (1978) noted that the reduced palpal segmentation common to doryline army ants also supports the hypothesized subterranean existence of the ancestral forms.

Although the mandibles of doryline workers exhibit a diverse morphology, those of the soldiers of *Eciton* and *Anomma* are morphologically similar and are celebrated for their ability to pierce human skin. In soldiers of both, the mandible is falcate and sharply pointed, and is clearly a piercing type considered best adapted to a defensive function (Wilson, 1971) (Fig.1D). In polymorphic species such as *D. (A.) nigricans* and *C. morosus,* the mandibles undergo a gradual but continuous transition in shape from the smallest to the largest workers (Hollingsworth, 1960; Gotwald and Kupiec, 1975). According to Gotwald (1978), this produces mandible morphologies that are variously adapted to different tasks. The mouthparts are morphologically distinctive at the tribal level and the Ecitonini, Aenictini and Dorylini can be separated on the basis of mouthparts alone (Gotwald, 1969). On the other hand, mouthpart morphology shows the Ecitonini and Cheliomyrmecini to be closely related (Gotwald, 1969). The number of palpal segments is significantly reduced from the primitive number in ants of six in the maxillary palpus and four in the labial palpus. In the dorylines, the maxillary palpus ranges from one to two segments and the labial palpus from two to three segments (Gotwald, 1978).

Reid (1941) found that there were two distinct types of worker alitrunk in the army ants, one typical of the Dorylini and the other of the Ecitonini. In the former, the alitrunk consists of two parts of approximately equal size, an anterior part consisting of the pronotum and a posterior part composed of the mesonotum, metanotum, and propodeum. In the latter, the alitrunk is a single, undivided structure in which the sutures are usually greatly reduced. Reid (1941) noted that the alitrunk of *Aenictus* most closely resembles that of the genus *Eciton*.

The worker waist has two segments or is binodal in the Ecitonini and Aenictini. In the Dorylini and Cheliomyrmecini the worker waist has one segment or is uninodal (Fig.2A,B). If the Dorylinae (*sensu lato*) are monophyletic, the segmental nature of the waist in this subfamily is certainly less conservative than it is in other subfamilies where the number of segments may be constant for an entire subfamily (e.g., the Formicinae) (Gotwald, 1978). Pullen (1963) suggested that the binodal waist facilitates stinging by making the gaster more maneuverable, and Schneirla (1971) added that this condition is important to surface-adapted species in subduing strong, fast moving prey. Schneirla also concluded that this flexibility is of some advantage in laying chemical trails and in carrying brood and prey (generally beneath the worker's body). These assumptions remain to be demonstrated empirically.

The doryline poison apparatus, including the sting sclerites, has been described by Whelden (1963), Hermann and Blum (1967), Hermann (1969), and Gotwald (1971). The soft parts of the poison apparatus include an elongate, pear-shaped or spherical, poison sac or venom reservoir with a conspicuous duct that terminates in the sting bulb (Hermann, 1969; Gotwald, 1971). Free poison filaments arise from the base of the poison sac or its duct. These filaments are sometimes branched and are distributed in the vicinity of the poison sac. An elongate Dufour's gland composed of cuboidal and/or columnar cells is present, but its function remains to be discovered (Hermann, 1969). Among the skeletal components, two are of special interest. The furcula, a sclerite located anterior to the sting bulb, is common to all ants thus far examined except the dorylines, cerapachyines, and one species of ponerine (Hermann, 1969). Its absence in the dorylines and *Simopelta oculata,* a ponerine with army ant lifeways, suggests that this sclerite is convergently lost in species adopting army ant behavior patterns (Hermann, 1968b). The sting, the other sclerite of interest, is broad and spatulate in the Dorylini, a morphological development that may be correlated with this group's inability to sting. The sting is slender in those doryline species that do sting (Hermann, 1969; Gotwald, 1978).

All New World army ants possess a functional sting. When attacking prey, these forms may bite and sting simultaneously. However, the Old World species of *Dorylus* are not known to sting, although they are ferocious, effective biters (Gotwald, 1978). Schneirla (1971) noted that the *Aenictus* workers that he observed in the Philippines had "potent stings," but Gotwald (1978) observed that African *Aenictus* may not sting. The absence of stinging in *Dorylus* may be a behavioral manifestation of the spatulate condition in the sting. Although Hermann (1969) found the sting of *Aenictus gracilis,* a species in which stinging is documented, to be slender as in New World stinging species, Gotwald (1978) noted that at least one apparently nonstinging species of African *Aenictus* possessed a spatulate sting.

Few studies have focused on the internal morphology of army ant workers. Mukerjee (1933) examined internal structes of *D. (Alaopone) orientalis,* Whelden (1963) published an extensive study of *E. burchelli* and *E. hamatum* and Gotwald (1971) and Gotwald and Kupiec (1975) produced anatomical descriptions of *C. morosus.* Other such observations are fragmentary at best. Bugnion (1930) described the pharynx of *Anomma* as reduced and hypothesized that the reduction is correlated with the absence of trophallaxis in the dorylines. Eisner (1957) found the army ant proventriculus to be degenerative and concluded that since the damming of this membranous valve is probably dependent on muscle contraction, crop storage in these ants may be of relatively short duration. The Malpighian tubules are probably histologically uniform throughout the army ants, although the numbers of tubules vary considerably. Even so, overlap in the ranges of tubule numbers between species is common, and the number of tubules per individual is so closely correlated with body size as to preclude their use taxonomically in polymorphic species (Gotwald, 1971). The rectal papillae, on the other hand, may be more constant in number, although Whelden (1963) reported a range of 3–6 in *E. burchelli* and *E. hamatum.*

Glands thus far identified in the workers include the mandibular glands, maxillary glands, pharyngeal glands, labial glands, metapleural glands, Dufour's gland, and the convoluted gland of the poison sac (Gotwald and Kupiec, 1975). Of these, only the convoluted gland has been studied in detail (Hermann and Blum, 1967).

The army ant nervous system has received some attention. In the brain, the corpora pedunculata or "mushroom-bodies," the size of which is often used as an indicator of mental capacity, are smallest in *Eciton hamatum* when compared to all other ants examined (Vowles, 1955; Bernstein and Bernstein, 1969). They are largest, for instance, in the formicine ant, *Formica rufa.* However, the relative size of the doryline brain is predictably smaller than in other similarly sized ants because, without eyes, the optic centers are greatly reduced (Werringloer, 1932; Vowles, 1955).

Ovaries composed of polytrophic ovarioles are found in the workers of *D. (A.) orientalis* (Mukerjee, 1933), *E. burchelli* and *E. hamatum* (Whelden, 1963), and *C. morosus* (Gotwald, 1971). In some species, each ovary consists of a single ovariole, while in others the number of ovarioles per ovary may range from one to three (Gotwald and Kupiec, 1975). Although ovaries are probably present in a majority of workers, Holliday (1904) failed to find them in *Neivamyrmex nigrescens.* Whether or not army ant workers actually lay these eggs is unknown, although the production of "trophic eggs" (i.e., eggs consumed by other colony members) among other ant species is not uncommon (Wilson, 1971).

b. Worker Functions. Wilson (1953) considered worker polymorphism, which is developed to some degree in all army ant genera except *Aenictus,* a special adaptive characteristic that results "in various types and degrees of worker labor." Schneirla (1971) noted that colony tasks in the monomorphic *Aenictus* are probably carried out by all workers at various times. Indeed, Topoff (1971) indicated that although "differences in sensory thresholds to chemical stimuli may exist among workers of different size groups" in *Eciton, Labidus,* and *Neivamyrmex,* workers of *Aenictus* all react similarly, for instance, to arousal stimuli.

Caste polyethism for worker subcastes has been documented in some cases and hypothesized in others. Rettenmeyer (1963b) found that major workers of *Eciton* participate in the formation of hanging clusters essential to assembling the bivouac (or temporary nest), in capturing prey, and in defense. However, both he and Schneirla (1971) pointed out that *Eciton* major workers are "automatically excluded from nearly all transport work since with their great double-fishhook jaws they cannot pick up, hold, or release objects." Topoff and Mirenda (1978) have shown that the callow workers of *Neivamyrmex nigrescens* do not participate in colony foraging excursions until 3–7 days after eclosion. Gotwald (1978) noticed that although some functions (e.g., the construction of soil particle walls) in *Anomma* correlate with body size, they are an adaptive consequence of mandible morphology, which in turn is allometrically determined.

A defensive function of major workers in *Eciton* is evident when a bivouac is torn apart. At the time of disturbance a large number of excited majors gather about the queen; in the undisturbed bivouac, the queen is surrounded by a "tight ball" of the smallest workers (Rettenmeyer, 1963b). Schneirla (1971) generalized about polymorphic species suggesting that media workers are less involved in "rough operations." Schneirla further noted that during emigration in *Eciton* there is a positive correlation between size of worker and load (i.e., brood) carried and that, in the nest, the smallest workers generally handle and feed the small larvae of a young brood. Among Old World species, a division of labor in prey retrieval in *Anomma* is apparent. Gotwald (1974a) observed that nearly all "preyless" foragers returning to the nest, i.e., workers not carrying visible pieces of prey, have liquid-filled crops. Statistical analysis revealed that larger workers tend to carry pieces of prey while smaller workers specialize in carrying prey liquids. Also in *Anomma,* the workers that construct the soil particle walls that often border their chemical trails are among the smallest in the colony (Kistner and Gotwald, 1982).

Although worker subcaste polyethism in polymorphic species of army ants is documented in some cases, a paucity of quantitative studies is clearly evident.

2. Queens

a. Morphology. Most observations of doryline queen morphology are limited to external features, specifically to those of taxonomic interest. For Old World species such observations have been reported by Emery (1887),

André (1900), Brauns (1903), Menozzi (1927), Arnold (1953), Raignier and van Boven (1955), van Boven (1967, 1968, 1972, 1975), Raignier *et al.* (1974), Gotwald and Cunningham-van Someren (1976), and Gotwald and Leroux (1980) to cite only a few. New World queens have been similarly described by Wheeler (1900, 1921, 1925), Bruch (1934), Weber (1941), and Rettenmeyer (1974). Studies that are entirely morphological in focus and include internal features are limited to those of Whelden (1963) and Hagan (1954a,b,c).

Army ant queens are dichthadiigynes, i.e., they possess a greatly enlarged gaster and petiole, are blind or nearly so, are permanently wingless, and have strong legs (Fig. 1A). Wilson (1971) pointed out that this unusual morphology adapts the queen to nomadic life in two ways: (1) the enlarged gaster contains ovaries composed of many ovarioles that enable the queen to "deliver large quantities of eggs during a short span of time," and (2) the queen is able to move under her own power from one temporary nesting site to the next. This specialized reproductive design is correlated with the acquisition of an expanded tracheal system and the ability to store large amounts of fat as a reserve energy source (Wheeler, 1928).

Since *Eciton* colonies alternate between statary and nomadic phases, an endogenously controlled rhythm, on a regular schedule, it is possible to observe certain external changes in the queen that are synchronized with these cyclic phases (Schneirla, 1971). As the colony enters the nomadic phase, the queen's ovaries undergo remarkable development, causing the gaster to become physogastric. The intersegmental membranes of the gaster stretch as they accommodate the hypertrophying ovaries, and the gastral sclerites are separated from one another (Schneirla, 1971). Although other doryline queens become physogastric, the regular alternation of statary and nomadic phases may be atypical of army ants in general. Rettenmeyer (1963b) noted, for instance, that Schneirla studied perhaps the most highly specialized species of *Eciton* (i.e., *E. burchelli* and *E. hamatum*) and that many of the generalizations offered by Schneirla about the army ant functional cycle are based on his knowledge of these species.

The mandibles of doryline queens differ in one conspicuous respect from those of the workers: they are devoid of subapical teeth (Gotwald, 1969) (Fig. 1B). They also depart from the triangular-shaped mandible typical of a majority of ants and are, instead, linear and slightly curved apically. The queen mandibles of the ponerine army ant, *Simopelta oculata,* are of a similar design (Gotwald and Brown, 1966). The maxillary palpus has two segments in the Aenictini, Dorylini and Ecitonini; the labial palpus has one segment in the Old World dorylines and two segments in the ecitonines (Gotwald, 1969). While compound eyes in the form of ocelluslike structures are common to the New World forms, they are entirely absent in Old World queens.

The alitrunk of the army ant queen is characterized by a general reduction in suturing, although the extent of reduction is far from universal. Even within a single genus the variation in suturing may vary considerably. For instance, queens of African *Aenictus* have a simplified (derived) thorax, whereas Asian species possess a more complex (primitive) suturing (Gotwald and Cunningham-van Someren, 1976). In fact, the differences between the Asian and African queens prompted Wheeler (1930) to suggest that the Asian species "would seem to belong to a distinct genus." However, a comprehensive, comparative survey of thoracic suturing in queens is wanting.

The petiole of doryline queens, at least in the Ecitonini and Dorylini, is enlarged and armed with caudally directed dorsal or lateral horns. These are most conspicuously developed in the ecitonines. In two *Eciton* matings reported by Schneirla (1949), each of the males involved grasped a petiolar horn in its mandibles during copulation. According to Rettenmeyer (1963b), this grasping behavior "suggests that contact between the mandibles and the queen's petiole may be important for mating and perhaps for preventing interspecies mating," since the petioles of the queen and the mandibles of the males differ so strikingly among the species. Schneirla (1971) assumed that these petiolar horns also serve a protective function.

The expansive gaster of the army ant queen is uniquely developed in the Dorylini where the bifurcated hypopygium extends conspicuously beyond the posterior of the pygidium, and the copulatory bursa is open, i.e., not covered by the pygidium (Raignier and van Boven, 1955; van Boven, 1967). In the Aenictini and Ecitonini, the hypopygium is not so extended and the bursa is closed by the pygidium. Hagen (1954a) noted that in *Eciton* during the nomadic phase, when the queen is contracted, the five visible gastral segments are strongly telescoped with the gastral sclerites greatly overlapping one another. During physogastry, of course, these sclerites are widely separated.

Internal features of interest include a proventriculus that is smaller than in the *Eciton* worker, although the structure is identical in both castes. The ventriculus of the *Eciton* queen is different from that of the worker both in shape and in the morphology of its component cells. For instance, the ventriculus of the queen is more or less cylindrical while that of the worker is pear-shaped. Whelden (1963) found approximately 30 Malpighian tubules in the *Eciton* queen, a number that is almost twice as many as in the workers. The number of rectal papillae, elliptical to nearly circular structures, in *Eciton* queens is "frequently six, infrequently three" (Whelden 1963).

Whelden (1963) described for the *Eciton* queen numerous glands. In the head there are mandibular glands, maxillary glands, pharyngeal glands, and three small glands (unnamed) that open through a membrane that extends from the mandibles to the bases of the mouthparts. The alitrunk contains the labial glands, metapleural glands, six small glands (one at the base of each leg), and a "moderately large" gland that is not present in the workers.

Although Whelden found the cells of this latter gland to be easily confused with those of the metapleural glands, their ducts open through the membrane that connects the alitrunk with the petiole. A gland is also present in the petiole. In the gaster, each segment contains a pair of large glands the cells of which each possess a duct that opens separately through the intersegmental membranes. Also present are the poison sac and the "alkaline" gland (Dufour's gland).

The brain or supraesophageal ganglion is larger in the *Eciton* queen than in the worker, but while the most conspicuous cranial nerves in the workers are those that innervate the antennae and mandibles, the optic nerves are most fully developed in the queen. The queen nervous system departs most noticeably from that of the worker in its possession of five large gastral ganglia instead of the single fused ganglionic mass of the worker.

The reproductive systems of the queens of *Eciton* and *Neivamyrmex* include a vagina and a median oviduct which bifurcates to form the paired lateral oviducts. These in turn lead to the ovaries which are composed of large numbers of polytrophic ovarioles (Holliday, 1904; Hagan, 1954a; Whelden, 1963). The lateral oviduct in *Neivamyrmex* extends for the entire length of the ovary; the ovarioles open into the duct (Holliday, 1904). In *Eciton* the lateral oviduct expands within the ovary to form a calyx and is capable of accommodating the vast numbers of oocytes that are discharged simultaneously into it by the ovarioles (Hagan, 1954a). There are approximately 1200 ovarioles in the ovary of the *Eciton* queen (Hagan, 1954a) and about 500 in the ovary of *Neivamyrmex* (Holliday, 1904).

Other reproductive structures in *Eciton* include an exceptionally long (probably longer than the gaster itself) spermathecal duct that originates on the dorsal wall of the vagina (Hagan, 1954a). This duct forms several tight coils that are peculiar in shape and position in the individual queens. This duct leads to a spherical or irregularly ovoid spermatheca (Whelden, 1963). Also present are a pair of irregularly twisted, tubular spermathecal glands that attach to the spermatheca directly above its junction with the spermathecal duct and a pair of accessory glands that arise anterior to the basal bulb of the ovipositor.

b. Queen Functions.   Above all else, the army ant queen is the reproductive core of the colony. So vital is she to the colony that it cannot survive as an integrated social unit without her. As Schneirla (1953) pointed out, "although the *Eciton* queen does not directly lead the activities of her colony, her role is critical in the functional pattern." In fact, Schneirla (1944) regarded the queen as a "pacemaker" in the colony behavior pattern of each species. There is but a single, functional queen per colony in the Ecitoninae and Dorylinae. The only known exception to this rule are colonies of *Neivamyrmex* *carolinensis* which regularly have 3 to 13 queens (Rettenmeyer and Watkins, 1978).

While the queen influences colony function by the types of eggs she produces (i.e., worker eggs versus sexual brood eggs), Schneirla (1971) concluded that the schedule of egg laying is determined by intracolony processes external to the queen. Central to this "colony-situation-feedback" hypothesis is the effect of brood-stimulative phenomena that regulate colony activities through the mediation of the queen's corpora allata. This in turn accounts for the egg-laying cycles in the queen. This is true, at least, for those army ants that exhibit a well marked functional cycle of regularly alternating nomadic and statary phases. Hagan (1954c) found that the oocyte cycle in *Eciton* "harmonizes" perfectly with colony behavior.

During the nomadic phase of those species of *Eciton, Neivamyrmex,* and *Aenictus* studied by Schneirla (1971), the queen remains contracted but as this phase ends and the statary begins, the queen's gaster swells, exposing the intersegmental membranes. The distension continues for about 1 week into the statary phase at which time the queen achieves full physogastry and egg laying commences. The adaptive advantage of cyclic physogastry that coincides with the statary phase is that is reduces the dangers that confront the queen during the emigrations of the nomadic phase. For instance, she must move under her own power along emigration routes that are often strewn with obstacles that could abrade or puncture her intersegmental membranes if exposed.

However, many army ant species, perhaps even a majority, do not have a well defined functional cycle, and emigrations occur as single events often separated by lengthy intervals. The queens of these species, e.g., of *Dorylus,* do not function on a precise reproductive schedule (Raignier and van Boven, 1955). The interval between emigrations of *D. (Anomma) molesta,* for instance, may vary from 3 to 67 days, and evidence suggests that during adverse conditions such as drought, a colony may remain at a single nest site for many months (W. H. Gotwald and G. R. Cunningham-van Someren, unpublished data). Long intervals may result from the inability of this species to maintain an elevated, optimal nest temperature (Raignier and van Boven, 1955). In some species of *Anomma,* Raignier and van Boven (1955) found that the queen is in a permanent but moderate state of physogastry, although egg laying may be discontinuous. In *D. (A.) wilverthi,* they observed that egg laying peaks occur at 20 to 25-day intervals and that it usually intensifies just after emigration. A period of physogastry longer than that in *Eciton* queens may also exist in *Labidus,* and certainly *Labidus* produces asynchronous broods or at least much less synchronous than *Eciton* (Rettenmeyer, 1963b).

The egg-laying capacity of the army ant queens is truly extraordinary. Schneirla (1971) estimated that a single queen of *Aenictus gracilis* produces as many as 240,000 eggs per year and that the annual yield of an *E. burchelli* queen may be as high as 2,400,000 eggs. But the queen of *D. (A.) wilverthi* may be unsurpassed. Raignier and van Boven (1955) estimated her egg production to be from 3 to 4 million per month.

If a colony of army ants loses its queen, it cannot ordinarily replace her unless a sexual brood is already present in the colony. Not only will the colony not flourish without its queen, but if it cannot join another conspecific colony, it faces certain extinction. Schneirla (1971) noted that a queenless colony of *Eciton* will within 1 or 2 hr of meeting a queen-normal colony begin merging with it. Workers of the queenless colony, with their brood, abandon their nest in favor of joining the columns of workers of the adoptive colony. The brood of the adopted colony is consumed within 1 or 2 days.

3. Males

a. Morphology. Because the army ant male represents a unique morphology (i.e., it is unusually large, has a long cylindrical gaster, highly modified mandibles, and uncommonly developed genitalia), Wheeler (1910), in his classification of caste variants, assigned to it the term "dorylaner" (Fig. 1E). Although this term is no longer used, it does point to the fact that these males are undoubtedly exceptional. However, in spite of their conspicuous appearance and the fact that they are commonly represented in museum collections, there is a dearth of information on army ant male morphology.

The males have well developed compound eyes and ocelli (Fig. 1F). The mandibles are distinctive enough to be of practical value in identifying some forms, e.g., in separating the subgenera of *Dorylus.* The maxillary palpus has two segments in all four tribes, while the labial palpus has one segment in the Aenictini and Dorylini, two segments in the Ecitonini, and three segments in the Cheliomyrmecini (Gotwald, 1969).

Tulloch (1935) noted that the alitrunk of *Dorylus helvolus* displayed the "greatest departure from the fundamental type of any of the [ant] subfamilies." He regarded the male doryline thorax as highly specialized. The male is the only doryline phenon with wings, and their wing venation is regarded as relatively primitive (i.e., unreduced) (Brown and Nutting, 1950). Within the army ants, *Cheliomyrmex* is the most primitively veined. Brown and Nutting (1950), in their analysis of formicid wing venation, placed great emphasis on the position of wing vein Mf1. Because this vein arises proximal to crossvein cu-a in the doryline specimens included in their study, they concluded that the dorylines arose from the main formicid line at an early date. However, within the genus *Dorylus,* the position of Mf1 varies considerably (Fig. 8E,G). Indeed, in the subgenus *Alaopone* it is distal to cu-a and in *Rhogmus,* the subgenus figured by Brown and Nutting, it may be found on either side of cu-a. Certainly wing venation patterns in the subfamily await careful analysis of large numbers of congeneric and conspecific specimens.

As in the queen, the waist of the male always has one segment, but unlike the queen, the male petiole is unarmed. The external genitalia of the male have been studied in greater detail than most other external features because of their potential taxonomic value. Even so, published comparative genitalic studies are nonexistent. The genitalic capsule of these males is retracted into a large cavity ventral to the rectum and anus in the last few gastral segments (Borgmeier, 1955; Forbes, 1958; Forbes and Do-Van-Quy, 1965). Of the genitalic sclerites, only the ninth sternum, or subgenital plate, projects beyond the tip of the gaster. As in other ants, the external genitalia consist of three pairs of valves that are surrounded anteriorly by a basal ring (lamina annularis) (Forbes and Do-Van-Quy, 1965). The outer valves are generally referred to as the parameres, the middle valves as the volsellae and the inner valves as the aedeagus. In *Dorylus* the basal ring is narrow and the parameres and volsellae are simple (Raignier and van Boven, 1955). Although the parameres and volsellae are also simple in the ecitonines, the basal ring is conspicuously wider (Forbes, 1958; Forbes and Do-Van-Quy, 1965).

The internal anatomy and histology of the army ant male remains virtually unexplored. Only the alimentary canal and the reproductive system have received attention.

Mukerjee (1926) noted that the alimentary canal of *D. labiatus* is straight and simple and lacks a crop and a well developed proventriculus. Shyamalanath and Forbes (1980) detailed the anatomy of the digestive system of the male of *Aenictus gracilis* and concluded that the unique features of this system lent support to the proposition that *Aenictus* arose independently of other army ants. It was pointed out by Brown (1968) that the metapleural gland is absent in army ant males and that these males must mate with apterous queens in alien conspecific colonies. Brown speculated that the metapleural glands might produce a substance that labels the individuals of one colony as aliens or enemies should they enter another colony of the same species. Without this labeling substance, the army ant male can enter an alien colony unmolested. As attractive as this hypothesis is, and Brown included in his observations a number of other instances in nondoryline ants where these glands are also absent, it does not explain the obvious exception when the workers of a queenless colony merge with a queen-normal colony.

The testes of *N. harrisi* and *E. hamatum* are composed of long, slender tubules that number 20–25 per testis (Forbes, 1958; Forbes and Do-Van-Quy, 1965). In *D. labiatus* the testicular tubules are quite small (Mukerjee, 1926). While a single capsule covers both testes in *N. harrisi,* each testis is invested with its own capsule in *E. hamatum* and in *D. labiatus* there is no capsule at all. Other structures in the system include the vasa deferentia, accessory glands, the bound accessory gland ducts, the ejaculatory duct, and the aedeagal bladder (Forbes, 1958). The accessory glands of *D. labiatus* are large and thick walled (Mukerjee, 1926), whereas those of *Eciton* and *Neivamyrmex* are formed of tightly coiled tubes (Forbes, 1958; Forbes and Do-Van-Quy, 1965). Curiously, in many studies of male ants, the accessory glands were not identified as such until Hung and Vinson (1975) examined males from five subfamilies including the army ants. They found the coiled accessory glands of the ecitonines to be unique among the ants, while the glands of *D. labiatus* are similar to those of the Myrmicinae and Ponerinae. They concluded that these differences gave credence to the polyphyletic hypothesis of doryline origins. Ford and Forbes (1980) described male reproductive anatomy of the Old World species *D. (A.) wilverthi* and *D. (A.) nigricans,* and Gotwald and Burdette (1981) compared the morphology of the male internal reproductive system in representative species of both New and Old World species. In this latter study, it was concluded that spermatogenesis and storage of newly formed sperm cells in the seminal vesicles occur during pupal development before emergence. In most species the testes greatly atrophy prior to the time of eclosion and little evidence of their existence can be found in the adult. Gotwald and Burdette (1981) further noted that the extraordinary differences in male internal genitalic morphology between New and Old World forms constitute further evidence of the polyphyletic origin of army ants.

b. Male Functions. Male ants do not contribute to the daily maintenance of a colony; they are, for all intents and purposes, little more than "flying sperm dispensers" (Wilson, 1971). In *Eciton* and other genera where colonies pass through alternating statary and nomadic phases, males appear periodically in large sexual broods (which include queens as well) that are coordinated with all-worker broods that precede and follow them (Schneirla, 1971). This precision in the appearance of males is probably absent in army ant species without the regular statary-nomadic functional cycle. For instance, in *Anomma,* sexual broods can appear during any season of the year (Raignier, 1959, 1972). Thus the male exodus from the nest following eclosion from the pupal stage is "relatively precise and genus typical" in species with a regular functional cycle and variable in other groups such as *Dorylus* (Schneirla, 1971).

The presence of sexual brood is prerequisite to colony division, and although males may fly from the nest site directly following eclosion, alate males may emigrate with their colonies (in *Eciton* this would be during the first nomadic phase of a new daughter colony following colony division) and may break away from the columns of workers and fly off. In this way they "literally seed the area through which they pass" (Schneirla, 1971). In *Anomma* the male brood, fully grown larvae and pupae, are left in the old nest with the new queen following colony division. The pupae then eclose and the adult males fly away (Raignier, 1972). *Anomma* males may also emigrate with the workers, but more often than not, these males are dealate and probably move with their adoptive colonies following their initial exodus flight (Savage, 1847; Donisthorpe, 1939; Raignier and van Boven, 1955).

When the exodus flight is ended, the males lose their wings, possibly as a physiological consequence of the flight itself, although in some cases they are torn off by workers. However, in order for fertilization to occur, these males must in some way find and gain entrance to another conspecific colony. By running on the substrate surface, the males may intersect with foraging or emigration trails of other colonies which they may then follow to the nest site; males that enter columns or nests of other species are killed (Schneirla, 1971). *Anomma* males, for example, can locate and follow the abandoned chemical trails of other colonies and follow these trails to the nest site (Raignier and van Boven, 1955). Schneirla (1971) conjectured that postflight *Eciton* males may attract conspecific workers to themselves by releasing a pheromonal attractant that is spread on the substrate surface by a brushlike collection of setae at the tip of the gaster.

Schneirla (1971) suggested that males of surface-active species leave their colonies around dusk and those of subterranean species leave in the evening or at night. However, males of D. (A.) *nigricans* and D. (A.) *wilverthi,* which are certainly surface-active species, fly in complete darkness only after 2000 hr, at least in Rwanda (J.K.A. van Boven, personal communication). Schneirla also noted that males of different species may respond differently to environmental stimuli once they have landed following their exodus flight. For example, surface-active species of *Neivamyrmex* have relatively small, flat eyes, react positively to doryline chemical trails and may on landing after their dusk flight, reach other conspecific colonies primarily through chemical stimuli. On the other hand, subterranean species have large eyes, react weakly to chemical trails and may rely on visual stimuli by moving toward moonlight silhouetted objects, such as rocks and logs, places where the trails of subterranean species are most likely to be located (Schneirla, 1971). Rettenmeyer (1963b) concluded that following eclosion New World males are initially negatively phototaxic and unable to fly for 1 to 3 days. He proposed that only after flight do they become positively phototaxic. Haddow *et al.* (1966) found that *Dorylus* males (of all five African subgenera) fly at all times of the year. Leston (1979) noted that there is a regular cycle in the timing of *Dorylus* male flight in which males are produced about every 30–32 days from March through September and around every 28–29 days from December through February. He also observed that this cycle is synchronous in the four most commonly trapped species and that males are produced ''more or less'' throughout the year, although with distinct seasonal peaks. Leston concluded that the syncronicity evident in these cycles was not related to climatic factors, as Schneirla suggested for ecitonine males, but was instead a mechanism for oversaturating an area with their numbers. In this way, the survival of at least some males from an onslaught of predators is enhanced. In a survey of New World army ant flights, Kannowski (1969) collected males of two species only during the dry season, of three other species during the dry season and well into the wet season, and of one species during the entire four months of the study. The males of nine remaining species were collected almost exclusively after the wet season began. Kannowski concluded that the rains may serve as triggering mechanism for these nine species. Seasonal flight periodicities in Nearctic army ant males have been reviewed in detail by Baldridge *et al.* (1980).

Haddow *et al.* (1966) also noted that the males of each species consistently peak, in total numbers of individuals trapped using light, at different times of night, starting at about 1800 hours and ending about 0600 hours. Because "dusk" is a phenomenon of higher latitudes, it is impossible to use the data of Haddow *et al.* to support or refute the Schneirla hypothesis that surface-active species fly at dusk. Kannowski (1969) examined the frequency distribution of ecitonine males during the hours of night and discovered that the species form two distinct groups: one which conducts "post sunset" flights and another that launches predawn flights. One species, *Neivamyrmex pilosus,* was found to be essentially "circum-nocturnal." Temporally spaced flights may in some way contribute to a species isolating mechanism. Apparently the males of several Nearctic species of *Neivamyrmex* have diurnal flights and this explains their absence in light traps used to determine seasonal periodicities (Baldridge *et al.,* 1980).

The frequency with which males may inseminate queens of their own mother colony,whether the queens be old or newly emerged, is not known. However, the genetic advantages of promoting gene flow between colonies may keep the frequency low. Once in an adoptive colony, a dealate male may remain there for days or even weeks before it mates and dies (Schneirla, 1971). Rettenmeyer (1963b) noted that such males probably live a few days but seldom more than 3 weeks.

One thing more is intuitively obvious: between the time that a male leaves its own colony and finds a foster colony, it is subject to many life-threatening risks. Predators and vagaries in the environment may, as risks, be so great that relatively few males ever achieve fertilization of a conspecific queen, which seems to be, after all, their only function.

C. Army Ant Brood

1. Morphology

Of the army ant brood, the larvae have been studied most thoroughly. The larvae were examined taxonomically by G. C. Wheeler (1943) and Wheeler and Wheeler (1964, 1974, 1976). Doryline larvae are "elongate, slender, subcylindrical'' with the anterior end slightly curved; 12 or 13 distinct somites are evident. Spiracles are minute. Vestigial legs (imaginal discs) are large and conspicuous (Wheeler, 1938). Setae are short, sparse to moderately abundant and mostly simple, except in *Neivamyrmex* where they are branched or plumose. The head is large, with short, simple setae and antennae composed of two sensillae each. The mandibles are weakly developed and of two types: "elongate, slender, slightly curved and denticulate" or "short, small, acuminate and feebly sclerotized". The maxillary palpi are either absent or represented by a slightly elevated group of sensillae. The trophorhinium ("the aggregate of roughened surfaces of the mouthparts which might be used in triturating food") is poorly developed or absent (Wheeler, 1943; Wheeler and Wheeler, 1964). Petralia and Vinson (1979) pointed out that the larvae of *Neivamyrmex nigrescens* do not show specializations for holding food on the ventral body region as do other ant species. This lack of specialized development they noted probably relates to the fact that prey food items are placed next to the larvae but never on them. Thus army ant larvae are not required to hold or manipulate their food items. Wheeler (1943) noted that the male larva in *Anomma* is "enormous" when compared to the worker larva and that the anterior portion is bent ventrally at an angle of 90°. Raignier (1972) observed that young male larvae of *Anomma* one week or older can be distinguished from worker larvae, because they are slender and they make curling movements. Those movements may provoke increased foraging activity in adult workers.

Lappano (1958) provided a detailed description of the external morphology of *E. burchelli* larvae. Schneirla *et al.* (1968) analyzed allometric growth in the larvae of *Eciton, Neivamyrmex,* and *Aenictus.* Their results confirmed the "empirical expectation" that the larval brood of *Eciton* and *Neivamyrmex* are polymorphic and that those of *Aenictus* are "quasi-monomorphic." Tafuri (1955) studied the larvae of *E. hamatum* and noted that several external features (such as the leg discs, shape of the head segment, and pilosity) are correlated with specific days in the nomadic phase. As a result, he was able to formulate a key for separating larvae according to the nomadic day represented in their stage of development.

Internal features of *E. burchelli* (Lappano, 1958) and *N. nigrescens* larvae (Wang and Happ, 1974) have been described. In the latter species, the labial gland was specifically targeted as a possible source of worker stimulating substances produced during the nomadic phase. The alimentary canal of *E. burchelli* is essentially a straight tube consisting of a foregut, midgut, and hindgut. The midgut is a blind sac, and the lumina of the midgut and hindgut are not continuous until late in the prepupal phase or in the pupa (Lappano, 1958). Wheeler and Bailey (1925) found the "stomach" of *E. burchelli* larvae to be unlike that in other known ant larvae because it is "very long and slender" and has "unusually thick, muscular walls." They assumed that the larva is fed at considerable intervals with large pellets of the "rolled up soft-parts of insects." Because these pellets are so compact, they retain their shape even in the lumen of the midgut. Four Malpighian tubules are present throughout larval development. The nervous system is composed of a central division, consisting of a brain and a ventral nerve cord with twelve paired ganglia, and a stomatogastric division. The *E. burchelli* larvae have a "secretory system" that includes the corpora allata and labial glands. The latter undergo striking morphological and histological changes during the course of larval development. Other *Eciton* features include the dorsal vessel and the ovaries (Lappano, 1958).

Larval cocoon spinning in the army ants varies remarkably from genus to genus, but the functional and phylogenetic significance of this variation remains a mystery. Cocoons are not spun in *Dorylus* and *Aenictus* but are present in *Eciton* and *Labidus*. In *Neivamyrmex* cocoons are spun only by sexual brood. It is not known whether cocoons are present in *Cheliomyrmex* and *Nomamyrmex* (Schneirla, 1971). Since it is assumed that the ant cocoon is inherited from a wasplike ancestor (Wheeler, 1915), the absence of a cocoon is a derived characteristic. While watching *Eciton* larvae spin cocoons, Beebe (1919) noted that, "I watched the very first thread of silk drawn between the larva and the outside world, and in an incredibly short time the cocoon was outlined in a tissue-thin, transparent aura within which the tenant could be seen skillfully weaving its own shroud."

2. Biology

Army ant colonies that are functionally normal always contain developing brood. In species with regularly alternating nomadic and statary phases, it is evident that the brood are produced on a periodic schedule. Larvae are present in the nomadic phase, a period characterized by elevated colony activity, and pupae are present in the statary phase when colony activity is low (Schneirla, 1971). In studies regarded by many as ethological classics, Schneirla (1933, 1938) determined that it was the larvae that somehow energized the colony to the activity level of the nomadic phase and that it was the inertness of the pupae that produced the depressed activity levels of the statary phase (Fig. 10).

All-worker broods that normally appear in a cyclic sequence are occasionally interrupted by a sexual brood that also has a profound influence on the colony via the process of colony division. In the family Formicidae, the dorylines certainly produce the largest all-worker broods per colony. These range from 30,000 individuals in *A. laeviceps* (Schneirla and Reyes, 1966) to 1,500,000 in *D. (A.) wilverthi* (Raignier and van Boven, 1955). The polymorphism evident in the brood of all genera except *Aenictus* is a function of

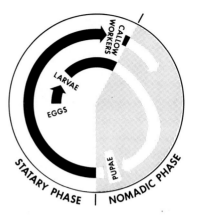

Fig. 10. Functional cycle of group A dorylines. Eggs are deposited by the queen during the statary phase. These hatch and develop as larvae during the latter one-half of this phase and the entire nomadic phase. Pupation of these larvae triggers the beginning of the statary phase. During the statary phase these pupae continue to develop. Their emergence as callow workers stimulates the start of the next nomadic phase. Note that two sets of brood are present during the statary phase, one consisting of eggs and young larvae and the other of pupae.

nongenetic factors that include trophogenic effects, stimulative effects of the workers on the larvae, "incubative" factors, and the physiological condition of the queen at the time of egg laying (Schneirla, 1971).

From his observations, Schneirla (1938) developed the concept of a brood-stimulative theory of army ant cyclic activity. Initially working with *Eciton,* he demonstrated that the nomadic phase begins with the emergence of callow workers from their cocoons which has a stimulatory effect on the colony. Colony activities in this phase are characterized by a nightly emigration from one nesting site to another preceded by almost frenzied foraging. During the nomadic phase, the queen is contracted and the larvae stimulate, on a sustained basis, the level of activity typical of the phase. This phase ends and the statary begins with maturation of the larvae. At this time of low brood stimulation, daily emigrations cease and the queen achieves full physogastry. At the middle of the phase, she deposits a single series of eggs that constitute a unitary population. Larvae hatch from these eggs prior to the emergence of the callow workers that will stimulate the next nomadic phase. Thus two brood populations overlap one another in time and space. These well-marked functional cycles have been identified in at least some species of *Eciton, Neivamyrmex,* and *Aenictus.*

However, in *Dorylus,* and perhaps in numerous species in the other genera, the functional cycles are irregular at best and probably not even homologous to those investigated by Schneirla. Although Raignier and van Boven (1955) found that emigrations in *D. (A.) wilverthi* can be initiated by the eclosion of callow workers, 6–40 day intervals separate emigrations. *Anomma* driver ants do not follow a functional cycle of alternating nomadic and statary phases. In fact, they appear to exist in successive "statary" phases separated by emigrations that may last for 2 or 3 days or even longer. Indeed, Raignier and van Boven (1955) recorded one intermigratory interval for *D. (A.)*

nigricans of 125 days. Thus the influence of brood on the nesting and emigration behavior of *Dorylus* may be far less significant than it is in *Eciton*.

The periodically produced sexual broods of army ants are evidenced by an occasional mass exodus of alate males from the nest. Male emergences of this kind have been recorded for *Dorylus* (Mayr, 1886; Brauns, 1901), *Eciton* (Rettenmeyer, 1963b; Schneirla, 1971), and *Neivamyrmex* (Gallardo, 1915; Bruch, 1923) among others. In two species of *Eciton,* Schneirla and Brown (1952) found that "fertile individuals" are produced in distinctive annual broods. In *E. hamatum* a mature sexual brood may contain, for example, only six new queens but as many as 1500–2000 males. Schneirla and Brown (1952) believed that the trophallactic relationship between the workers and the sexual brood is more intensive than it is between workers and an all-worker brood.

In *Eciton, Neivamyrmex,* and *Aenictus,* the onset of dry weather in areas with distinct seasonal changes may stimulate the production of sexual broods (Schneirla, 1971). Although Schneirla (1971) convincingly argued his "dry-impact" hypothesis for the initiation of sexual broods, Rettenmeyer (1963b) pointed out that all colonies of *E. hamatum* in the same locality do not produce sexual broods at the same time nor in the same dry season. In fact, he noted that some species of *Eciton* actually produce sexual broods in the rainy season. Schneirla (1971) surmised that the abrupt environmental change represented by the onset of the dry season radically affects the reproductive physiology of the queen. The physiological changes thus wrought can inhibit the process of fertilization and establish the pattern for sexual broods. However, the first group of eggs in a sexual brood series are deposited before the mechanism inhibiting fertilization is fully operative. These eggs are fertilized and are destined to become new queens. A second set of eggs in the series goes unfertilized and will produce the males which are haploid. Schneirla (1971) postulated that a third series of eggs, also unfertilized, might be laid and later used to feed the newly hatched queen larvae. The number of eggs laid in a sexual brood series may initially approximate the number produced in an all-worker series, but perhaps for a variety of reasons the number of sexual brood eggs is substantially reduced (Schneirla, 1971). Therefore sexual broods are small and individuals are assured greater attention and heavier feeding from the workers than is the case in all-worker broods where the sheer numbers of individuals guarantee intense competition for food. Flanders (1976) hypothesized that the physiological change induced by the dry season and responsible for the inhibition of fertilization is "simply the lack of sperm gland stimulation," i.e., the failure of sperm-activating secretions to be produced by the spermatheca. To explain the presence of sperm that fertilizes the first set of eggs, he suggested that some sperm remains in the coils of the spermathecal duct between broods. This residual sperm is reproductively important only when egg deposition occurs during the dry season and spermathecal activity is inhibited (Flanders, 1976).

In *Anomma,* sexual broods can appear during any season of the year, although they appear with greater frequency during the last third of the dry season (Raignier and van Boven, 1955; Raignier, 1959, 1972). Male brood can number from 200 to 3400 individuals per nest, while the maximum number of queens ever observed in one nest was 56 (Raignier, 1959). As in *Eciton,* the *Anomma* male and queen larvae enjoy greater attention from the workers which includes over-feeding. Developmentally, however, the queens are approximately 10 days ahead of the males and are already adults when the males are reaching the pupal stage (Raignier, 1959). Males may appear in worker broods.

V. BEHAVIOR OF ARMY ANTS

A. Pheromonal Communication

1. Trail Pheromones

Given the fact that a majority of army ant species are blind and that the remainder are nearly so, pheromonal communication along with tactile stimuli assume added importance in the biology of these ants (Topoff and Lawson, 1979). In no instance is this more pertinent than in trail following where visual orientation, except perhaps to light intensity, is at best of little significance. As Blum (1974) noted, odor trails are an effective method of coordinating the movements of groups and individuals. Furthermore, in few colonies of ants are there more reasons than in the dorylines, with their extraordinarily large colonies, to effectively organize the workers into a cohesive, moving force. After all, other ant species do not combine group predation and nomadism to the extent that the army ants do, and both activities require an exceedingly sophisticated system for group organization and orientation.

Watkins (1964) concluded that trail substances in *Neivamyrmex* may be contained in the feces, perhaps even added to the feces by some hitherto undescribed glands. In *Eciton,* the hindgut was similarly implicated as the source of trail pheromone (Blum and Portocarrero, 1964). However, the specific source is unknown. Although the pheromone in *Eciton* may be produced by specialized gland cells in the digestive tract, circumstantial evidence indicates that this substance may in fact be a digestive product. There may be more than a single source for trail substances, at least in *Eciton.* For example, *E. burchelli* workers whose gasters have been removed and whose petioles are sealed with wax are still capable of laying a trail that elicits trail-following behavior. However, the substance responsible for this trail, perhaps a "footprint" secretion, is probably not a true pheromone since it is of short duration (Torgerson and Akre, 1970a,b). After examining trail following in *Neivamyrmex nigrescens,* Topoff and Mirenda (1975) concluded that "it is not surprising that the ants can utilize not only the non-volatile and volatile components of the chemical trail deposited from the hindgut, but other chemicals secreted from the surface of the ants' bodies." The sources of trail substances in *Aenictus* and *Dorylus* have not been determined.

The functional duration or stability of army ant trails, at least in the genus *Eciton,* depends on four factors: (1) the nature of the substrate, (2) the kind of trail laid (foraging versus emigration), (3) the amount of precipitation that falls on the trail, and (4) the lifeway of the ant producing the trail (i.e., whether the ant is surface-adapted or subterranean). Trails are more stable when established on "porous substrates with many adsorptive sites such as roots, logs and lianas" and less persistent on soil and leaf-litter surfaces. Emigration trails are more stable than are foraging trails but this stability may result from the fact that a greater number of workers are involved in establishing emigration trails and that emigration trails are used for a longer period of time. Trails are less persistent when subjected to rain, and once a trail is wet, the workers encounter considerable difficulty in sensing the trail. Surface-adapted species apparently establish trails that are significantly more durable than those created by subterranean species (Torgerson and Akre, 1970a). Increased trail stability certainly seems advantageous to surface-adapted species, since their trails are more vulnerable to the effects of air currents, humidity, rain, and solar light and heat.

The chemistry of doryline trail substances is unknown, except that, for the social Hymenoptera in general, trails are "usually generated with a mixture of compounds" (Blum, 1974). Laboratory experiments on the *Eciton* trail substance have demonstrated its stability, low vapor pressure, and water insolubility.

A lack of trail odor specificity is apparently common among the ecitonines. Watkins (1964) demonstrated that three species of *Neivamyrmex* followed each other's trails. In fact, two of the species followed trails prepared from 70% alcohol extracts of whole workers of *Eciton dulcius.* However, this promiscuity in trail following did not extend to extracts prepared from nondoryline ants. Males of *N. opacithorax* followed the trails of all three *Neivamyrmex* species, and *Neivamyrmex* queens of two species also followed the trails of other congeneric species (Watkins, 1964). Although Watkins *et al.* (1967a) showed experimentally that the trails produced by four species of *Neivamyrmex* and *Labidus coecus* elicited trail following by all five species, they did note that some species preferred trails laid by conspecific workers. This indicates that trail substances vary from species to species and that workers can detect these differences. However, the lack of specificity among army ant trail substances is not universal. In the laboratory studies of Watkins *et al.* (1967a), *Neivamyrmex pilosus* did not readily follow the trails of other species. Torgerson and Akre (1970b) also noted that although army ants can detect interspecific trails, the trail-following behavior elicited is variable. They hypothesized that trail following involves two chemical components, one which serves as a general releaser and to which all ecitonines respond, and another which is genus- or species-specific and results in sustained trail following.

Intraspecific subcaste and age differences in trail following behavior were investigated by Topoff *et al.* (1972a,b, 1973). Although callow workers of *Eciton* have the same relative ability to detect and follow conspecific trails as do the mature adult workers, their running speed along the trails is significantly slower (Topoff *et al.*, 1972b). Topoff *et al.* (1973) proposed that major workers of *Eciton* are less responsive than small workers to trail substance and that in the field major workers are more prone to leave trails than are smaller workers, particularly when disturbed. Experimental data revealed that differences between major and intermediate-sized workers in trail following performance are not significant. Topoff *et al.* (1972b) hypothesized that the readiness with which the major workers left a trail and approached sources of disturbance is attributable to the differences in sensitivity of major workers to other compounds, in this case, alarm substance.

A veritable menagerie of arthropods are known to follow army ant trails (Rettenmeyer, 1962a), and in fact, interordinal trail following includes vertebrates as well, since certain species of blind snakes have also found an adaptive advantage in detecting and following the pheromonal trails of doryline ants (Watkins *et al.*, 1967b, 1969, 1972; Gehlbach *et al.*, 1968, 1971; Brown *et al.*, 1979). Myrmecophilous species of millipeds (Rettenmeyer, 1962b), mites (Akre and Rettenmeyer, 1968), thysanurans (Rettenmeyer, 1963a; Torgerson and Akre, 1969), phorid flies (Rettenmeyer and Akre, 1968), sphaerocerid flies (Richards, 1968), beetles of the families Histeridae (Akre and Rettenmeyer, 1968), Staphylinidae (Wasmann, 1904, 1917; Patrizi, 1948; Paulian, 1948; Seevers, 1965; Akre and Rettenmeyer, 1966, 1968; Akre and Torgerson, 1968, 1969; Kistner, 1976) and Carabidae (Plsek *et al.*, 1969; Topoff, 1969), and diapriid wasps (Masner, 1976, 1977) are all capable of following the chemical trails of army ants.

Blum (1974) pointed out that although ants use "subtle blends of exocrine products" to insulate their trails from a majority of other ant species, they have not been able to deter unrelated animals from using their trails. Furthermore, many myrmecophiles are proficient in detecting species-specific differences in doryline trails and apparently respond more frequently to these differences than do the army ants themselves (Akre and Rettenmeyer, 1968).

The ability of myrmecophiles to follow chemical trails is, for the most part, of some detriment to the army ants. Although dorylophilous millipeds and phorids function primarily as scavengers on organic debris in and around the ant nest (Rettenmeyer, 1962b; Rettenmeyer and Akre, 1968) and the thysanurans appear to feed mainly on surface secretions and particles scraped from the bodies of the adult ants and their brood (Rettenmeyer, 1963a), the coleopterous dorylophiles feed on army ant prey and brood. For example, in laboratory colonies staphylinid beetles feed on both prey and brood (Akre and Rettenmeyer, 1966; Akre and Torgerson, 1969). However, although the staphylinid genus *Diploeciton* is recorded in the field to feed on army ant brood, Akre and Torgerson (1969) surmised that "in nature," staphylinids may feed almost exclusively on prey returned to the nest. Carabids of the genus *Helluomorphoides* also prey on army ant brood and occasionally on the adults (Plsek *et al.,* 1969). Watkins *et al.* (1967b) theorized that the blind snake *Leptotyphlops dulcis* enters raiding columns of *N. nigrescens* and is subsequently able to locate the nest and feed on the brood and possibly prey. Curiously, the indigenous peoples of the Usambara Mountains of Tanzania report that blind snakes of the genus *Typhlops,* which they call mkonko, are often seen slithering along in the columns of *Anomma* driver ants (Loveridge, 1949). They refer to the snake as the "cow" of the ants that the ants keep until there is a food shortage, whereupon they kill and devour it. It would seem that once the snake reaches the *Anomma* nest the reverse is true.

In all of these instances, trail following is essential for the myrmecophiles either to gain entry to the ant nest or to maintain a more or less permanent association with a particular colony. It would not be imprudent to conjecture that army ants that can insulate their trails from myrmecophiles are being selected for.

2. Recruitment Pheromones

Recruitment of sister workers to a food source by foragers that discover the source is common among ants. However, until recently this phenomenon was not sufficiently documented in army ants. Chadab and Rettenmeyer (1975) discovered that *E. hamatum* and *E. burchelli* workers lay down recruitment trails that are qualitatively different from foraging trails. After being exposed to a food source, a recruiting worker returns to the foraging column, intermittently dragging her gaster on the substrate as she goes. When the recruiter reaches the foraging column, she runs 5–10 cm in each direction in the column, contacting workers with her antennae and body. She returns to and runs along the recruitment path periodically, and in doing so contacts new workers and reinforces the trail. Within 30 sec, workers from the column are diverted to the recruitment trail, and in the first 5 min, 50–100 workers are recruited. Some of the recruited workers in turn recruit new workers to the scene in a process Chadab and Rettenmeyer called "secondary recruitment." Although there is a recruitment message in the recruitment trail substance, the effectiveness of the message is enhanced by recruiter behavior. Field experiments led Chadab and Rettenmeyer to conclude that (1) recruitment pheromone is either an entirely different substance from foraging trail substance or it is a combination of hindgut material and a substance from some other source; (2) the army ant recruitment system, "in terms of gathering large numbers of workers quickly," is among the most efficient in the ants; and (3) the "combination of continuous foraging columns, a recruitment trail that attracts and orients workers, secondary recruitment, and persistent recruiters results in the efficient gathering of the large attack force essential for army-ant raiding."

Laboratory experiments with *Neivamyrmex nigrescens* (Topoff *et al.*, 1980b) have confirmed the existence of a recruiting trail pheromone that is qualitatively different from the ants' exploratory trail pheromone. This recruitment pheromone alone is apparently sufficient enough to initiate mass recruitment equal to that brought about by recruiting workers that interact tactually with nestmates.

3. Alarm Pheromones

Releasers of alarm behavior in the doryline ants have not been chemically isolated or identified, although the existence of such substances is not in doubt. In the Formicidae, alarm pheromones are generally produced by the mandibular glands (Blum, 1974). Brown (1960) first demonstrated that alarm pheromone in *Eciton, Nomamyrmex,* and *Labidus* is produced in the head. The crushed head of a worker dropped into a column of workers elicits attack behavior during which the workers bite and attack the head, while crushed, headless bodies of workers attract only momentary attention. In *Eciton* and *Nomamyrmex,* the crushed worker heads emit a "meaty" odor that Brown (1960) assumed to be associated with the alarm substance. Inanimate objects, such as twigs, rubbed against the crushed heads also elicit attack behavior when placed among the workers. Brown speculated that workers emitting alarm pheromone are not themselves attacked by sister workers, either because the workers normally give off lesser amounts of alarm substance than is released by the crushed heads or because the worker also secretes, from its alitrunk or gaster, a substance that neutralizes the attack behavior of its nestmates. In the latter case, the substance might be an "identification pheromone" or a "nest odor" (Brown, 1960).

Torgerson and Akre (1970b) repeated some of Brown's field experiments and found that ecitonine alarm pheromones are more specific than is the case, for instance, in the Formicinae. While the crushed heads of *E. burchelli* elicited alarm behavior from *E. hamatum,* and those of *E. hamatum* stimulated such behavior in *Nomamyrmex esenbecki,* reciprocal tests did not; *E. mexicanum* and *E. vagans* did not react at all interspecifically.

4. Queen Odor

Army ant workers are no doubt attracted to their queen and recognize her odor. Although the source of the odor remains to be identified, evidence of this attraction can be seen in *Eciton* where the queen is surrounded by a cluster of small workers in the undisturbed nest or bivouac and by larger workers when the nest is disturbed. In an emigration column, she is always accompanied by a retinue of workers (Rettenmeyer, 1963b; Schneirla, 1971). In the laboratory, if an *Eciton* worker is taken from a dish with a queen and added to a dish containing workers without a queen, the latter workers become greatly excited, apparently because the added worker carries with it some queen odor (Rettenmeyer, 1963b). Specifically, the workers are attracted to the anterior portion of the queen's gaster and rarely to her mouthparts or anal region (Schneirla, 1949; Rettenmeyer, 1963b). Laboratory experiments with five species of *Neivamyrmex* and one species of *Labidus* demonstrated that workers prefer secretions of their own queen over those of queens from other conspecific colonies and that although workers are attracted to the secretions of queens of other species, they are most attracted to conspecific queens (Watkins and Cole, 1966).

B. Mating Behavior

Mating in army ants has seldom been observed or recorded. Smith (1942) reported that a mating pair of *Neivamyrmex carolinensis* was found in a nest in 1932 and constituted the first such observation for army ants in the United States. Mating in *E. hamatum* has been observed twice (Schneirla, 1949) and in *E. burchelli* once (Rettenmeyer, 1963b). Matings in Old World dorylines have not been described. Schneirla (1971) suspected that in *E. hamatum* "the leading callow queens" (i.e., those that eclose first from the pupa and around which a new colony, as a result of fission, forms) are inseminated in or near the nest of the parent colony within a few days after they eclose or in the early part of the first nomadic phase of their new colony. Raignier and van Boven (1955) noted that queens of *Anomma* are fertilized many times during their lives and that they may even be inseminated by males of different subgenera. Intersubgeneric copulation and the subsequent "loss" of male gametes is reproductively inefficient and difficult to explain. Schneirla (1971), for instance, hypothesized that the male must be dealate before mating is possible and that once dealate, it must live for a period of time in its adoptive colony and acquire the colony's odor before being accepted by the queen and that, after becoming habituated to the host-colony odor, it will mate only with a queen bearing that odor.

Because both matings in *E. hamatum* observed by Schneirla (1971) occurred after the ants were captured and consequently excited, he speculated that matings in nature may occur not in the nest but rather during emigration since it "presents intervals of excitement with opportunities for union when a dealate male nears the queen as she pauses at an obstacle or enters a new nest." In the first coupling he observed, the male was oriented over the queen's dorsum with its mandibles securely locked about one of her petiolar horns. Insertion of the male genitalia in the queen occurred with such penetration that the queen's gaster was considerably deformed as a result. While the male appeared "lethargic," the queen moved about the enclosure carrying the male with her. They remained in copula for 2 hr. The second coupling, between a different pair, lasted 10 hr and ended with the male apparently entering the first stages of death. An examination of the queen's spermatheca revealed the presence of several "ball-like" masses of sperm, but the ovarioles contained only immature eggs.

The *E. burchelli* mating occurred between a previously fertilized queen that was more than 1 year old and a dealate male, both of which were taken from the same emigration column and placed together in a petri dish. They remained in copula for 1 hr. The male held the queen with its middle and hind legs and grasped the queen's petiole with its mandibles posterior to the horns. It did not grasp the horns as was observed in *E. hamatum*. The *E. burchelli* mating lends support to Raignier and van Boven's claim that army ant queens mate more than once during their lives.

C. Foraging Behavior

1. Group Predation

Group predation enables species to exploit as food sources other social insects and large arthropods (Wilson, 1958a). In each case it is the collective efforts of cooperating foragers that permits these species to enter and ravage the vigorously defended nests of termites, wasps, and ants or to physically overwhelm prey larger than themselves. These are prey not ordinarily available to solitary foragers. Group predation also increases the efficiency with which a colony can flush and capture prey. This is particularly important for army ant species that function as general predators. Group predation combined with nomadism, which permits colonies to periodically shift trophophoric fields, makes possible exceptionally large colonies like those achieved in some species of *Anomma* (Raignier and van Boven, 1955). As defined by Wilson (1958a), group predation must include two components: group raiding and group retrieving of living prey.

2. Group Raiding

a. Column Raids.   Schneirla (1934) discerned two types of raiding (or foraging) patterns in army ants: column raids and swarm raids. These patterns are species specific, although gradations between the two patterns are evidenced in some species such as *Labidus coecus* (Rettenmeyer, 1963b). A column raid consists of a system of branching columns of foraging workers that diverge from a single base or trunk column that connects the nest with the foraging sector. Each of the branching columns usually terminates in an advancing group of workers that push forward into new territory with their antennae actively playing over the substrate and objects before them. At the same time, the trails over which the advancing columns move are chemically laid and augmented by the workers as they touch their gasters to the substrate. Membership in the advance guard of the terminal groups constantly changes as workers push forward and then retreat, only to be replaced by other workers that behave in the same way (Schneirla, 1934). In other words, there are no trailblazer specialists. The direction that each of these terminal groups takes is determined by the momentum of newly arrived workers and by topographical features.

As raiding continues, trails to prey-depleted areas are abandoned. However, one advancing trail is usually maintained, and this gives rise to new branches farther along as other previously established trails are vacated. Thus the one continuing base column grows progressively longer. "In this manner the typical fan-shaped complex of moving ant columns is moved forward" (Schneirla, 1934). Terminal raiding groups can advance rapidly, as much as 20 m/hr in *E. hamatum,* and to considerable distances, as far as 350 m in the same species (Schneirla, 1971).

The organization brought to column raiding ranges widely from the simplest of doryline systems, as exemplified by *Aenictus,* to the complex, specialized raids of *E. hamatum,* with species like *Neivamyrmex nigrescens* and *Nomamyrmex esenbecki* organizationally somewhere in between (Schneirla, 1971). In the Asian species *A. laeviceps,* the base column is 1–5 workers wide and often extends as far as 20 m from the nest. The terminal groups may range in width from a few centimeters to a few meters (Schneirla and Reyes, 1966). Foraging columns in some African species of *Aenictus* are weakly developed, consisting of small groups of 3 or 4 workers, running together in single file. These groups are often widely separated but all follow precisely the same trail, revealing its chemical basis (Gotwald, 1976). Rettenmeyer (1963b) noted that the column raider *Neivamyrmex pilosus* travels over its foraging trails in a similar manner, i.e., its workers move along in "spurts" with gaps between individuals and groups of individuals. The complexity of *E. hamatum* raids is evinced by the fact that this species usually develops three systems of trails on any one foraging expedition (Schneirla,

1971). The complexity and "strength" of a raid vary with the functional cycle of a colony. That is, in species with alternating nomadic and statary phases, such as *E. hamatum,* the workers conduct their heaviest, most complex raids during the nomadic phase when workers are most excitable (Schneirla, 1971).

b. Swarm Raids. In a swarm raid, the base column divides in the raiding area into a series of anastomosing columns that coalesce to form a single, advancing swarm of workers. These swarms in *E. burchelli* are commonly over 5 m in width and occasionally surpass 15 m (Schneirla, 1971), while in some species of *Anomma* they are likely to be 20 m or more in width (Kistner and Gotwald, 1982). In *E. burchelli,* the advancing swarm proceeds by alternate flanking movements, i.e., the swarm swings in its movement first to one side and then to the other. This is responsible for the meandering course that the base column takes as it lengthens behind the advancing foragers (Schneirla, 1934). Schneirla (1971) considered these flanking movements to be superior in organization to the advances of other swarm raiders and that they permit *E. burchelli* to better hold a single direction of advance than can *Anomma* and *Labidus praedator* (for a description of foraging behavior in *L. praedator,* see Fowler, 1979). Swarm behavior was considered by Schneirla (1940) to be a "highly complicated and variable phenomenon," although the organization of swarm activity rests on certain stereotypical responses of individual workers during raiding. He listed three individual behavior patterns that occur during swarm activity: (1) a "track phase" during which a worker runs rapidly over a chemically-saturated trail, (2) a "pioneering phase," in which the worker encounters chemically unsaturated terrain and reverses direction after a short advance, and (3) a "retreat phase" when the worker returns to the swarm and is directionally reoriented.

In *Anomma,* the raiding column swarm advances as rapidly as 20 m or more per hour. The raiding columns become so crowded with workers that they may achieve a density of 13 individuals per cm^2 (Raignier and van Boven, 1955) (Fig. 11A). The average length of the raiding column in *D. (A.) wilverthi* and *D. (A.) nigricans* is 125 m (Raignier and van Boven, 1955), although Leroux (1975) found that 63% of the raids of *D. (A.) nigricans* that he observed measured between 26 and 75 m. Raiding expeditions in *Anomma* commonly range from 9 to 27 hr in length (Raignier and van Boven, 1955; Leroux, 1975, 1977b).

A comparison of the two types of raiding systems reveals two strategies, one in which a series of terminal raiding groups forage over narrow strips of substrate (column raiders) and the other in which a single mass of workers sweeps across a wide area of substrate (swarm raiders). Intuitively it would appear that the latter strategy is most adaptively significant for the general predator, since it provides a mechanism for flushing simultaneously a taxonomically diverse group of prey.

c. Search Activities.   While the discovery of prey by foragers is, in part, fortuitous, workers do respond positively to the movements of potential prey, perhaps through the detection of substrate vibrations. Prey odors are no doubt also important cues, but so are the odors of the excited foraging workers themselves, possibly via recruitment and/or alarm pheromones. Schneirla (1971) noted that column raiders are not as responsive to prey movements as are swarm raiders, and that they respond first to prey odors. Indeed, Rettenmeyer (1963b) noted that the swarm-raider *E. burchelli* was so responsive to movement that the raiding workers would attack even a blade of grass, if it moved. He found as well that some arthropods escaped being killed by these workers when they remained motionless.

In *E. hamatum,* the advancing workers "probe the surface and low vegetation, prying beneath surface cover and into insect burrows and other niches" (Schneirla, 1971). Rettenmeyer (1963b) found that *E. hamatum* workers run primarily on the ground and leaf surfaces, and although they climb vegetation and buildings, they rarely go beyond 2 m above the ground. This contrasts markedly with some *Anomma* species that are known to forage in trees to heights of more than 3 m (Raignier and van Boven, 1955; Gotwald, 1972a) and with *E. burchelli* which may raid into the tops of trees (Rettenmeyer, 1963b). In *Anomma,* at least, workers foraging on low vegetation to 1 m or more in height habitually drop to the substrate at the conclusion of their search activities instead of climbing back down. This creates a sound that resembles a light rain shower, as the workers strike the substrate surface. Their behavior probably functions to reduce the energy expended on search activities (Kistner and Gotwald, 1982). General search activities of some African *Aenictus* workers involves an apparent random meandering and an exploration of the soil surface as well as crevices and holes in the soil (Gotwald, 1976).

d. Swarm "Followers."   Swarm raiders are not alone as they advance across the forest floor in their quest for prey. Indeed, they are frequently accompanied by animals that benefit from the ability of the raiders to flush from cover large numbers of arthropods. Most commonly, these swarm followers are insectivorous birds and parasitic flies.

Almost 150 years ago Lund (1831) noted, when observing army ants in Brazil, that "ces troupes des fourmis voyageuses sont constamment suivies par une bande d'oiseaux. . . " He further noted that one bird species "announce au loin par son crimonotone et lugubre la presence de ces troupes."

Fig. 11. (A) Raiding column of *D. (Anomma) nigricans;* (B) a major and media worker of *D. (A). nigricans* returning to the nest with prey (note that the workers straddle the prey); (C) an *Anomma* worker returning to the nest with prey. (Photographs by W. H. Gotwald, Jr.)

Thus detecting the presence of army ant swarms through the calls of attendant "antbirds" has long been a part of tropical forest lore. As Johnson (1954) noted,

> On a walk through the forest in tropical America, long periods may pass without the glimpse of a bird. Then, suddenly, all about one hears the chirring, twittering and piping of birds, and sometimes a dim murmur, as if a light [rain] shower were striking the leaves of the forest floor. This gentle pattering—it soon becomes clear—is caused by the frantic fluttering and hopping of countless insects trying to escape a swarm of raiding army ants. . .

The behavior of birds that follow or attend the swarm raids of the New World species *E. burchelli* and *L. praedator* was studied by Johnson (1954) and Willis (1960, 1966, 1967). These birds, e.g., the red-throated ant-tanager (*Habia gutturalis*) and the bicolored antbird (*Gymnopithys bicolor*), forage on the arthropods that so predictably flee the advancing swarm. Seldom do the birds take the army ants themselves. The presence of "professional ant-followers," such as bicolored antibirds, may also attract other birds to the swarming site, consequently a number of bird species are clearly opportunistic in taking ant-flushed prey (Willis, 1967). In fact, Willis (1966) found that even migrant bird species from North America (e.g., Acadian Flycatchers and Wood Thrushes) attend army ant swarms in the tropics. Noting the presence of "professional followers" and opportunists on Barro Colorado Island, Johnson (1954) defined the bird associations in the following way: (1) birds of the "feeding aggregations, whose association seemed wholly dependent on the feeding opportunities afforded by the army ant raids and which remained with the ants for long periods of time" (e.g., *Gymnopithys leucaspis, Hylophylax naevoides,* and *Formicarius analis*), and (2) birds of the "social aggregations, whose association was independent of army ant raids, but which would attend raids for varying periods of time" (e.g., *Dendrocincla fuliginosa, Dendrocolaptes certhis,* and *Microrhopias quixensis*). Willis and Oniki (1978) have reviewed what is known about the New World birds that follow army ant swarms.

Anomma driver ants attract a similar group of avian attendants (Bequaert, 1922). DuChaillu (1861) wrote of the behavior of one species:

> . . . they fly in a small flock, and follow industriously the bashikouay [*Anomma*] ant in their marches about the country. The bird is insectivorous; and when the bashikouay army routes before it the frightened grasshoppers and beetles, the bird, like a regular camp-follower, pounces on the prey and carries it off. I think it does eat the bashikouay.

Some of these birds are from the genera *Alethe, Neocossyphus,* and *Bleda* (Bequaert, 1922).

Conopid flies of the genus *Stylogaster* and tachinid flies of the genera *Calodexia* and *Androeuryops* are commonly associated with the raiding swarms of *E. burchelli* and *L. praedator* (Rettenmeyer, 1961). *Stylogaster* flies are parasitic on cockroaches driven from cover by the foraging workers. Darting at the arthropods fleeing the advancing swarm, these flies deposit eggs on the exposed cockroaches. *Calodexia,* on the other hand is larviparous and deposits larvae on its hosts. The *Androeuryops* flies are also parasitic, probably on arthropods flushed out by the ants (Rettenmeyer, 1961). In Africa, calliphorid flies of the genus *Bengalia* accompany the swarm raids of *Anomma* and steal prey directly from the workers. Lamborn (1913–1914) described the behavior of *B. depressa* as follows:

> Suddenly the fly rushed forward, and it must have driven its proboscis, which seems to me armed with strong bristles, into the pupa, for the ant was brought to a standstill with a sharp jerk. Then ensued a tug-of-war between ant and fly fastened on at opposite ends of the pupa, but neither had the advantage till, as it seemed to me, the ant must have got annoyed and loosening its hold rushed towards the fly, which of course instantly flew off with the pupa, and this it proceeded to suck on the ground about a foot away from the ants.

Stylogaster is also present around *Anomma* swarms and is parasitic on fleeing cockroaches, although it also parasitizes calypterate Diptera (Smith, 1967, 1969).

Certainly among the most surprising of swarm followers are several species of ithomiine butterflies. Drummond (1976) was the first to observe this puzzling association. He witnessed in the field six butterflies "flying low over the leading edge" of a swarm of *Eciton burchelli* but did not discover any functional reason for this association. He further noted the presence of antbirds foraging at the head of the swarm, and this observation was later to be seen as relevant to understanding this unusual association. Ray and Andrews (1980) have since shown that these butterflies feed on the bird droppings deposited by the antbirds. That is, the antbirds "provide a predictable source of droppings, an otherwise sparsely distributed resource."

Swarm raiders, their myrmecophiles and their "followers" constitute a complex and intricate balance of relationships that represents one of the most elegant coevolved systems in the tropics.

3. Hypogaeic and Epigaeic Foraging

A majority of army ants are subterranean foragers, although many will also forage beneath forest litter. Few chance to forage on the substrate surface exposed to solar light and heat and the evaporative effects of the air. Schneirla (1971) applied the term hypogaeic to species he regarded as subterranean and epigaeic to those that are surface adapted. However, he usually utilized the nesting behavior of species as a measure of their adaptation to a particular lifeway, essentially ignoring extranidal activities. In doing so, Schneirla failed to characterize accurately the composite nature of any one species' behavior.

In order to portray the biology of army ants precisely, the terms hypogaeic and epigaeic must be applied independently to the three basic components of army ant behavior: nesting, foraging, and emigration (Gotwald, 1978).

Although a majority of Old World dorylines are hypogaeic foragers, a significant number of species are not. Chapman (1964) considered five species of *Aenictus* in the Philippines to be epigaeic foragers. Raiding in two of these species, *A. gracilis* and *A. laeviceps,* was studied in detail by Schneirla and Reyes (1966). Most of the remaining Asian species and a clear majority, if not all, of the African species forage either hypogaeically or beneath forest litter, although raiding workers must sometimes surface to cross hardpacked footpaths and roads (Gotwald, 1976). Only one African species observed thus far may commonly forage exposed on the substrate surface (Gotwald, 1976).

Dorylus, in its numerous forms is also primarily a hypogaeic forager, although several species of *Anomma* are conspicuous exceptions and owe their sometimes exaggerated reputations to the fact that they commonly forage on the substrate surface. Epigaeic foraging in New World species appears to be less extraordinary than it is among Old World forms, but the literature is often vague in defining the nature of foraging in some of the New World species. Certainly the best known epigaeic foragers are *E. burchelli, E. hamatum,* and *N. nigrescens.*

In an analysis of nesting and foraging behavior in Old World dorylines, Gotwald (1978) placed species into one of three categories: (1) species in which nests and foraging are both hypogaeic, (2) those in which nests are hypogaeic and foraging is essentially epigaeic, and (3) species in which the nests are surface phenomena and foraging is epigaeic. He concluded that surface foraging from a hypogaeic nest is a relatively recent derivation from a totally hypogaeic lifeway and that surface foraging from a surface nest is the most recently derived army ant adaptation. This totally epigaeic life mode has been attained by only a few species and is especially evident in *A. gracilis, A. laeviceps, E. burchelli,* and *E. hamatum.* Because these species are so accessible, they are also the most thoroughly studied and understood, and yet, as Rettenmeyer (1963b) pointed out, these represent "the most highly specialized or atypical" army ant species. Therefore to extrapolate from our knowledge of these species conclusions about the biology of the more cryptic forms is unsound. Epigaeic foraging from a hypogaeic nest is strongly developed in *Anomma,* moderately so in *Labidus,* and weakly developed in numerous species of *Aenictus.* The primitive hypogaeic mode is maintained in some species of *Aenictus,* many species of *Dorylus,* in some species of Ecitonini, and most certainly in *Cheliomyrmex.* Gotwald (1978) speculated that selective pressures are operating on hypogaeic army ants to become surface foragers.

Raiding schedules are unrelated to whether or not a species is an epigaeic or hypogaeic forager. Surface-adapted species of *Eciton,* for example, have a distinct diurnal routine, raiding from dawn until dusk, while *N. nigrescens,* also an epigaeic forager, raids from dusk to dawn. However, diurnal surface foragers are sensitive to daily rhythms in temperature change and exhibit a midday lull in foraging that Schneirla (1949) termed the "siesta effect." Common in *Eciton,* this effect has also been documented by Rettenmeyer (1963b). Raiding schedules may be quite flexible. Surface-foraging *Aenictus* may initiate raids at any time of day or night (Schneirla and Reyes, 1966). Raignier and van Boven (1955) reported that *D. (A.) nigricans* and *D. (A.) wilverthi* mount foraging expeditions at any time of day or night but show a preference for beginning in early evening and ending toward the middle of the following day. Leroux (1975) found that 68% of the raids of *D. (A.) nigricans* that he observed began during the cooler hours, between 1800 and 0800 hours. The hypogaeic foragers *L. praedator* and *Nomamyrmex* forage by day or night (Schneirla, 1971).

4. Group Retrieval of Prey

a. Prey Immobilization and Sectioning. After prey are captured by the foraging workers, they are immobilized, dismembered and sectioned (if they are large prey), transported back to the nest, and finally distributed to the nestmates.

Foraging workers bite their prey and in many instances also sting them. When large prey, such as scorpions, offer strong resistance to attack, they are "first pinned down by raiders anchored firmly by their tarsal hooks," and then are "spread-eagled by oppositely pulling groups" and torn apart (Schneirla, 1971). Earthworms are attacked in this manner by *Anomma* driver ants. While some workers anchor the earthworm in place (which is not an easy task), others tear small pieces of tissue from the captive (Gotwald, 1974b). New World species both bite and sting their prey, although the swarm raiders, *E. burchelli* and *L. praedator,* are considered more potent than column raiders in both respects (Schneirla, 1971). Even though *Dorylus* army ants possess a morphologically complete sting apparatus, they do not sting and instead rely exclusively on their ability to bite when capturing prey (Gotwald, 1978). Their sharp, cutting mandibles plus their great numbers permit them to kill and dissect even vertebrate prey that are not often attacked by New World species (Schneirla, 1971). The ability to sting is variable in *Aenictus.* While Schneirla (1971) found that Asian *Aenictus* possess potent stings and strong bites, Gotwald (1978) observed that African species do not readily sting.

Large prey organisms are usually sectioned before being transported to the nest, but even the smallest of prey may have its appendages torn from it, especially by *Anomma* (Gotwald, 1974a). A survey of prey unit size in *D. (A.) nigricans* conducted by Gotwald (1974a) revealed that the measured prey units fell into two nonoverlapping size categories. All prey units from annelids, arachnids, diplopods, insects, and snails measured between 0.1 and 1.5 cm along their longest axes, while chilopod prey units measured between 2.0 and 3.0 cm. He observed in the field that the long centipede prey units are usually carried by two or more cooperating workers running in tandem. Other less linearly shaped prey units are most often carried by individual workers (Fig. 11B,C). Cooperating *Anomma* foragers straddle their centipede prey just as do the workers of *Eciton* (Rettenmeyer, 1963b).

b. Prey Transport to the Nest.   As a raid begins and progresses during its early stages, worker traffic on the base trail and its branches is unidirectional away from the nest. When the foragers encounter increasing numbers of prey, traffic on the trails becomes bidirectional as prey-laden workers return to the nest. Finally, as the raid diminishes in momentum and comes to a close, traffic on the trails again becomes unidirectional, only this time toward the nest. This temporal shift in the directional flow of workers has been documented for *Eciton* (Schneirla, 1971), *Anomma* (Raignier and van Boven, 1955), and *Aenictus* (Gotwald, 1976). In highly productive raids, particularly in *Eciton* and *Aenictus,* large caches of prey are sometimes deposited at points where columns branch from one another near the raiding area. This prey is later returned to the nest (Rettenmeyer, 1963b; Schneirla and Reyes, 1966; Schneirla, 1971).

Labidus and *Anomma* commonly form walls and arcades that border and cover their foraging columns. These are composed either of clustered workers or of soil particles and pellets (Cohic, 1948; Raignier and van Boven, 1955; Rettenmeyer, 1963b). Although this phenomenon may occur to an even greater extent along emigration trails (Schneirla, 1971), these walls and arcades, whether constructed of the workers themselves or of soil, probably serve the same functions (Fig. 12A,B). Shade and perhaps increased humidity are created by these structures, especially for the columns that cross open, exposed surfaces. Further, the walls formed of clustered workers may serve to keep itinerant arthropods and small vertebrates from accidentally straying into the midst of the column and interrupting the flow of traffic. This author has witnessed *Anomma* workers repel insects this way. Major workers often assume a "defensive" posture when positioned at trail margins (Fig. 12C).

The number of foraging workers returning to the nest with visible pieces of prey is small in proportion to the total number of individuals that participate in a raid. Raignier and van Boven (1955) calculated that 6–22% of the returning workers of *D. (A.) wilverthi* carry prey and only 0.8–10% of the *nigricans* workers do so. Gotwald (1974a) examined the crops of returning "preyless" workers of *D. (A.) molesta* and found that most had liquid-filled crops. A statistical analysis of prey-carrying and preyless workers disclosed a division

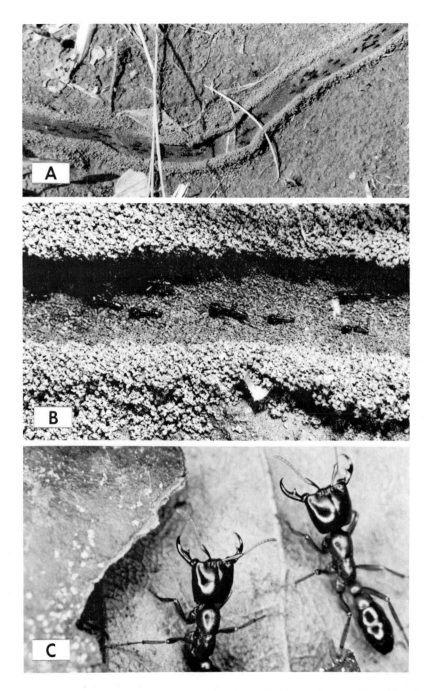

Fig. 12. (A) *Dorylus (Anomma) nigricans* foraging trail with conspicuous soil particle walls; note the workers on the trail; (B) as in A but at closer range; (C) *D. (Anomma) gerstaeckeri* soldiers in "defensive" posture. (Photographs by W. H. Gotwald, Jr.)

of labor corrlated with size: large workers most often carry prey units while smaller workers transport prey liquids.

   c. Trophic Relationships within the Nest.   In terms of trophic relationship, the larvae are no doubt the focus of attention when foraging workers arrive at the nest with prey. However, it is not clear when, how, and what the adult workers themselves eat. Wheeler and Bailey (1925) suspected that *E. burchelli* larvae are fed at considerable intervals with large pellets composed of "the rolled up soft-parts of insects." They found these pellets so compact that they retained their shape even in the larval gut where they formed an "irregular longitudinal series." Because only occasional, minute, hard fragments occur in these pellets, Wheeler and Bailey (1925) concluded that the workers, when preparing these pellets, must trim away the hard, chitinous materials. They further surmised that the workers consumed the liquids expressed from the prey tissue when it was formed in the pellets. However, as the larvae grow larger, they also feed with their mouthparts applied directly to whole pieces of prey (Schneirla, 1971). Although army ant workers possess ovaries with maturing ova, it is not known whether the larvae consume worker produced eggs as happens in other ant species (Gotwald, 1971; Gotwald and Schaefer, 1982).

   Schneirla (1971) noted that the larvae, through their movements and odors, attract the workers that both feed the larvae and "stimulate" them. In *Eciton*, as the larvae mature the workers become increasingly attentive and more frequently drop food on the larvae. At the same time, the larvae are more frequently carried and dropped by the workers on caches of prey in the nest (Schneirla, 1971).

5. Diet

   a. Types of Prey Taken.   Dietary observations for army ants are often anecdotal, incomplete, and rarely quantified. Although some species may utilize plant nutrients, army ants are decidedly carnivorous. For the New World dorylines, ants rank as the most important and commonly taken food items. However, ecitonines do not take other ecitonines as prey (Rettenmeyer, 1963b). Only a few species appear to be generalists. One of these, *Labidus coecus,* "perhaps feeds on a greater variety of substances than any other species of Ecitonini" and takes ants, orthopterans, adult moths, homopterans, beetles, amphipods and spiders (Rettenmeyer, 1963b). Lenko (1969) observed *L. coecus* even attacking a Brazilian fresh water crab, *Trichodactylus argentinianus,* as it rested in its nonaquatic burrow. This species also takes a variety of plant products and is known to be attracted to walnut and pecan kernels (Wheeler, 1910), seeds, and cooked rice (Borgmeier, 1955). Another generalist, *E. burchelli,* takes, in decreasing order of importance, ants, wasps (vespids of the subfamily Polistinae), Orthoptera, spiders, and scorpions (Rettenmeyer, 1963b). Most New World dorylines are specialized predators; *E. hamatum* capture ant and wasp brood and a few adult insects, and some species, such as *E. vagans, Nomamyrmex esenbecki,* and *Neivamyrmex pilosus* take almost exclusively ant brood (Borgmeier, 1955; Rettenmeyer, 1963b). *Neivamyrmex nigrescens* takes both termites and ants, with a decided preference for ants of the genus *Pheidole* (Mirenda *et al.,* 1980). Only one species, *Nomamyrmex hartigi,* consistently takes large numbers of termites as prey. *Neivamyrmex opacithorax* is unusual in that it feeds solely on ants and carabid beetles (both larvae and adults) (Rettenmeyer, 1963b). Wheeler and Long (1901) found a colony of this species in which the "workers had stored their nest with a considerable number of small carabid beetles that had evidently been captured on one of their marauding expeditions." Although Schneirla (1956b) noted that *E. burchelli* occasionally kills snakes, lizards, and nestling birds, Rettenmeyer (1963b) supposed that none of the New World dorylines kills vertebrates on a frequent basis. Rettenmeyer also noted that all of the New World army ants studied thus far kill many more arthropods than they can eat. The feeding habits of many of the more cryptic, hypogaeic species, particularly those belonging to the genus *Cheliomyrmex,* are unknown.

Dietary information on the Old World dorylines is so scattered and fragmentary that only a few of the more interesting observations can be cited here. Savage (1847) provided the first systematic description of *Anomma* behavior and listed as food sources for the driver ants, domestic fowl, lizards, snakes, fresh meat, and fresh oil, particularly of the oil palm, in addition to the usual bill of fare of assorted arthropods. Swynnerton (1915) investigated the food preferences of *D. (A.) molesta* experimentally but offered the foraging workers prey not normally available to driver ants. Some of the experimental prey he selected repelled the workers with defensive secretions. As for the "natural" diet of this species, Swynnerton (1915) listed such prey items as millipeds, ticks, and insects of the orders Orthoptera, Hemiptera, Coleoptera, Lepidoptera, and Diptera. The same east African species was recorded by Loveridge (1922) to capture grasshoppers, crickets, and pentatomoid bugs and to feed on a chamelion, geckos, a caged crocodile, and "eagle flesh" distributed as bait. Raignier and van Boven (1955) collected the prey of such general predator species as *D. (Anomma) burmeisteri, D. (A.) sjoestedti, D. (A.) terrificus,* and *D. (A.) wilverthi.* An analysis of their data (Gotwald, 1974a) revealed that 64% of the prey units collected were of insect origin, 16% from arachnids, 9.7% unidentifiable, 4.5% plant seeds, 3.6% from isopods, 0.7% from diplopods, and 0.5% from mollusks. Cohic (1948) noted that *D. (A.) nigricans* takes spiders, cockroaches (including oothecae), grasshoppers, crickets, and dipterous and ant pupae. In a quantitative study of *Anomma* diet, Gotwald (1974a) determined that although these ants take a wide variety of invertebrates, insects comprise the largest portion of their diet. However, he concluded that this bias is more a reflection of insect abundance than it is an indication of food preference on the part of the ants. Earthworms and arachnids were the next most common sources of prey items. Much less is known about the hypogaeic species of *Dorylus*. While most of these species forage almost exclusively through subterranean galleries and beneath litter, they are attracted to vertebrate carcasses and meat baits on the soil surface (W.H. Gotwald, unpublished data). Certain of these species forage in compost heaps (Wheeler, 1910) and village trash deposits (W.H. Gotwald, unpublished data) for insect larvae. Some species of *Dorylus* are reportedly termitophagous. The subgenera *Typhlopone* and *Rhogmus* attack termites of the genera *Acanthotermes* and *Basidentitermes* respectively (Wheeler, 1936). Bodot (1961) observed *D.* (*Typhlopone*) *dentifrons* raiding the termitaria of *Bellicositermes natalensis* with devastating effectiveness. Some epigaeic foragers, such as *D.* (*A.*) *nigricans* also attack termites, though probably not often (Bequaert, 1913) and *D.* (*A.*) *wilverthi* destroys large numbers of swarming, alate males (Burgeon, 1942a).

Some *Dorylus* species also feed on plant materials, and Schneirla (1971) suggested that army ants that do so are usually hypogaeic species. Green (1903) reported that *D.* (*Alaopone*) *orientalis* in India feeds on potatoes, the tubers of dahlias, and the roots of the common sunflower and insisted that these ants are "confirmed vegetarians." Supporting Green's conclusion, Roonwal (1975) argued that *D.* (*A.*) *orientalis* is sometimes a serious pest of "vegetables, tubers, bulbs, shrubs, trees and . . . cash crops such as sugarcane, coconut palm, citrus and groundnut." *Anomma* driver ants commonly forage on fallen palm nuts (*Elaeis guineensis*) from which they remove the pithy outer covering, leaving the seed and numerous fibers behind (Gotwald, 1974a). *Dorylus* (*A.*) *nigricans* was observed to gather pieces of corn cob in a village refuse heap (Gotwald, 1974a) and *D.* (*A.*) *molesta* to take banana (Swynnerton, 1915).

African species of *Aenictus* studied thus far feed exclusively on ant brood. Prey collected by Gotwald (1976) from seven foraging columns of *Aenictus* yielded only ant larvae (36%) and pupae (64%). In fact, two *Aenictus* colonies that he observed displayed little or no hostility toward the adults of brood that the foragers carried away unchallenged. The prey adults abandoned their nest to the *Aenictus* foragers, although they remained close to the nest opening, slowly milling about the external features of the nests. Crawley and Jacobson (1924) made a similar observation of the Asian species, *Aenictus aratus,* when it attacked a nest of *Pheidole*. The *Pheidole* adults offered no resistance, but instead fled "with as many of their brood as they were able to save." Weber (1943) noted the prey of one colony of the African species *A. rotundatus* to be ant brood of the subfamily Myrmicinae. Twenty-nine percent of the *Aenictus* prey collected by Gotwald (1976) belonged to the genus *Pheidole,* also a myrmicine ant. Chapman (1964) recorded that the Asian species *A. gracilis* and *A. laeviceps* feed primarily on ants (they take adults as well as brood) but also take a wide variety of other prey. Schneirla and Reyes (1966) confirmed Chapman's findings for these species, stating that the prey "may be described as almost any invertebrate that the ants can find and overcome, hunted down in crannies from the depths of the soil to the tops of tall trees." *Aenictus* species that are trophic specialists generally take prey that are approximately the same size as or smaller than the foraging workers themselves, whereas the general predators take prey of any size, subdividing larger prey into transportable pieces (Gotwald, 1976).

Army ants at times consume their own brood, but the extent to which cannibalism occurs and how important the brood are as a protein source are not known. Cannibalism of brood is common in ants, and the brood may in fact serve as an emergency food supply in times of food shortages (Wilson, 1971). After comparing the size of egg and young larval broods with the size of mature larval and pupal broods in *E. hamatum,* Rettenmeyer (1963b) noted a significant decrease in numbers of individuals over the time period during which maturation occurs and theorized that cannibalism of worker brood may be extensive. The cannibalism of sexual brood, particularly of potential queens, appears to be an important factor in the development and production of males and queens, particularly since it helps reduce the total number of immatures that must share the worker-supplied food (Schneirla, 1971).

The use of plant liquids as nutrient sources by doryline ants is problematic. Schneirla and Reyes (1966) observed instances when workers of *A. gracilis* and *A. laeviceps* gathered on plants, "clustering" near nectaries. Although nectar attraction has not been noted for other dorylines, Arnold (1915) recorded a report that the African species *D. (Typhlopone) fulvus* tended immature membracids feeding on the roots of maize. Similarly, Santschi (1933) noted that *Aenictus eugeniae,* an East African species, was once collected while tending a species of *Pseudococcus.*

b. Prey Determinants. A variety of factors combine to determine what the potential prey of army ants will be and whether or not these prey will actually be captured. Certainly army ant foragers, even the swarm raiders, are not as efficient in gathering prey as their reputations would have them be, i.e., that they capture and consume all living things in their path. Rettenmeyer (1963b) estimated that 30–50% of the arthropods attacked by *E. burchelli* escaped being killed and Schneirla (1945) noted that even when an *E. burchelli* colony stayed at a nest site for 23 days, the colony could not deplete the surrounding area of prey. He observed that (1) the colony incompletely covered

some sectors during foraging, (2) it entirely missed some sectors when foraging, and (3) areas cleared of prey were quickly repopulated. Schneirla (1971)
reasoned that the dietary range of army ants was influenced primarily by (1)
the habitat, (2) worker structures, such as potency of the sting, (3) the
workers' chemotactic thresholds, (4) the species' raiding pattern and level of
excitability, and (5) the colony population itself, including colony conditions
and the degree of polymorphism present.

In a study of food preferences in three species of *Anomma,* Gotwald
(1974a) proposed that the habitat was the single most important factor in
determining the composition of *Anomma* diet. For example, earthworms
constitute a larger portion of *Anomma* diet in forest habitats than they do in
savanna, and this is directly related to the relative availability of earthworms
in these two ecosystems, not to the preferences of the *Anomma* foragers. The
influence of the habitat extends as well to the kind of insect prey taken. Gotwald (1974a) further noted that prey vulnerability to *Anomma* attack must
strongly affect diet composition. A minimum of 61.8% of the insect prey
units collected from the foraging columns of 11 *Anomma* colonies were from
immature stages; 63.9% of the insect prey collected from 12 foraging
Anomma colonies were from holometabolous forms. The larvae of Diptera,
Coleoptera, Lepidoptera, and Hymenoptera (particularly the Formicidae)
are especially vulnerable to *Anomma* attack. Probably most arthropods that
escape army ant attack avoid capture by running, flying, or hopping away
from the foragers or by climbing vegetation (Rettenmeyer, 1963b; Schneirla,
1971). Adult polistine wasps detect foraging army ants by sight and odor and
flee their nest. Although the wasp brood cannot be rescued from the ensuing
attack, the adults can at least recolonize (Chadab, 1979). Some species of the
ant genus *Camponotus* evacuate their nests, carrying their brood with them,
when attacked by *Neivamyrmex nigrescens,* while other species defend their
nests through the recruitment of the major caste (LaMon and Topoff, 1981).
Some potential prey may employ specialized escape mechanisms. Ants of the
genus *Cryptocerus* (*Cephalotes*), for example, are so heavily armored that
they seem to be immune to attack (Schneirla, 1971). Another method of
escape effectively separates the prey from the predator by means of a slender
cord, a "bridge" that the foraging workers are unable to negotiate.
Lepidopterous larvae escape *Anomma* attack in this way, by dropping down
from the leaves of low vegetation, suspended on threads of silk (Swynnerton,
1915; Gotwald, 1972a). Gotwald (1972a) discovered that slugs (mollusks of
the order Pulmonata) employ an analogous method of escape from *Anomma*
foragers by suspending themselves from vegetation on a cord of integumental
slime. Spiders also drop down from vegetation on silken escape threads
(Swynnerton, 1915; Rettenmeyer, 1963b). Although poorly documented,
there can be little doubt that defensive secretions confer protection on a wide

variety of potential prey. Carpenter (1914–1915) observed a snail escape a driver ant attack when it withdrew into its shell and "produced a mass of bubbles of mucus which so completely surrounded it, shell and all, by a barrier about half an inch thick, that the ants could not get at any part of its body." Carpenter (1920) also witnessed an hemipterous bug repel driver ants, that tugged at its antennae and legs, with a "powerful odour" that "may have been disagreeable to the ants."

c. Trophic Specialists and Generalists.   With the exception of the totally epigaeic forms, especially *A. gracilis* and *A. laeviceps, Aenictus* species are monophagous or oligophagous, specializing as predators of the immature stages of other ants (Brauns, 1901; Crawley and Jacobson, 1924; Sudd, 1959; Chapman, 1964; Gotwald, 1976). Most of these specialist species are, in turn, hypogaeic column raiders. Hypogaeic foragers of *Dorylus* are also specialists, primarily termitophagous and myrmecophagous, although some dietary departures have been recorded (Green, 1903; Lamborn, 1913–1914; Forel, 1928). All of these specialists are also column raiders and maintain hypogaeic nests. Although the nests of the social insects on which these *Aenictus* and *Dorylus* species prey are scattered about the trophophoric field, requiring a greater search investment on the part of the predator, they are an "exceptionally concentrated food source" and are well worth the energy expended in search activities (Carroll and Janzen, 1973). On the other hand, *A. gracilis* and *A. laeviceps* and other surface-adapted *Aenictus* are column raiders, frequently form surface-exposed nests, and are general predators, although they display a predilection for other ants (Schneirla and Reyes, 1966). Even though *Anomma* driver ants maintain hypogaeic nests, a number of species are epigaeic swarm raiders and general predators (Raignier and van Boven, 1955). Gotwald (1978) noted that these data suggest that epigaeic foragers become trophic generalists while hypogaeic foragers remain specialists. He also pointed out that it may be advantageous in tropical habitats for army ants to become epigaeic foragers since a concomitant development may include the qualitative and quantitative expansion of the diet. Since a taxonomically diverse prey is more uniformly distributed in the trophophoric field, less energy may be required of the predator in search activities. The polyphagous predator not only capitalizes on a diverse prey on a daily basis but can also exploit prey sources that are periodically abundant, such as emerging male termites (Gotwald, 1978).

In the New World army ants, the positive correlation between epigaeic foraging and dietary expansion is not as clear. *Eciton burchelli, L. coecus,* and *L. praedator* are evidently the most polyphagous of the New World species. *Eciton burchelli* forms a surface-exposed nest and is an epigaeic column raider. In contrast, *Labidus coecus* and *L. praedator* are both hypogaeic

nesters and both are swarm raiders, although Rettenmeyer (1963b) noted that
L. coecus is somewhat intermediate between column and swarm raiding.
Labidus praedator will raid on the substrate surface while *L. coecus* is more
negatively phototaxic. These two species behaviorly resemble some species of
Anomma. The epigaeic nester *E. hamatum* is an epigaeic column raider and a
specialist predator of ants and wasps. However, it does take other insects
(Rettenmeyer, 1963b). Other specialists such as *Nomamyrmex esenbecki* and
Neivamyrmex pilosus are column raiders and negatively phototaxic, although
they will forage on the soil surface.

From these facts and other data assembled by Rettenmeyer (1963b) and
Schneirla (1971), an evolutionary scenario can be constructed. Primitive
dorylines (i.e., those that have retained the ancestral life mode) are hypogaeic
nesters, hypogaeic column raiders, and specialist predators of social insects,
especially ants, wasps, and termites. There is a tendency for army ants to
become surface foragers and in doing so to expand their diet. However, if the
epigaeic forager retains column raiding, its dietary expansion is limited,
whereas, if the epigaeic forager becomes a swarm raider, its diet expands to include a wide variety of arthropods and even vertebrates. In some instances,
swarm raiding, at least to some extent, may evolve in species that often forage
hypogaeically. The most derived army ant lifeway is that in which the species,
in addition to being an epigaeic forager, also becomes an epigaeic nester. A
species that becomes totally epigaeic also develops a functional cycle of alternating statary and nomadic phases. This has been achieved in so few species as
to be truly atypical for the army ants in general. With the acquisition of swarm
raiding and polyphagous feeding, plus occasional emigrations to new nesting
sites, extraordinarily large colonies, like those of *Anomma,* are possible.

6. Army Ants and Pest Control

The manipulation of ant populations in tropical agriculture is receiving attention as a possible approach to integrated pest control, particularly in
tropical tree crops (see Leston, 1973). Army ants, especially those species that
are polyphagous, are obvious candidates for study in this respect.

Wellman (1908) considered *D.* (*A.*) *nigricans* of economic importance in
Angola, but primarily because as he explained, "careful housewives . . . wel-
come the approach of the ants and joyfully vacate for them the bungalow,"
for "after a column of 'army ants' has minutely explored a dwelling not a bug,
beetle, cockroach, mouse, rat, snake, or other pest remains behind." Dutt
(1912) remarked that *D.* (*Alaopone*) *orientalis* was a beneficial ant in India,
since it attacked and killed in large numbers *Pheidole indica* which he described as an occasional nuisance. Burgeon (1924a) published some biological
observations on *D.* (*A.*) *wilverthi* and submitted that, "il est certain que ces
Fourmis rousses sont des utiles pour l'agriculture, detruisant foule d'insectes

nuisibles.'' Both Alibert (1951) and Strickland (1951) implied that driver ants might be beneficial in cocoa farming but offered no corroborative evidence. Gotwald (1974b) studied the foraging behavior of two species of *Anomma* in Ghana cocoa farms and discovered that these ants do not forage in the cocoa canopy, although they readily climb trees in other habitats. They are apparently repulsed in any attempt they might make to climb cocoa trees by dominant ant species in the canopy, such as *Oecophylla longinoda*. Although the effect of driver ants on the ground stratum fauna was not investigated, Gotwald concluded that driver ants were of little potential value in ant manipulation schemes to control cocoa pests, since they fail to have any visible effect on the composition of the canopy fauna.

In the New World, *Labidus coecus* is an important predator of the screwworm, *Cochliomyia hominivorax,* and other carcass-infesting, dipterous larvae. In fact, Lindquist (1942) concluded that the animals dying from screwworm infestations ''are the source of only a small emergence of flies during the warmer seasons of the year, chiefly because of the predatory actions of several species of ants.'' Ants effectively reduced the number of adult flies emerging from carcasses from about 93% to 4%, and *L. coecus* figured most importantly in this predatory control. Larvae dropping from the wounds of relatively healthy, ambulatory animals were also subject to intense ant predation but in this case, the role of *L. coecus* was not clearly delineated.

D. Emigration Behavior: Nomadism

1. Army Ant Nests

Theoretically, all army ant nests are temporary, since all army ants are presumably nomadic, but the frequency with which nest sites are abandoned by many species is unknown. Surely a majority of species occupy subterranean quarters, while a few maintain surface or above surface nests. Schneirla (1971) applied the term ''bivouac'' to all army ant nests and suggested that a bivouac was more ''the state of a colony'' than it was a physical place. However, many nondorylines also occasionally move from one nesting site to another (Wilson, 1971). What distinguishes army ants in this respect is that many of the better known species undertake emigrations with great frequency and regularity. However, recurrent and regular movement is by no means universal in the army ants, because even some surface-active species (e.g., *Anomma*) tend to move at irregular intervals (Gotwald and Cunningham-van Someren, 1982). Only in those species that have a distinct functional cycle of alternating statary and nomadic phases (e.g., *A. gracilis* and *E. hamatum*) are surface nests formed, and because these are hardly nests in the conventional sense, the name bivouac is appropriate. However, it is more descriptively accurate to restrict the term bivouac to these atypical surface formations and to continue to apply the term nests to all situations where the nest indeed is a place, i.e., housed within the soil. This dichotomy in doryline nest formation was evident in Forel's (1896) classification of ant nests. He noted that some army ants have "earth nests" and others "migratory nests" and was unsure how to reconcile the two in what he considered to be a homogeneous and monophyletic group.

Raignier and van Boven (1955) found the subterranean nests of *Anomma* to be of two general types, one exemplified by *D. (A.) wilverthi* and other by *D. (A.) nigricans*. In the *D. (A.) wilverthi* type the entire colony population is clustered together in a central cavity, forming a compact mass in which a higher, presumably optimal, temperature can be maintained. The nest is 1–2 m deep. In the second or *D. (A.) nigricans* type, there is no central chamber; the colony members are dispersed in deep (2–4 m down) galleries and chambers, apparently without much control over nest temperature. The majority of both types of nests were located at the bases of trees, both living and dead, associated with the root systems, although the nests of *D. (A.) wilverthi* were also found in treeless, loose earth. No one species of tree was preferred for nest site locations, although nests were most often associated with living trees. In both types there may be surface manifestations of the nest in the form of craters composed of excavated soil particles. Leroux (1977a) noted, in a study of *D. (A.) nigricans* in savanna and gallery forest in Ivory Coast, that although a majority of nests are constructed at the bases of trees, 27 percent of the nests he examined were located in soil without a supporting network of roots. He observed that during the first week that a nest is constructed, as much as 20 kilograms of soil a day may be excavated. The location of the *D. (A.) wilverthi* queen in the nest is variable (e.g., she may be situated in the central group or in a lateral gallery), whereas the *D. (A.) nigricans* queen is typically lodged at great depth in the nest. The nests themselves do not appear to be highly organized structures, in fact driver ants use existing cavities in the soil where possible. Over a century ago, Savage (1849) wrote that the interior of the *Anomma* nest "exhibits no mechanical contrivance for the depositing of food, or hatching of eggs; for these purposes, spaces between the stone, sticks, etc., found within, are adopted." A nest of *D. (A.) nigricans sjoestedti* described by van Boven and Levieux (1968) departed from the *D. (A.) nigricans* type in three respects: it was shallow, only 60–80 cm deep; it had a large central cavity which was occupied by an enormous mass of workers; and it was not situated at the base of a tree or associated with a root system. The queen was found in the central cavity, to one side. Kistner and Gotwald (1982) also found that the *D. (A.) nigricans* nest had a central chamber. A nest of *D. (Typhlopone) fulvus,* described briefly by Normand (1931), consisted of a vaulted chamber, about 50 cm below the soil's surface, with numerous galleries leading from it in all directions, some extending deeply in the soil.

Normand communicated to Santschi (1931) that the chamber was filled with larvae, pupae, and workers, and that the queen was situated in the midst of the workers.

Most *Aenictus* species in both Africa and Asia are hypogaeic nesters, although their nests have yet to be described in detail (Gotwald, 1976). Menozzi (1936), for instance, located a colony of a new species, which he named *A. rhodiensis,* on the Island of Rhodes at a depth of 60–70 cm beneath a stone.

Some New World dorylines live in subterranean nests at some times and form bivouacs at other times. For example, *E. vagans* may form more or less exposed bivouacs, but more often than not lodges itself in underground or well sheltered nests (Rettenmeyer, 1963b). Rettenmeyer (1963b) found one colony of *E. vagans* clustered on small roots that traversed a subterranean cavity measuring 30 cm in diameter. This cavity, most likely not constructed by the ants themselves, was positioned 30 cm below the soil surface and gave rise to four principal galleries that descended into the soil to an unknown depth. Another colony formed a surface bivouac that was suspended from several support objects including a log, branch, and large rock (Rettenmeyer, 1963b). The nests of *L. coecus* may be located at great depths in the soil or they may be close to or on the soil surface but well sheltered. One nest found by Rettenmeyer (1963b) was fashioned in a decaying log. The eggs were concentrated in three long chambers, the cocoons situated primarily in a single, separate chamber, and the larvae were scattered throughout the nest in numerous chambers. The nests of *Neivamyrmex opacithorax* are hypogaeic as are those of *N. nigrescens.* In the *N. nigrescens* nest, a daily vertical migration takes place in which a portion of the colony's brood is brought to near the surface to a position beneath sun-exposed stones (Rettenmeyer, 1963b).

The nature of the bivouac or surface-exposed formation depends on the physiological state of the colony, i.e., on whether it is in the statary or nomadic phase. In the former phase, the nesting cluster is assembled in an enclosed or sheltered space where it remains for many days. In the latter, a new, exposed cluster is formed each night in a new location (Schneirla, 1971). Bivouacs in *E. hamatum* are seldom more than 1 m above the ground, whereas those of *E. burchelli* may be formed as high as 30 m above the ground in trees (Schneirla, 1971). A bivouac is formed essentially of the bodies of the ants themselves, suspended from a support object and from each other (for a discussion of the nesting behavior of *E. burchelli,* see Teles da Silva, 1977b). Fundamental to bivouac formation is a clustering group response. In *E. hamatum* during the nomadic phase, bivouac formation begins at dusk with the creation of clusters of workers that hang from the support object (e.g., a log) near a booty cache [note that *Anomma* workers can also produce hanging clusters (Raignier, 1959b)]. Other workers are attracted to the clusters and attach themselves, usually by means of interlocking tarsal claws.

As Wheeler (1900) picturesquely explained, in "forming these chains, which remind one of the pictures of prehensile-tailed monkeys crossing a stream, the insects make good use of their long legs and hooked claws." First strands, then "ropes" of workers are formed that ultimately fuse into a "heavy fabric" (Schneirla, 1971). Workers fastened into the bivouac wall usually hang head downward, a phenomenon that Wheeler (1900) attributed to the "positively geotropic" nature of the workers, but which Schneirla (1971) explained in terms of the physical stresses exerted by the placement of the tarsal claws and the weight of the ants. The bivouacs of surface-adapted species of *Eciton* are certainly the most specialized of all nesting formations in the army ants. In *E. hamatum* they may take the form of cylinders suspended between the support object and the substrate or of a curtain between the buttressed roots of a tree (Schneirla, 1971). In *Aenictus gracilis* and *A. laeviceps,* the bivouac during the nomadic phase is little more than a disc-shaped cluster of workers on the soil surface, usually beneath litter. The cluster in *A. laeviceps* is between 15 and 18 cm in diameter and 6 and 9 cm in height (Schneirla and Reyes, 1966).

The bivouac of such species as *E. hamatum* permits colony mobility and yet at the same time provides protection for the brood from the vagaries of the external environment. Temperature variations within the bivouac, for instance, are significantly less than ambient conditions and provide a constancy that optimizes conditions for the developing brood (Schneirla *et al.,* 1954; Jackson, 1957). In *E. hamatum* the general intrabivouac temperature is $1°-2°C$ above that of the environment and even higher in the compact brood mass (Jackson, 1957). Bivouac temperatures fall in early morning, rise in late morning, fall again in the afternoon and rise in the evening despite fluctuations in ambient temperature (Jackson, 1957).

Refuse deposits, some diffuse and others rather circumscribed are probably formed by most army ants and are located a short distance from the nest or bivouac. While the fate of refuse, such as the remains of prey and carcasses of dead workers, is difficult to determine in hypogaeic nests, it is fairly easy to study in surface-exposed formations. There is little refuse around the bivouac of *E. hamatum,* reflecting that the diet consists primarily of soft-bodied prey (Rettenmeyer, 1963b). However, the refuse deposits of *E. burchelli* are extensive and include the sclerotized parts of dead prey, dead workers, and empty cocoons. This refuse is placed in more or less distinct areas by workers that move along short trails in what Rettenmeyer (1963b) called "refuse columns." Workers defecate in the refuse deposits and along the trails with such frequency that the feces may form a distinct white trail during the statary phase. These refuse deposits usually contain thousands of living arthropods, most of whom are scavengers on the discarded refuse. Reference to refuse deposits associated with colonies in subterranean nests are rare. Bruch (1923), in one such reference, found refuse deposits in three peripheral chambers of a leaf cutter ant nest that had been occupied by a colony of *E. dulcius*. He noted that the refuse was "repletas de incontables fragmentos, cabezas y alas de otras hormigas, formas sexuales que correspondian principalmente a *A. (M.) Heyeri* y *Trachymyrmex,* mezclados, a trozos quitinosis, élitros, miembros, etc. de coleópteros y otros artrópodos."

2. Origin of Nomadism: Evolutionary and Ontogenetic Considerations

Wilson (1958a, 1971) reasoned that group predation permits the exploitation of other social insects and large arthropods as food sources. Since these sources are more widely dispersed than other types of prey, nomadic behavior was a natural consequence of this new dietary adaptation. That is, nomadism either developed concurrently with group predation or it was added soon afterward as a means for colonies to shift trophophoric fields in order to find new food supplies. Some nondoryline ants, especially of the genus *Cerapachys,* have embarked on group raiding without a concomitant disposition toward nomadism (Wilson, 1971). In fact, there may be some army ants that emigrate infrequently; the biology of a majority of hypogaeic species is unknown. Species that maintain a pattern of precisely spaced emigrations are exceptional. Gotwald (1978) suggested that the selective pressures for the establishment and maintenance of regularly spaced emigrations may be significantly reduced for species that become epigaeic foragers and general predators via swarm raiding. This conclusion is a reasonable extension of Wilson's observation that large arthropods and social insects are widely spaced. Certainly the prey of the most general predator is more numerous and more uniformly distributed in the trophophoric field and less likely to be totally exhausted by intense foraging. Thus the adaptive value of emigrating to new trophophoric fields may consequently be reduced. While the ultimate evolutionary grounds for nomadism may seem obvious, the proximate reasons are still a matter of debate for some species. Schneirla (1938, 1957, 1971) determined that the initiation of the nomadic phase, during which time an *Eciton* colony emigrates daily, is the result of brood stimulative factors. In particular, the statary phase ends when the colony is aroused by the emergence of callow workers from their cocoons. Although the callow effect soon diminishes, the nomadic phase continues to be maintained, stimulated by the developing larval brood. When this brood matures and enters the prepupal and pupal stages, the colony settles into the statary phase at a single nest site (Fig. 10). With his discovery that brood stimulative cues are operating to provoke emigration, Schneirla (1971) concluded that while food shortage may have been the ancestral basis for nomadism, it is no longer so. In this author's opinion he obscured the distinctions between the ultimate and proximate reasons for nomadism. Brood originated cues for colony movement constitute only the proximate, ontogenetic stimuli that effectively achieve the ultimate end of increasing food supply for the maturing brood. The foregoing conclusion would no doubt be refuted by Topoff (1972) who pointed out that Schneirla emphasized ontogenetic rather than evolutionary processes in his studies of army ant behavior, and that this author is therefore guilty of confusing these approaches and the answers they supply.

Brood stimulative factors appear to be operating in *E. burchelli* and *E. hamatum* (Schneirla, 1971), *Neivamyrmex nigrescens* (Schneirla, 1958, 1963), and *Aenictus gracilis* and *A. laeviceps* (Schneirla and Reyes, 1969), all of which evidence a functional cycle of alternating nomadic and statary phases. However, in *Aenictus* the nomadic phase appears to be initiated by the larvae rather than by the callow workers (Schneirla and Reyes, 1969). Even in army ants that do not pass through such cycles but instead emigrate irregularly, brood stimulative factors operate. In *Anomma,* for example, the emigration of a colony from one nesting site to another is not the direct result of food shortage or other environmental factors. Instead, Raignier and van Boven (1955) concluded that, "Le rythme des exodes est determine par le rythme du couvain." Specifically, emigrations are correlated with the emergence of callow workers. However, if this is the case, the irregularity of movement suggests that eggs are laid in unsynchronized bursts, and indeed, Raignier and van Boven came to such a conclusion. Additionally they found that when male larvae are present in *Anomma* nests, the colonies will remain at a single nest site for an average of 56 days. Colony emigrations in some *Anomma* are so irregular as to suggest that other factors may supersede the stimulative effects of the brood (Gotwald, 1978b). Gotwald and Cunningham-van Someren (1982) recorded the number of days between 38 emigrations observed for a colony of *D. (A.) molesta* in Kenya. In chronological order the intervals were: 9, 9, 19, 4, 14, 35, 11, 33, 7, 44, 5, 23, 4, 45, 14, 24, 10, 10, 7, 9, 5, 19, 23, 11, 12, 13, 18, 7, 8, 5, 14, 11, 12, 16, 3, 5, and 6 days. Rettenmeyer (1963b) believed that Schneirla went too far in discounting the influence of the amount of prey taken on the emigration activities of army ants. Since Rettenmeyer (1963b) demonstrated that the amount of prey in an area affects the "direction, distance, and duration of raids and emigrations," he presumed that the amount of food in the nest or bivouac must influence the larvae that in turn stimulate the adults. He speculated that as the larvae get larger and require more food, they stimulate the workers to raid and emigrate at greater distances.

Nevertheless, in some species brood stimulative factors in the role of proximate initiators of nomadism and other behavior patterns are of paramount importance. Observations of *Eciton* led Schneirla (1971) to expand Wheeler's concept that trophallaxis, or oral food exchange, constituted the social cement of the colony, and included under the term all communicative relationships between workers and brood. Because these relationships, whether they be tactile or pheromonal, "entail a bilateral arousal basic to colony function," Schneirla (1971) termed them "reciprocal stimulation." This phenomenon is no more evident anywhere than in the army ant functional cycle, particularly as manifested in the callow arousal factor.

3. Army Ant Population Dynamics

Little is known about the number of army ant colonies that can be supported by any given unit of area and space. Unfortunately, colony censuses locate primarily those colonies whose behavior brings them to the soil surface, while the hypogaeic species, for the most part, go undetected. In African environments, one can intuitively conclude that species diversity and the number of colonies are greater in forest habitats than they are in drier, savanna localities (Kistner and Gotwald, 1982).

Leroux (1975, 1977a) calculated the density of D. (A.) nigricans colonies at the Laboratoire d'Ecologie Tropical, Lamto, Ivory Coast. The reserve at Lamto encompasses approximately 580 hectares of forest and 2120 hectares of savanna. Leroux reckoned that there were 3.16 colonies per 10 hectares of forest and 0.79 colonies for the same area of savanna, or for every 13 hectares there were 4 colonies in forest and 1 in savanna. Since his calculations did not consider numerous other dorylines that are also present, mostly hypogaeic species of Aenictus and non-Anomma Dorylus, the carrying capacity of these habitats for such predators remains undetermined.

Although similar calculations for New World species have not been made, Barro Colorado Island, because it is a water-bound unit of land, would seem to be a logical place to begin such an investigation.

4. Emigration

a. Process of Colony Movement

The purpose of these expeditions of *Eciton* is, without doubt, multiple, for the circumstances that these *sorties* . . . coincide more often with a change of season, hardly permits one to consider them exclusively as simple razzias undertaken at the expense of other insects. One can believe them to be sometimes expeditions of pillage, sometimes changes of domicile, veritable migrations.

With this declaration about New World ecitonines, Sumichrast (1868) recognized the dualism of army ant behavior. Especially did he note that army ants emigrate from one nesting site to another, a behavior pattern which was to be the focus of attention for much of Schneirla's exhaustive research (e.g., 1938, 1944, 1945). But long before this research, other observers noted that army ants emigrate periodically. In Africa, Savage (1847), Brauns (1901) and Vosseler (1905) did so, Vosseler offering the explanation that colony emigrations were stimulated by food shortage. In the New World in addition to Sumichrast's (1868) observations, Müller (1886) and Wheeler (1900) noted that the army ant colony was prevented from moving at times by reproductive processes.

In *Eciton,* colonies always conduct day-long foraging expeditions before emigrating (Schneirla, 1938, 1945, 1971). As raiding progresses in the afternoon, at least in *E. burchelli* and *E. hamatum,* three conditions that preclude emigration are evident: (1) a high level of excitement in the colony is maintained which guarantees a continued exodus of foraging workers from the bivouac, (2) traffic complications among workers on the raiding trails prevent a general return of foragers to the bivouac, and (3) environmental changes at dusk trigger a behavioral shift in the raiding workers from foraging activities to actions essential to emigration (Schneirla, 1971). In *E. hamatum,* one of the three main raiding trails serves as the emigration route and teems with workers leaving the bivouac and with prey-laden foragers returning from the raiding system. These returning workers are intercepted and "sucked" into the outward moving stream (Schneirla, 1938). This exodus becomes an emigration when workers begin carrying brood from the bivouac. While some workers are still leaving the old bivouac, others form the new bivouac, a process that often begins between 1800 and 2000 hours in *E. hamatum* and 1930 and 2130 hours in *E. burchelli* (Schneirla, 1971). When the transport of brood is more than half completed, the queen passes with her retinue of workers. In the early nomadic phase this occurs between 1900 and 2000 hours in *E. hamatum* and 2000 and 2200 hours in *E. burchelli* (Schneirla, 1971). Rettenmeyer (1963b) found that in *E. hamatum* the size of the queen's retinue depends primarily on the speed with which the queen travels. If her progress is unimpeded, the retinue is small. The largest retinue seen by Rettenmeyer (1963b) included between 25 and 50 major workers and a larger number of smaller workers that stayed within 15 to 30 cm of the queen. The queen's retinue, at least in *Eciton,* is not a fixed group of individuals that remains with the queen throughout an emigration. Instead, membership in the retinue constantly changes. Even so, major workers comprise a greater percentage of the retinue than they do of the colony in general. Probably the retinue functions to protect the queen against predators and various environmental hazards and consequently is largest in army ant species that nest and forage epigaeically (Rettenmeyer *et al.,* 1978).

In *Neivamyrmex nigrescens* the general pattern of emigration is similar to that of *E. hamatum,* except that it may forego emigrating on some nights during the nomadic phase. Schneirla (1958) observed 60 emigrations in this species. All of the colony movements occurred at night and all grew out of raiding activities. Usually the emigrations began before midnight about 6 hr following the beginning of raiding, which is also crepuscular and nocturnal. An emigration required about 6–8 hr to complete, thus most were over at dawn. Schneirla (1958) found that the "emigration always occurs over a principal trail developed from the bivouac in raiding." In some instances, when interrupted by such things as heavy rain, an emigration may require a second night to complete. The distances covered varied considerably and ranged from 2.5 to 76 m. The queen of *N. nigrescens* seems to leave the nest earlier than does the queen of *E. hamatum* and appears in the column of march at about the time when one-third of the emigration is completed (Schneirla, 1958).

Less is known about the emigrations of the more cryptic, subterranean species of New World army ants. For instance, *Labidus praedator* was observed emigrating by Schneirla (1971) on only four occasions. The emigration columns were commonly 4–10 ants wide and included thousands of callows. Large numbers of worker cocoons were carried. This species also constructs walls of earthern pellets that flank the columns and sometimes form arcades completely shielding the emigrating workers (Rettenmeyer, 1963b; Schneirla, 1957). Rettenmeyer (1963b) observed an emigration of this species that required more than 1 day to complete, but only a short portion of the column (about 4 m) moved on the soil surface and it was shielded by soil particle walls and arcades and by "guard" workers. Wheeler (1921) observed an emigration of *Cheliomyrmex megalonyx*. He described the workers as

running along in dense, orderly columns under leaves, sticks or boards, wherever such cover was available, but where they had to cross open spaces, they had built covered galleries about four-fifths of an inch wide, of small particles of earth.

He noted that of the workers moving in the procession, the smallest were carrying larvae "tucked under their bodies."

In *A. gracilis* and *A. laeviceps* emigrations are usually initiated along a major raiding column that has been in progress for a considerable length of time. However, emigrations in these two species can begin within 20 min of the first signs of excitement in the bivouac and may occur without the raiding precondition evident in *Eciton*. Furthermore, emigrations may begin at any time of day or night and may start "early or late in raiding, as actions overlapping previous emigrations, or as actions ending a quiescent interval without extra-bivouac group operations" (Schneirla and Reyes, 1969). The fact that emigration in these two species could arise from raiding or other colony activities prompted Schneirla and Reyes (1969) to propose that this may represent a "generalized, primitive condition of colony organization."

During the first emigrations of the nomadic phase in these two species of *Aenictus,* the brood, which are quite small, are carried in packets by the workers and consequently the entire brood may be transported out of the old bivouac in less than 20 min. Later in the nomadic phase when the larvae are larger, each is carried by an individual worker, increasing the time required to remove the larvae from the bivouac. Generally the queen leaves the old bivouac in the second half of the emigration and is accompanied by a retinue of workers (Chapman, 1964; Schneirla and Reyes, 1969). In *A. laeviceps,* the queen's entourage is usually 5–8 cm at its widest point and as long as 1 m. Curiously, both of these species may average more than one emigration per day. Early in the nomadic phase the emigrations last from 2 to 3 hr, later they may take 4–7 hr.

Emigrations of African *Aenictus* were described by Brauns (1901) and Gotwald (1976). Brauns noted that the workers carry the larvae slung beneath their bodies. The emigration recorded by Gotwald was discovered, in progress, at 0950 hours crossing a path between plots planted with cassava. Except for an occasional worker, the column was unidirectional and approximately six individuals wide. All workers moving toward the new bivouac carried larvae; pupae were not seen. Workers did not assume "guard" positions along the column; nor did they construct earthen borders. At 1010 hours the queen, whose gaster was contracted, passed with a small entourage of workers.

Of the extensively studied army ants, the *Anomma* driver ants are most irregular in their emigrations. Savage (1847) noted, when observing *D. (A.) arcens,* that from "its locomotive habits the impression . . . has obtained, that it has no fixed habitation." Raignier and van Boven (1955) observed that *Anomma* emigrations often follow previously used foraging trails and that emigration columns are more often subterranean than are foraging columns. An emigration in *Anomma* is a single episode that may take 2 or 3 days (or more) to complete (mean duration: 56 hr) and may not be followed by another emigration for as long as 125 days, although this was an extreme case seen in *D. (A.) nigricans* (Raignier and van Boven, 1955). The mean distance covered by *Anomma* is 223 m. Raignier and van Boven noted that emigration columns can move along more rapidly than foraging columns and calculated some to move as fast as 155 m in 5 hr. Gotwald and Cunningham-van Someren (1982) discovered that some *Anomma* colonies emigrate only short distances and reuse emigration trails and nest sites. In fact, different conspecific colonies may utilize the same trails and nests at different times. For example, they found that one nest was occupied on 15 separate occasions by four different colonies.

b. Army Ant Functional Cycle.   Schneirla (1933, 1938, 1945, 1949, 1971) categorized army ants as either group A dorylines or group B dorylines, by virtue of the type of cyclic phenomena evident in their behavior. In the former group he placed species that exhibit a well defined cycle of alternating nomadic and statary phases that are conditioned by brood stimulative factors (Fig. 10, p. 189). Surface-active species of *Eciton, Neivamyrmex,* and *Aenictus* belong to this group. In group B he placed species in which there is an absence of these alternating phasic events. This group is characterized by variable nomadism: emigrations occurring as single events separated by intervals of nonnomadic behavior.

The nomadic period in Group A dorylines is one of elevated activity. For *E. hamatum* this means a large daily raid that begins at dawn and which, at dusk, is converted into an emigration to a new nesting site. Daily raiding and nightly emigrations typify this phase. At the very beginning of the phase there are two broods present in the colony, one which emerges from cocoons as callow workers and energizes the phase to begin, and another that consists of young worker larvae that hatched from eggs laid in the immediately preceding statary phase (Fig. 12). During the nomadic phase, which lasts 16–18 days, the queen does not lay eggs and her gaster remains contracted. Initially the young larvae are maintained en masse at the center of the bivouac, which is generally exposed, but as they mature, the larger individuals are kept at the bivouac periphery. The nomadic phase ends when the larvae reach maturity and spin their cocoons; as they do this their stimulative effect on the workers wanes and the colony enters the statary phase.

Lasting for a period of 18–21 days in *E. hamatum,* the statary phase is best characterized by subdued colony activity. Emigrations cease and raids, although still executed on a daily basis, are small and often feeble. The bivouac is established in a sheltered location, such as a hollow tree or log, and the workers are less excitable and active. During the second week of the phase, the queen, who has by then become fully physogastric, delivers a new cluster of eggs. These hatch and larval development ensues. The phase ends when the pupating brood from the previous statary phase eclose. These phases alternate regularly throughout a colony's existence, except when sexual brood is present. At that time the nomadic phase is shortened.

Eciton burchelli has a similar cycle, although its nomadic phase of 11–16 days is considerably shorter and more variable than that of *E. hamatum* (Schneirla, 1945). The statary phase is 19–22 days long. An equivalent nomadic–statary cycle exists in *Neivamyrmex nigrescens,* although under Nearctic conditions, the functional cycle is completely interrupted during the winter months (Schneirla, 1958, 1961). Interestingly the first emigration of the nomadic phase, which is triggered as in *Eciton* by the emergence of callow workers, is not followed by a succeeding emigration until about the fourth night of the phase. Furthermore, the colony remains in the nomadic phase until the larvae enter the early pupal stage. Schneirla (1958) postulated that nomadism continues past the point of larval activity as a result of some "stage-specific" secretory or metabolic products or functions. It could be that because the pupae are not enclosed in cocoons (i.e., there are no silken barriers between pupae and workers) their stimulative cues wane less abruptly. The nomadic phase for this species in Arizona lasts for 20–31 days, the statary phase for about 18 days. During the statary phase, *N. nigrescens* workers become more photonegative and exhibit a stronger tendency to cluster together, changes which correspond to the decrease in the excitability of workers during this phase (Topoff, 1975).

Schneirla (1971) concluded that of the Old World army ants, *Aenictus,* at least *A. gracilis* and *A. laeviceps,* most closely resemble *Eciton* and *Neivamyrmex* in their *functional* cycle. When nomadic, *A. gracilis* and *A. laeviceps* are capable of raiding and emigrating at any time of day or night. During the nomadic phase, the bivouac is a disc-shaped cluster of workers, either exposed or beneath litter, although near the end of the phase and throughout the statary phase the bivouacs are well sheltered or hypogaeic. The duration of the nomadic phase is about 14 days and thus similar to that of the New World group A army ants. However, the statary phase lasts about 28 days which is strikingly longer than in the New World forms. The statary phase begins when the advanced brood enters the prepupal stage, but the emergence of this brood as callows does not elevate the colony's activity level into a new nomadic phase. Instead, this phase begins some days later.

Schneirla's exhaustive studies of group A dorylines have been supplemented by numerous field and laboratory investigations. Studies of *Eciton burchelli* by Teles da Silva (1977a) have confirmed the endogenous nature of the nomadic-statary cycles in this species. Extensive observations of *Neivamyrmex nigrescens* support Schneirla's theory that brood stimulation is a proximate cause of the nomadic phase in group A dorylines. However, his theory does not account for characteristics such as frequency, direction, and distance of emigrations within the nomadic phase (Mirenda and Topoff, 1980; Topoff *et al.,* 1980a). Topoff and Mirenda (1980a, b) have explored the relationship between food supply and emigration frequency in *N. nigrescens* and have concluded that the "amount and location of food strongly influence the frequency and direction of emigrations." They found in laboratory studies that overfed colonies emigrate far less frequently during the nomadic phase than do underfed colonies.

Of the group B dorylines, *Anomma* is best known. Although the driver ants exhibit some regularity in raiding and emigration, they do not pass through alternating nomadic and statary phases (Raignier and van Boven, 1955), nor for that matter do a majority of dorylines. However, Schneirla (1957) attempted to homologize the functional cycle of *Eciton* with components of *Anomma* behavior, starting with the questionable premise that they are "similarly adapted on a nomadic, predatory basis probably by virtue of common ancestry." He concluded that the first part of the *Eciton* nomadic phase and the single emigrations of *Anomma* involve "homologous reproductive processes as essential causes," i.e., the excitatory effects of eclosing callow workers. According to Schneirla, it then followed that each emigration in *Anomma* should be regarded as a nomadic phase and the interval between emigrations as a statary phase. "Consequently," Schneirla stated, "*Anomma* may be said to have an identifiable nomadic-statary functional cycle equivalent to that of *Eciton.*" However, it is just as valid, and perhaps even more empirically demonstrable to invoke the process of convergent evolution as an explanation for the presence of brood stimulative cues for nomadic behavior.

E. Colony Division

1. Colony Founding in Nondoryline and Nonecitonine Ants

Typically, the founding of new colonies in a majority of ants, in which the males and females are initially alate, is preceded by a nuptual flight (Wheeler, 1933). The emergence of conspecific males and females in a specific locality is often synchronized for a majority of colonies, a phenomenon that no doubt favors cross fertilization. Following the nuptual flight, the newly fecundated queen descends to the substrate and either removes her wings with her legs and mandibles or rubs them off against a readily available object. She then excavates a small burrow, lays eggs, and cares for the larvae once the eggs hatch. She feeds the growing larvae salivary secretions that are metabolically derived from her fat bodies and flight muscles. Once these larvae pupate and eclose as adult workers, the queen does little more than ingest food gathered by her offspring and lay eggs. The rearing of new brood becomes the preoccupation of overlapping generations of sister workers (Wheeler, 1933; Wilson, 1971).

2. Colony Founding in Doryline and Ecitonine Ants

Army ant colonies are founded by the subdivision of existing colonies into daughter colonies. Since the queens are apterous, there are no nuptual flights and queens are not left to found new colonies by themselves. New colonies come "ready made."

In *E. burchelli* and *E. hamatum,* worker broods are produced throughout most of the year, to the exclusion of sexual broods, on a 33–36 day cycle of development, and colony division does not take place. Sexual broods are prerequisite to division. In Central America, sexuals are produced in the first one-third of the dry season (Schneirla, 1956a). With a sexual brood present in the *Eciton* colony, a "bipolar organization arises" in which some workers become greatly attracted to the brood, while others remain fixed in their attachment to the functional queen. This organization is basic to the actual process of division itself, which does not begin until the sexual brood emerges from their cocoons. The young queens eclose before the males, the first about 3 days earlier. The order in which the queens emerge is critical to queen selection, for usually only one callow queen will survive. In the hierarchy of attractiveness of callow queens to workers, the first one or two young queens to emerge normally have the advantage and are the leading candidates for survival in a new daughter colony. The first queens to eclose each attract an entourage of hundreds of workers. Queen number one leaves the bivouac, accompanied by her entourage and situates herself at a location usually within 1 m of the bivouac. The workers cluster about her. The next one or two virgin queens to emerge behave similarly, although they usually remain closer to the nest and have smaller clusters of affiliated workers. Generally these clusters and their queens gradually move farther away from the bivouac. Later eclosing queens are not greatly attractive to the workers and are confined by the workers to the bivouac wall. Thus the "period of a few days preceding appearance of the males . . . is one in which the young queens normally become consolidated in very different relations of acceptance or rejection with the worker population" (Schneirla, 1956a).

The eventual eclosion of the males galvanizes the statary colony to conduct a major foraging expedition that ends in an emigration. So vigorous is the raid that two or more base or trunk columns are established. This development is imperative for the division to take place. Once the raid is well established, about midmorning, the functional queen and her entourage leave the bivouac on one of the base trails. The leading virgin queen and her retinue then set out on another trail. Sometimes the functional queen is superseded by one of the virgin queens, but Schneirla (1956a) considered this a secondary form of the main pattern. Superfluous virgin queens are "sealed-off" by the workers and are eventually abandoned. In the final stage of division in *Eciton,* the newly eclosed males, the brood, and the workers are divided nearly equally, one group following the functional queen and the other the new queen. In most cases, fission is completed by nightfall.

Little is known about colony division in other New World species except *N. nigrescens.* Colony division in this species is not unlike that of *Eciton,* although the males, instead of being divided into two equal groups, all emigrate with the callow queen (Schneirla, 1971).

In *Aenictus,* the flight of newly eclosed males from the parent colony occurs prior to colony division, not following the process as occurs in *Eciton* and *Neivamyrmex* (Schneirla, 1971). While Schneirla (1971) found similarities between the patterns in *Aenictus* and the ecitonines, he concluded that the process of colony division in *Aenictus* is the most generalized of the known species.

Although the presence of male brood is necessary for colony division in *Anomma,* male larvae and pupae may be found in colonies that lack queen brood (Raignier, 1959, 1972). This means that the appearance of male brood cannot always be interpreted to signify that colony fission is imminent. However, Raignier (1959) observed colony fission when only male brood was present. In fact, this kind of division occurs more frequently than does division with a mixed brood present. However, in this case, most of the colony emigrates with the functional queen, leaving behind the male brood and some workers. After the males eclose and fly from the old nest, the queenless workers die.

While *Anomma* males may be present and colony division can take place at any time of the year, division occurs with greatest frequency during the last third of the dry season. The new queen or queens eclose approximately 10–14 days before the males and actually achieve adulthood at about the time the males enter the pupal phase. Division begins with the exodus from the nest of the functional queen and about half the workers and worker brood, an exodus which is identical to a normal emigration (Raignier, 1959). A bipolar organization in the colony prior to division, as can be seen in *Eciton,* is evident. Even though the exodus appears to take place after a large scale emergence of callow workers, the sexual brood may be in various stages of development. The males may be larvae, pupae, or adults and the new queens may be callow or fully pigmented. The virgin queens and male brood are always left behind in the original nest, and once exodus of the functional queen takes place, the workers kill all but one of the remaining queens. Raignier (1959) found as many as 56 virgin queens in a single nest and pointed out that the first queen to eclose is not necessarily the one to be spared. Once the males eclose, they fly from the nest.

VI. ARMY ANT LONGEVITY

A. Caste-Related Determinants of Army Ant Life Spans

1. Workers

Army ant workers suffer enormous losses when attacking prey, therefore much of the developmental energy expended in the production of brood is devoted to the replacement of workers lost in foraging. That raiding is hazardous for workers is easily observed in *Anomma* when numbers of injured stragglers return to the nest from raiding forays long after most other foragers have reached the nest. Even these stragglers stand little chance of continued survival, since on their laborious return they often fall prey to other ants, especially those belonging to the genus *Crematogaster* (W. H. Gotwald, unpublished data). Schneirla (1971) speculated that swarm raiders must lose many workers to the defensive secretions of some potential prey and that hypogaeic foragers must suffer high mortality from the secretions of nasute termites. He further presumed that large numbers of foragers are crushed in the mandibles of such ants as *Atta* and *Odontomachus* and that others succumb to the bites, stings and repellents of even less formidable ants. The actual turnover rate of workers in a colony has not been determined, nor would it seem reasonable that the rate could be easily calculated.

Environmental factors also play a role in determining the lifespans of workers. Increasing atmospheric dryness, for instance, may take its toll, particularly among the smallest workers since they appear to be the most susceptible to the fatal consequences of decreases in relative humidity (Schneirla *et al.,* 1954) (as one might predict given their surface area to volume ratio). Certainly workers are lost through excessive exposure to solar heat and many must drown when caught in torrential rains that characterize many of the tropical localities where army ants abound.

Watkins and Rettenmeyer (1967) determined that worker army ants live longer when in the presence of their queen, most likely because of certain secretions that are "licked" from the body of the queen.

2. Queens

Captured, marked, and recaptured queens of *Eciton* indicate that army ant queens may function in their colonies in excess of 4 years (Rettenmeyer, 1963b; Schneirla, 1971). While Schneirla (1971) assumed that some *Eciton* queens succumb to disease, parasites, and fatal accidents, he concluded that most queens probably die after being superseded by virgin queens during colony division.

3. Males

The male lifespan is relatively short and probably does not extend much beyond the time of copulation, if indeed copulation is achieved. It is not known whether such males expire at that time of "natural" causes or if they are exiled or killed by perhaps increasingly intolerant workers.

B. Predators of Army Ants

1. Invertebrate Predators

Army ants are themselves not without a myriad of predators. As noted previously (Sections V, A, 1 and V, C, 2, d), dorylines play host to a large number of inquilines and followers, many of which prey on army ant brood. In addition to the pressures brought about by these associates of army ants, the army ant lifespan is often shortened by the intervention of predators of a more general nature. For instance, it must be assumed that newly emerged males are subject to intense predation by invertebrates and vertebrates alike. I have witnessed evidence in both the New and Old World that doryline males are commonly attacked by spiders (Fig. 13). That males are attracted, often in

Fig. 13. Unidentified spider with captured army ant male of the subgenus *Dorylus*. The male was attracted to light between 2030 and 2200 hours which led to his capture (taken at Lamto, Ivory Coast). (Photograph by W. H. Gotwald, Jr.)

substantial numbers, to light at night no doubt increases their vulnerability to the attack of opportunistic predators. In Ivory Coast at Lampto, for example, I observed light-attracted males being captured by the ponerine ant *Megaponera foetens*.

Invertebrate predators of army ants, outside of those that could be classified as guests or followers, are, however, not numerous. Other ants appear to be the most common of these predators. Lamborn (1913–1914) observed workers of *Camponotus sericeus* on the earthenworks of a *D. (A.) nigricans* column, reaching down occasionally to grasp in their mandibles minor workers. However, he did not see any of the *Anomma* workers carried off for all were eventually released. *Crematogaster* workers will attack and drag away injured *Anomma* foragers and *Paltothyreus tarsatus* may attack *Anomma* brood under unusual circumstances (Gotwald, 1972b). Where *Anomma* columns cross the territories of the red tree ant, *Oecophylla longinoda,* this formicine ant may be a formidable predator of the driver ants (Gotwald, 1972b). In fact, in areas where these species are sympatric, *O. longinoda* may be the single most important insect predator of *Anomma.* An *O. longinoda* attack consists of individual workers reaching into an *Anomma* column, each seizing a driver ant worker in its mandibles and pulling it quickly from the column. The *Anomma* workers thus removed are then immobilized through prolonged stretching by numerous cooperating *O. longinoda* workers and transferred to the *O. longinoda* nest.

2. Vertebrate Predators

Vertebrates probably constitute the greatest predatory threat to army ants. Bequaert (1922) reported that of 1815 ants found in 194 stomachs of five species of Congo toads, 8% were dorylines. Three species of African frogs were also found to take dorylines of the subgenus *Anomma*. Certain forest species of African skinks of the genus *Mabuya* actually follow the columns of driver ants and feast on the workers (Bequaert, 1922). In India, *Dorylus* is among the ants taken by some species of birds (Bequaert, 1922). Chapin (1932) reported that driver ants are eaten by several species of African birds, including the Guinea fowl, *Phasidus niger*. Gotwald (1972b) noted that even the domesticated chicken can be counted among the predators of army ants that forage in village refuse heaps. Of the mammals, pangolins or scaly anteaters of the genus *Manis* probably take large numbers of dorylines. Lang

Fig. 14. (A) Chimpanzee inserting ant dipping tool into a nest of *D. (Anomma) molesta.* (B) As the disturbed driver ant workers climb the dipping tool, the chimpanzee monitors their progress. (C) The "pull-through"; the left hand, in this case, slides up the tool, catching the ants in a mass which accumulates as the hand advances. The mouth is open ready to receive the ants. (D) The ants are ingested at the end of the "pull-through." (Photographs by Caroline Tutin.)

(in Bequaert, 1922) observed the pangolin "lashing its sticky tongue through the confused crowds," and noted that it lost "no time in moving back and forth along the . . . [*Dorylus*] column as quickly as the dense clusters vanished into its mouth." Patrizi (1946) estimated that among the stomach contents of a female aardvark there were in excess of 100,000 individuals of *D. (D.) helvolus*. That these individuals were taken in all stages of development indicated that the aardvark probably plundered the *Dorylus* nest. *Dorylus* workers (of a non-*Anomma* species) have even been found in fecal samples of the aardwolf, *Proteles cristatus*, who normally is a termite feeder (Kruuk and Sands, 1972). No doubt an even greater list of army ant predators could be assembled for the New World species.

Perhaps the most interesting of all army ant predators, at least from the standpoint of the predator's behavior, is the chimpanzee. Goodall (1963) first described the use by east African chimpanzees of modified sticks in harvesting driver ants, a tool-using behavior referred to by McGrew (1974) as "ant dipping." Van Lawick-Goodall (1968) noted that chimpanzees at Gombe Stream Reserve in Tanzania frequently eat *Anomma* driver ants in January and occasionally do during September through November and February through May. Chimpanzees apparently find the driver ants visually, either by seeing the moving columns of workers or by recognizing the earthen works of the nest. In ant dipping the chimpanzee fashions a tool from living, woody vegetation, inserts the slender end of this stick into the *Anomma* nest, and waits until the workers swarm about three-quarters of the way up the stick (Fig. 14). At this point, it withdraws the stick and holds it in a vertical position with the distal end just below the mouth. The chimpanzee then slides its free hand up the stick, catching the ants in a mass the size of a "hen's egg" and shoves this mass into its mouth. "The chimpanzee's mouth closes and the jaws gnash frantically and exaggeratedly, audibly crushing the ants between the teeth" (McGrew, 1974). Driver ants seem to constitute an important protein source for chimpanzees, probably second only to termites for the females (McGrew, 1974).

VII. PHYLOGENY OF ARMY ANTS

A. Zoogeography

There are approximately 127 species of New World army ants arrayed among five genera. Five of the species belong to the monogeneric tribe Cheliomyrmecini, the remainder to the Ecitonini (Watkins, 1976). Because many of the species are based on descriptions of unassociated phena, the actual number of species may be considerably lower. *Cheliomyrmex* has a much more restricted distribution than the ecitonine genera and is decidedly more tropical. *Cheliomyrmex morosus* occurs as far north as San Luis Potosi, Mexico and *C. audas* as far south as La Paz, Bolivia (Watkins, 1976). Even though the workers of *Cheliomyrmex* are hypogaeic and behaviorly cryptic, the males are taken commonly enough at light to suggest that the range of the genus is truly restricted and not just a collecting artifact.

Some ecitonine species are wide ranging while others are severely restricted in distribution. *Labidus coecus,* for example, is found from Argentina to Oklahoma, whereas *Neivamyrmex baylori* is known only from Texas. *Labidus coecus* and some species of *Neivamyrmex,* including *N. nigrescens,* quite clearly are adapted in parts of their ranges to temperate climates. *Neivamyrmex* species extend to 40° latitude on either side of the equator, and *N. nigrescens,* for instance, has been collected in Iowa, Illinois, and West Virginia, all areas that are subjected to harsh winters. Curiously, *N. nigrescens* is the only group A doryline that maintains a subterranean nest throughout its functional cycle, and this may, in fact, be a secondary adaptation to temperate conditions.

The Old World tribe Aenictini is represented by 34 species in the Indo–Australian region and by approximately 15 species in Africa (Wilson, 1964). Continental Asian species are known from India, Thailand, Burma, Malaya, Sumatra, and southern China. *Aenictus* is also found in Australia (New South Wales and Queensland) and on such islands as the Philippines, Ceylon, New Guinea, Borneo, Java, Aru, and Taiwan (Wilson, 1964). A single specimen of *Aenictus* has been collected on Iriomote in the Ryukyu Islands (Onoyama, 1976). Although *Aenictus* ranges through the Middle East into Africa, the two populations do not appear to share any species. As noted by Wheeler (1930), Asian and African *Aenictus* queens differ from one another in a number of significant morphological details. Menozzi (1936) collected a previously undescribed species of *Aenictus* on the Mediterranean island of Rhodes and believed it to have certain affinities with the Indo–Australian species. In Africa, except for the most arid of areas, *Aenictus* is rather ubiquitous, although it is infrequently collected and does not exhibit the diversity apparent in Asia. It is not found on Madagascar, nor for that matter is *Dorylus.*

The Dorylini are well established in Africa with about 50 species, but poorly represented in Asia by only four species (Wilson, 1964; Gotwald, 1979). The Asian forms include two species of *Alaopone,* one of *Typhlopone,* and one of the endemic subgenus *Dichthadia.* Three subgenera, *Anomma, Dorylus,* and *Rhogmus,* are endemic to Africa and not present in Asia. Conversely, *Dichthadia* has not dispersed toward the African continent. The Asian species range, on the continent, from Nepal and India to southern China. As far as it can be ascertained *D. (Dichthadia) laevigatus* is the only species recorded from Java, Borneo, and the Celebes. Although all five of the *Dorylus* subgenera in Africa are widespread, *Anomma* is least successful in xeric and cool habitats. Driver ants, therefore, are not found in northern Africa or South Africa. As in *Aenictus, Dorylus* appears continuous in distribution across the Middle East, connecting African and Asian populations but without these populations sharing any species. Surely the Middle East should be carefully searched for evidence of sympatry between Asian and African forms. Donisthorpe (1950) reported the African species *D. (D.) affinis* from Turkey, and Wilson (1964) expressed misgivings about maintaining *D. (Typhlopone) labiatus,* a species from India, as separate from *D. (T.) fulvus* of North Africa.

B. Origin of Army Ants

1. Phylogenetic Scenarios and Hypotheses

Wilson (1958a) formulated an hypothesis for the origin of army ant behavior in terms of major adaptive steps. First, he proposed that group predation arose because it permitted feeding on large arthropods and on other social insects; second, that nomadism developed concurrently with or shortly after group predation to permit shifting of the trophophoric field; third, that prey preference was secondarily expanded to include smaller nonsocial insects and arthropods, making general predators of at least some species; and, fourth, that further refinements in nomadic and group predatory behavior permitted large colony size.

However, even if this scenario is accurate, the question must still be posed: From what group or groups of ants did the army ants arise? Throughout the literature, the cerapachyines are most often regarded to be somehow ancestrally related to the army ants. Emery (1895, 1901, 1904) presumed that the cerapachyines linked the army ants to the Ponerinae. Initially, Emery (1895) even placed the cerapachyines as a tribe within the Dorylinae and was most impressed with the fact that the males of both groups have retractile genitalia. Later he relented, most likely falling sway to the arguments of Forel (e.g., 1901) and placed the cerapachyines in the section Prodorylinae of the Ponerinae. Wheeler (1902, 1920) discussed the affinities of the cerapachyines and was inclined to regard them as ponerines, while admitting to their army ant characteristics, especially in the larvae. Although Brown (1954) refused to accept a cerapachyine origin for the army ants in his review of ant phylogeny, he most recently (1975) admitted that "Emery's notion that the Cerapachyini gave rise to the Dorylinae may have something to it still." However, he speculated that, because the army ants might be polyphyletic, "*Eciton* and relatives in the New World and/or *Aenictus* in the Old World arose separately from cerapachyines through the genus *Acanthostichus,*" a genus that Brown (1954) considered aberrant and on which he felt too much emphasis had been placed because of this group's dichthadiiform queens. Pullen (1963) proposed that the army ants "passed through an intermediate termitophagous condition characteristic of some modern Ponerinae" and was convinced that the army ants were "amongst the first ants to exploit termites for food on a large scale." If the army ants are triphyletic, as is suggested in further discussion, then should one look to the cerapachyines as ancestral candidates for all three groups? If indeed diet provides clues to these phylogenetic relationships as Pullen would have one believe, then it should be noted that the tribe Cerapachyini, which is strongly developed in the Old World, raid the nests of other ants, while the Cylindromyrmecini and Acanthostichini [now regarded as distinct ponerine tribes and not cerapachyines by Brown (1975)], which are endemic to the New World, are termite predators. The myth persists, in the absence of quantitative, corroborative evidence, that termitophagy is prevalent among the army ants. To be sure, termites do constitute an important prey item for some hypogaeic species, but it may be misleading to make too much of this in phylogenetic interpretations.

The study of dorylophilous faunas, especially the Staphylinidae, may provide at least circumstantial evidence bearing on the phylogenetic relationship of the army ants. Seevers (1965), for instance, maintained that the pantropical staphylinid tribe Dorylomimini is monophyletic. If his conclusion were true, a strong case for monophyly in the army ants could also be argued. However, Seevers revealed his uncertainty by admitting that if it could be demonstrated conclusively that the army ants are polyphyletic, then retention of the Dorylomimini as a monophyletic group would be "indefensible." Kistner (1972) proposed, based on the relationship of Old and New World myrmecophilous staphylinids, that *Neivamyrmex* and *Aenictus* may share a common ancestry.

Wheeler (1928) supposed that the army ants were monophyletic, believing that *Cheliomyrmex,* with its uninodal waist, linked the New and Old World faunas. Schneirla (1971) felt strongly that the "concept of a monophyletic origin best fits available functional and behavioral evidence." Above all, he was convinced that the phasic behavior of the group A dorylines could be homologized with the cyclic activities of the group B species. Because Old World doryline workers and queens lack eyes, Schneirla (1971) hypothesized that the ancestors of *Aenictus* and *Dorylus* were subjected to harsh surface conditions and were "forced" to adopt a hypogaeic lifestyle. Retention of reduced eyes in the workers and queens of most New World forms indicated, according to Schneirla, that the ancestors of the New World forms separated early from *Aenictus*-like stock in Asia and dispersed to the New World over the northern route, which he assumed to be tropical at the end of the Cretaceous. His evolutionary scenario implies a common origin for the army ants in Laurasia.

However, even Borgmeier (1955) doubted that the Old and New World forms were closely related, and in a personal communication to Seevers (1965), he admitted to believing that the two groups arose independently. Brown (1954) too professed that the army ants possibly were diphyletic, and Gotwald (1969) noted that mouthpart morphology could be used to support a triphyletic hypothesis. Gotwald and Kupiec (1975) stated that geographic, morphological, and behavioral evidence indicates a triphyletic origin and that the three lineages composing the Dorylinae are the (1) Ecitonini–Cheliomyrmecini, (2) Dorylini, and (3) Aenictini. They proposed, contrary to Schneirla's hypothesis, that the two New World tribes arose from a common ancestor that possessed a one-segmented waist; that the Ecitonini and Cheliomyrmecini diverged from one another quite early, and that the genus *Cheliomyrmex* retained the primitive, uninodal waist, while the ecitonines independently evolved a two-segmented waist.

2. The Fossil Record

Doryline army ants are completely absent from the fossil record. Although some ant genera in the Florissant Shales show Neotropical affinities and suggest that the Nearctic fauna was once rich in genera now restricted to the neotropics, army ants are not among them (Carpenter, 1930). The Baltic Amber contains a diverse ant fauna of which 56% of the genera represented are extant. Wheeler (1914) noted the absence of dorylines from the amber and hypothesized that they were either restricted to the tropics during the Oligocene or were so hypogaeic as to preclude their entrapment in resin. One can only speculate about the time of origin for the army ants.

Until the discovery of the Mesozoic ant *Sphecomyrma freyi,* the earliest known ant fossils were of Eocene age (Carpenter, 1929). Found in the amber of the Magothy Formation, *S. freyi* can be dated with reasonable certainty to the lower part of the Upper Cretaceous. Thus sociality probably developed in the ants prior to the mid-Cretaceous, although Wilson *et al.* (1967a,b) concluded that social life in the Hymenoptera in general might not be much older than *S. freyi* itself. Gotwald (1977, 1979) concluded that the primitive nature of *Sphecomyrma* and the diversity of the Oligocene ant fauna suggest a late Cretaceous or early Tertiary (and perhaps even later) origin for the doryline ants. Schneirla (1971) was of a similar opinion. Because the two Old World tribes have distinct endemic elements in Asia and Africa, Wilson (1964) proposed that this faunal differentiation occurred since Miocene times.

3. Plate Tectonics

The biogeography of some organisms can be explained, in part, as a consequence of continental drift. In turn, phylogenetic information can sometimes be gleaned from the geological data. For instance, plate tectonics can often account for the distribution of tropical disjuncts, i.e., tropicopolitan organisms that are related but are now separated by oceanic barriers (Keast, 1972).

According to continental drift theory, today's continents once formed a single land mass, Pangaea. By the late Triassic to mid-Jurassic, Pangaea began to split into a northern cluster of continents called Laurasia and a southern cluster called Gondwana (Dietz and Holden, 1970). The clusters, in turn, fragmented to form the northern and southern continents. By the end of the Cretaceous, Africa and South America were well separated and the South Atlantic Ocean had widened to 3000 km. In fact, at the conclusion of the Cretaceous, the three tropical areas in which true army ants are currently distributed were all separated by substantial ocean barriers.

Although true army ants superficially resemble tropical disjuncts, faunal exchange between Asia and Africa has occurred in relatively recent times so that there are some shared species groups. Since army ants are notoriously poor dispersers, mainly because the queen is apterous and new colonies are produced by colony fission, and since they most likely arose following the breakup of Gondwana and Laurasia, Gotwald (1977, 1979) postulated that the army ants are indeed triphyletic. In other words, the probability is low that the army ants dispersed from a single place of origin across significant oceanic barriers. As Brown (1973) noted in his zoogeographical analysis of Hylean and West African ant faunas, the Atlantic Ocean has been a "formidable barrier" to even those genera that most likely could have rafted across it. Less formidable is the barrier between the Ethiopian and Oriental regions. Although separated during much of the Mesozoic and Tertiary by the pre-Mediterranean Tethys Sea (Cooke, 1972), and now extensive xeric habitats, there is a great deal of sharing of species groups between the regions (Brown, 1973). These are patterns to which the army ants also conform.

Gotwald (1979) pointed out that land bridges connecting the continents periodically since the end of the Cretaceous cannot explain the current distribution of army ants. Even the North Pacific bridge, which sometimes permitted intense faunal exchange of a variety of taxa and which Schnerila (1971) favored as a dispersal route overwhich the progenitors of the New World army ants dispersed from Asia, can probably be ruled out as playing a role in army ant dispersal. Indeed, Darlington (1957) noted that the groups exchanged across this bridge probably belonged to cool environments; it was not a tropical route as Schneirla (1971) supposed.

Thus Gotwald (1979) concluded the geological data indicated that (1) the army ants arose convergently on three separate occasions in three separate tropical loci, (2) the genus *Aenictus* arose in tropical Laurasia, possibly in the early Tertiary, and dispersed into Africa between the late Oligocene and late Pliocene, (3) *Dorylus* evolved in Africa during the early Tertiary but did not disperse into Asia until late in the Tertiary or even in Quaternary time, and (4) the ecitonines and *Cheliomyrmex* arose from a common ancestor in tropical South America, they underwent extensive diversification during a long period of geographic isolation, and they did not disperse into North America until the end of the Tertiary.

4. Polyphylogeny and the Ascendancy of the Army Ant Adaptive Syndrome

The true army ants may yet become a classic example of what convergent evolution can accomplish. Morphological, behavioral, and zoogeographical evidence point to a triphyletic origin for the army ants, but perhaps even more importantly a number of ponerine species can be judged as being or becoming army ants. *Leptogenys* and *Simopelta* are but two such genera containing some species that not only behave as army ants but show as well a concomitant morphological convergence toward the army ant habitus. In *Simopelta oculata,* for instance, the queen is dichthadiiform and the workers do not possess a furcula, the sting sclerite characteristically absent in the army ants. All of this means that army ant behavior must have arisen independently a minimum of seven times (Wilson, 1958a). Although it remains convenient to refer to the army ants as a single group, particularly when comparing them to other ants, it is not phylogenetically accurate. In fact, some investigators (e.g., Brown, 1973; Chadab and Rettenmeyer, 1975) have begun referring to the ecitonines as a separate subfamily, Ecitoninae, and Snelling (1981) has formally proposed this change in taxonomic status.

Clearly, army ant behavior in tropical environments confers considerable selective advantage on species that "earn" their living in this way. Army ants are exceedingly successful organisms and selective pressures for at least some ant species to evolve toward the army ant adaptive syndrome must be significant. In particular, ground stratum ants of the subfamily Ponerinae, ants that are committed predators and have not yet developed an obligatory "thirst" for plant liquids (especially via Homoptera), appear ready to become army ants where circumstances dictate. No doubt, the advantage in being an army ant lies in a qualitative and quantitative expansion of the diet. Army ants have access to a wide range of prey not available to the solitary forager.

C. Army Ant Role in Tropical Ecosystems: Some Reflections

Army ants cannot be dismissed merely as bizarre, albeit interesting, tropical creatures whose greatest claim is to have inspired breathtaking, fictionalized accounts of fearsome, ravaging hexapods on the loose. Instead, they must be accorded their very special place in tropical ecosystems. They constitute a dominate form of tropical life. Their predatory effect on the total biomass cannot be overestimated for their numbers are prodigeous. A host of other organisms have coevolved with the army ants, inquilines and followers alike, all bound up in the delicate fabric of tropical life. Army ants are the evolutionary center of a coevolved system that, when fully understood, may be dazzling because of its complexity and its omnipresence in tropical habitats. There are few tropical animals that are not affected, either directly or indirectly, by army ants.

Army ants also assume an aesthetic place in tropical life and should be valued, along with all other tropical organisms, for their diversity and for the lessons in biology they have to teach. This aestheticism of living things (and the shared concern of scientists for the practical problems produced by a reduction in species diversity) must be translated into tangible efforts to halt the accelerating destruction of tropical habitats, before army ants and myriads of other tropical organisms become but a wistful memory in our collective conscience.

ACKNOWLEDGMENTS

Much of my own research described in this chapter was made possible through the generous assistance of individuals who are too numerous to list here. Their contributions are nevertheless acknowledged with gratitude. The research, which included four field trips to Africa, was supported by National Science Foundation grants GB–22856, GB–39874, DEB 77–03356, and DEB 79–05835. I am grateful to Dr. William L. Brown, Jr. (Cornell University) for reviewing the keys and to Prof. Dr. J. K. A. van Boven (Universiteit Leuven, Belgium), Mr. Barry Bolton (British Museum-Natural History), and Mr. Roy R. Snelling (Los Angeles County Museum of Natural History) for critically reading the manuscript. Dr. William C. McGrew (University of Stirling, Scotland) kindly provided the photographs of the chimpanzees feeding on driver ants. I thank Ms. Virginia Marsicane and Ms. Joanna R.M. Gotwald for typing the drafts of the manuscript. Finally, I am especially grateful to Prof. van Boven and affectionately thank him for kindly assisting and encouraging me over the years in my frenetic pursuit of Old World army ants.

REFERENCES

Akre, R. D., and Rettenmeyer, C. W. (1966). Behavior of Staphylinidae associated with army ants (Formicidae:Ecitonini). *J. Kans. Entomol. Soc.* **39**, 745–782.

Akre, R. D., and Rettenmeyer, C. W. (1968). Trail-following by guests of army ants (Hymenoptera:Formicidae:Ecitonini). *J. Kans. Entomol. Soc.* **41**, 165–174.

Akre, R. D., and Torgerson, R. L. (1968). The behavior of *Diploeciton nevermanni,* a staphylinid beetle associated with army ants. *Psyche* **75**, 211–215.

Akre, R. D., and Torgerson, R. L. (1969). Behavior of *Vatesus* beetles associated with army ants (Coleoptera:Staphylinidae). *Pan-Pac. Entomol.* **45**, 269–281.

Alibert, H. (1951). Les insectes vivant sur les cacaoyers en Afrique occidentale. *Mem. Inst. Fr. Afr. Noire* **15**, 1–174.

André, E. (1900). Sur la femelle probable de *l'Anomma nigricans* Ill. (Hyménoptère). *Bull. Mus. Natl. Hist. Nat.* **6**, 364–368.

Arnold, G. (1914). Nest-changing migrations of two species of ants. *Proc. Rhod. Sci. Assoc.* **13**, 25–32.

Arnold, G. (1915). A monograph of the Formicidae of South Africa. *Ann. S. Afr. Mus.* **14**, 1–756.

Arnold, G. (1953). Notes on a female *Dorylus (Anomma) nigricans* Ill. taken with workers. *J. Entomol. Soc. South. Afr.* **16**, 141–142.

Ashmead, W. H. (1906). Classification of the foraging and driver ants, or family Dorylidae, with a description of the genus *Ctenopyga* Ashm. *Proc. Entomol. Soc. Wash.* **8**, 21–31.

Baldridge, R. S., Rettenmeyer, C. W., and Watkins, J. F., II. (1980). Seasonal, nocturnal and diurnal flight periodicities of Nearctic army ant males (Hymenoptera:Formicidae). *J. Kans. Entomol. Soc.* **53**, 189–204.

Baroni Urbani, C. (1977). Materiali per una revisione della sottofamiglia Leptanillinae Emery (Hymenoptera:Formicidae). *Entomol. Bras.* **2**, 427–488.

Barr, D., and Gotwald, W. H., Jr. (1982). Phenetic affinities of males of the army ant genus *Dorylus* (Hymenoptera:Formicidae:Dorylinae). *Can. J. Zool.* (in press).

Bates, H. W. (1863). "The Naturalist on the River Amazons." Everyman's Library, Dutton, New York (reprinted in 1910).

Beebe, W. (1919). The home town of the army ants. *Atl. Mon.* **124**, 454–464.

Belt, T. (1874). "The Naturalist in Nicaragua." Murray, London.

Bequaert, J. (1913). Notes biologiques sur quelques fourmis et termites du Congo Belge. *Rev. Zool. Afr.* **2**, 396–431.

Bequaert, J. (1922). The predaceous enemies of ants. *Bull. Am. Mus. Nat. Hist.* **45**, 271–331.

Bernstein, S., and Bernstein, R. A. (1969). Relationships between foraging efficiency and the size of the head and component brain and sensory structures in the red wood ant. *Brain Res.* **16**, 85–104.

Blum, M. S. (1974). Pheromonal sociality in the Hymenoptera. *In* "Pheromones" (M. C. Birch, ed.), pp. 222–249. Am. Elsevier, New York.

Blum, M. S., and Portocarrero, C. A. (1964). Chemical releasers of social behavior. IV. The hindgut as the source of the odor trail pheromone in the Neotropical army ant genus *Eciton*. *Ann. Entomol. Soc. Am.* **57**, 793–794.

Bodot, P. (1961). La destruction des termitieres de *Bellicositermes natalensis* Hav., par une fourmi, *Dorylus (Typhlopone) dentifrons* Wasmann. *C. R. Hebd. Seances Acad. Sci.* **253**, 3053–3054.

Bolton, B. (1973). The ant genera of West Africa: A synopsis with keys (Hymenoptera:Formicidae). *Bull. Br. Mus. (Nat. Hist.), Zool.* **27**, 317–368.

Borgmeier, T. (1950). A fêmea dichthadiiforme e os estadios evolutivos de *Simopelta pergandei* Forel), e a descrição de *S. bicolor*, n. sp. (Hym. Formicidae). *Rev. Entomol.* **21**, 369–380.

Borgmeier, T. (1953). Vorarbeiten zu einer Revision der neotropischen Wanderameisen. *Stud. Entomol.* **2**, 1–51.

Borgmeier, T. (1955). Die Wanderameisen der Neotropischen Region (Hym. Formicidae). *Stud. Entomol.* **3**, 1–717.

Borgmeier, T. (1958). Nachtraege zu meiner Monographie der Neotropischen Wanderameisen (Hym. Formicidae). *Stud. Entomol.* **1**, 197–208.

Brauns, J. (1901). Uber die Lebensweise von *Dorylus* and *Aenictus*. *Z. Syst. Hymenopterol. Dipterol.* **1**, 14–17.

Brauns, J. (1903). Ueber das Weibchen von *Dorylus (Rhogmus) fimbriatus* Shuck. (Hym.). *Z. Syst. Hymenopterol. Dipterol.* **3**, 294–298.

Brown, C. A., Watkins, J. F., II, and Eldridge, D. W. (1979). Repression of bacteria and fungi by the army ant secretion:skatole. *J. Kans. Entomol. Soc.* **52**, 119–122.

Brown, W. L., Jr. (1954). Remarks on the internal phylogeny and subfamily classification of the family Formicidae. *Insectes Soc.* **1**, 21–31.

Brown, W. L. Jr. (1960). The release of alarm and attack behavior in some New World army ants. *Psyche* **66**, 25–27.

Brown, W. L., Jr. (1968). An hypothesis concerning the function of the metapleural glands in ants. *Am. Nat.* **102**, 188–191.

Brown, W. L., Jr. (1973). A comparison of the Hylean and Congo-West African rain forest ant faunas. *In* "Tropical Forest Ecosystems in Africa and South America: A Comparative Review" (B. J. Meggers, E. S. Ayensu, and W. D. Duckworth, eds.), pp. 161–185. Smithson. Inst. Press, Washington, D.C.

Brown W. L., Jr. (1975). Contributions toward a reclassification of the Formicidae. V. Ponerinae, Tribes Platythyreini, Cerapachyini, Cylindromyrmecini, Acanthostichini, and Aenictogitini. *Search: Agric.* **5**, 1–116.

Brown, W. L., Jr., and Nutting, W. L. (1950). Wing venation and the phylogeny of the Formicidae (Hymenoptera). *Trans. Am. Entomol. Soc.* **75**, 113–132.

Bruch, C. (1923). Estudios mirmecológicos con la descripción de nuevas especies de dipteros ("Phoridae") por los RR. PP. H. Schmitz y Th. Borgmeier y de una araña ("Gonyleptidae") por el Doctor Mello-Leitao. *Rev. Mus. La Plata* **27**, 172–220.

Bruch, C. (1934). Las formas femeninas de *Eciton*. Descripción y redescripción de algunas especies de la Argentina. *An. Soc. Cient. Argent.* **108**, 113–135.

Bugnion, E. (1930). Les pièces buccales, le sac infrabuccal et le pharynx des fourmis. *Bull. Soc. Entomol. Egypte* **14**, 85–210.

Burgeon, L. (1924a). Les fourmis "siafu" du Congo. *Rev. Zool. Afr.* **12**, 63–65.

Burgeon, L. (1924b). Quelques observations sur les *Dorylus. Rev. Zool. Afr.* **12**, 429–436.

Carpenter, F. M. (1929). A fossil ant from the Lower Eocene (Wilcox) of Tennessee. *J. Wash. Acad. Sci.* **19**, 300–301.

Carpenter, F. M. (1930). The fossil ants of North America. *Bull. Mus. Comp. Zool.* **70**, 1–66.

Carpenter, G. D. H. (1914–1915). Observations on *Dorylus nigricans,* Illig., in Damba and Bugalla Islands. *Trans. R. Entomol. Soc. London* pp. 107–111.

Carpenter, G. D. H. (1920). "A Naturalist on Lake Victoria with an Account of Sleeping Sickness and the Tse-tse Fly." Unwin, London.

Carroll, C. R., and Janzen, D. H. (1973). Ecology of foraging by ants. *Annu. Rev. Ecol. Syst.* **4**, 231–257.

Chadab, R. (1979). Early warning cues for social wasps attacked by army ants. *Psyche* **86**, 115–123.

Chadab, R., and Rettenmeyer, C. W. (1975). Mass recruitment by army ants. *Science* **188**, 1124–1125.

Chapin, J. P. (1932). The birds of the Belgian Congo. I. *Bull. Am. Mus. Nat. Hist.* **65**, 1–756.

Chapman, J. W. (1964). Studies on the ecology of the army ants of the Philippines genus *Aenictus* Shuckard (Hymenoptera: Formicidae). *Philipp. J. Sci.* **93**, 551–595.

Cohic, F. (1948). Observations morphologiques et écologiques sur *Dorylus (Anomma) nigricans* Illiger (Hymenoptera Dorylidae). *Rev. Fr. Entomol.* **14**, 229–275.

Cooke, H. B. S. (1972). The fossil mammal fauna of Africa. *In* "Evolution, Mammals, and Southern Continents" (A. Keast, F. C. Erk, and B. Glass, eds.), pp. 89–139. State University of New York Press, Albany.

Crawley, B. A., and Jacobson, E. (1924). Ants from Sumatra. With biological notes. *Ann. Mag. Nat. Hist.* [9] **13**, 380–409.

Cros, A. (1939). *Dorylus (Typhlopone) fulvus* Westw. St. *juvenculus* Shuck.: Etude biologique. *Bull. Soc. Hist. Nat. Afr. Nord.* **30**, 205–220.

Darlington, P. J. (1957). "Zoogeography: The Geographical Distribution of Animals." Wiley, New York.

Darwin, C. (1859). "The Origin of Species by Means of Natural Selection or the Preservation of Favoured Races in the Struggle for Life" (6th ed., 1872) (reprinted in 1958 by Mentor Books, The New American Library of World Literature, Inc., New York).

Dietz, R. S., and Holden, J. C. (1970). Reconstruction of Pangaea: Breakup and dispersion of continents, Permian to Present. *J. Geophys. Res.* **75**, 4939–4956.

Dinesen, I. (1937). "Out of Africa." Random House, New York.

Donisthorpe, H. St. J. K. (1939). On the occurrence of dealated males in the genus *Dorylus* Fab. (Hym. Formicidae). *Proc. R. Entomol. Soc. London* **14**, 79–81.

Donisthorpe, H. (1950). Two new species of ants, and a few others from Turkey. *Ann. Mag. Nat. Hist.* [12] **3**, 638–640.

Drummond, B. A., III (1976). Butterflies associated with an army ant swarm raid in Honduras. *J. Lepid. Soc.* **30**, 237–238.

DuChaillu, P. B. (1861). "Explorations and Adventures in Equatorial Africa; with Accounts of the Manners and Customs of the People, and of the Chase of the Gorilla, the Crocodile, Leopard, Elephant, Hippopotamus, and Other Animals." Harper, New York.

Dutt, G. R. (1912). Life histories of Indian insects: (Hymenoptera). Memoirs Department Agriculture in India. *Entomol. Ser.* **4**, 183–267.

Eisner, T. (1957). A comparative morphological study of the proventriculus of ants (Hymenoptera: Formicidae). *Bull. Mus. Comp. Zool.* **116**, 437–490.

Emery, C. (1887). Le tre forme sessuali del *Dorylus helvolus* L. e degli altri Dorilidi. *Boll. Soc. Entomol. Ital.* **19**, 344–351.

Emery, C. (1895). Die Gattung *Dorylus* Fab. und die systematische Eintheilung der Formiciden. *Zool. Jahrb., Abt. Syst.* **8**, 685–778.

Emery, C. (1901). Notes sur les sous-familles des Dorylines et Ponérines (Famille des Formicides). *Ann. Soc. Entomol. Belg.* **45**, 32–54.

Emery, C. (1904). Le affinità del genere *Leptanilla* e i limiti delle Dorylinae. *Arch. Zool.* **2**, 107–116.

Emery C. (1910). Subfam. Dorylinae. *In* "Genera Insectorum" (P. Wytsman, ed.), Fasc. 102, pp. 1–34. V. Verteneuil & L. Desmet, Brussels.

Fabricius, J. C. (1793), "Entomologia systematica." Hafniae, Proft.

Flanders, S. E. (1976). Revision of a hypothetical explanation for the occasional replacement of a unisexual with a bisexual brood in the colonies of the army ant *Eciton. Bull. Entomol. Soc. Am.* **22**, 133–134.

Forbes, J. (1958). The male reproductive system of the army ant, *Eciton hamatum* Fabricius. *Proc. Int. Cong. Entomol. 10th, 1956* **1**, pp. 593–596.

Forbes, J., and Do-Van-Quy, D. (1965). The anatomy and histology of the male reproductive system of the legionary ant, *Neivamyrmex harrisi* (Haldeman) (Hymenoptera: Formicidae). *J. N.Y. Entomol. Soc.* **73**, 95–111.

Ford, F. C., and Forbes, J. (1980). Anatomy of the male reproductive systems of the adults and pupae of two doryline ants, *Dorylus (Anomma) wilverthi* Emery and *D. (A) nigricans* Illiger. *J. N.Y. Entomol. Soc.* **88**, 133–142.

Forel, A. (1896). "Ants' Nests," Smithson. Rep. 1894, pp. 479–505. Smithson. Inst. Press, Washington, D.C.

Forel, A. (1901). A propos de la classification des fourmis. *Bull. Ann. Soc. R. Entomol. Belg.* **45**, 136–141.

Forel, A. (1928). "The Social World of the Ants Compared with that of Man," Vol. II, Putnam, London.

Fowler, H. G. (1979). Notes on *Labidus praedator* (Fr. Smith) in Paraquay (Hymenoptera: Formicidae:Dorylinae:Ecitonini). *J. Nat. Hist.* **13**, 3–10.

Gallardo, A. (1915). Algunas hormigas de la República Argentina. *An. Mus. Nac. Hist. Nat.* **27**, 1–35.

Gallardo, A. (1929). Notas sobre las dorilinas Argentinas. *An. Mus. Nac. Hist. Nat.* **36**, 43–48.

Gehlbach, F. R., Watkins, J. F., II, and Reno, H. W. (1968). Blind snake defensive behavior elicited by ant attacks. *BioScience* **18**, 784–785.

Gehlbach, F. R., Watkins, J. F., II, and Kroll, J. C. (1971). Pheromone trail-following studies of typhlopid, leptotyphlopid, and colubrid snakes. *Behaviour* **40**, 282–294.

Goodall, J. (1963). Feeding behaviour of wild chimpanzees. A preliminary report. *Symp. Zool. Soc. London* **10**, 39–48.

Gotwald, W. H., Jr. (1969). Comparative morphological studies of the ants, with particular reference to the mouthparts (Hymenoptera: Formicidae). *Mem.—N.Y., Agric. Exp. Stn. (Ithaca)* **408**, 1–150.

Gotwald, W. H., Jr. (1971). Phylogenetic affinities of the ant genus *Cheliomyrmex* (Hymenoptera: Formicidae). *J. N.Y. Entomol. Soc.* **79**, 161–173.

Gotwald, W. H., Jr. (1972a). Analogous prey escape mechanisms in a pulmonate mollusk and lepidopterous larvae. *J. N.Y. Entomol. Soc.* **80**, 111–113.

Gotwald, W. H., Jr. (1972b). *Oecophylla longinoda,* an ant predator of *Anomma* driver ants (Hymenoptera:Formicidae). *Psyche* **79**, 348–356.

Gotwald, W. H., Jr. (1974a). Predatory behavior and food preferences of driver ants in selected African habitats. *Ann. Entomol. Soc. Am.* **67**, 877–886.

Gotwald, W. H., Jr. (1974b). Foraging behavior of *Anomma* driver ants in Ghana cocoa farms (Hymenoptera:Formicidae). *Bull. Inst. Fondam. Afr. Noire, Ser. A* **36**, 705–713.

Gotwald, W. H., Jr. (1976). Behavioral observations on African army ants of the genus *Aenictus* (Hymenoptera:Formicidae). *Biotropica* **8**, 59–65.

Gotwald, W. H., Jr. (1977). The origins and dispersal of army ants of the subfamily Dorylinae. *Proc. Int. Congr. Int. Union Study Soc. Insects, 8th, 1977* pp. 126–127.

Gotwald, W. H., Jr. (1978). Trophic ecology and adaptation in tropical Old World ants of the subfamily Dorylinae (Hymenoptera:Formicidae). *Biotropica* **10**, 161–169.

Gotwald, W. H., Jr. (1979). Phylogenetic implications of army ant zoogeography (Hymenoptera:Formicidae). *Ann. Entomol. Soc. Am.* **72**, 462–467.

Gotwald, W. H., Jr., and Barr, D. (1980). Quantitative studies on major workers of the ant genus *Dorylus* (Hymenoptera:Formicidae:Dorylinae). *Ann. Entomol. Soc. Am.* **73**, 231–238.

Gotwald, W. H., Jr., and Brown, W. L., Jr. (1966). The ant genus *Simopelta* (Hymenoptera: Formicidae). *Psyche* **73**, 261–277.

Gotwald, W. H., Jr., and Burdette, A. W. (1981). Morphology of the male internal reproductive system in army ants: phylogenetic implications (Hymenoptera:Formicidae). *Proc. Entomol. Soc. Wash.* **83**, 72–92.

Gotwald, W. H., Jr., and Cunningham-van Someren, G. R. (1976). Taxonomic and behavioral notes on the African ant, *Aenictus eugenii,* with a description of the queen (Hymenoptera: Formicidae). *J. N.Y. Entomol. Soc.* **84**, 182–188.

Gotwald, W. H., Jr., and Cunningham-van Someren, G. R. (1982). Emigration behavior of the East African driver ant, *Dorylus (Anomma) molesta* Gerstaecker (Hymenoptera: Formicidae: Dorylinae). (In preparation).

Gotwald, W. H., Jr., and Kupiec, B. M. (1975). Taxonomic implications of doryline worker ant morphology: *Cheliomyrmex morosus* (Hymenoptera: Formicidae). *Ann. Entomol. Soc. Am.* **68**, 961–971.

Gotwald, W. H., Jr., and Leroux, J. M. (1980). Taxonomy of the African army ant, *Aenictus decolor* (Mayr), with a description of the queen (Hymenoptera:Formicidae). *Proc. Entomol. Soc. Wash.* **82**, 599–608.

Gotwald, W. H., Jr., and Schaefer, R. F., Jr. (1982). Taxonomic implications of doryline worker ant morphology: *Dorylus* subgenus *Anomma* (Hymenoptera: Formicidae). *Sociobiology* (in press).

Green, E. E. (1903). Note on *Dorylus orientalis* West. *Indian Mus. Notes* **5**, 39.

Haddow, A. J., Yarrow, I. H. H., Lancaster, G. A., and Corbet, P. S. (1966). Nocturnal flight cycle in the males of African doryline ants (Hymenoptera: Formicidae). *Proc. R. Entomol. Soc. London* **41**, 103–106.

Hagan, H. R. (1954a). The reproductive system of the army-ant queen, *Eciton (Eciton)*. Part 1. General anatomy. *Am. Mus. Novit.* **1663**, 1–12.

Hagan, H. R. (1954b). The reproductive system of the army-ant queen, *Eciton (Eciton)*. Part 2. Histology. *Am. Mus. Novit.* **1664**, 1–17.

Hagan, H. R. (1954c). The reproductive system of the army-ant queen, *Eciton (Eciton)*. Part 3. The oöcyte cycle. *Am. Mus. Novit.* **1665**, 1–20.

Haley, A. (1976). "Roots." Doubleday, Garden City, New York.

Hermann, H. R., Jr. (1968a). Group raiding in *Termitopone commutata* (Roger) (Hymenoptera: Formicidae). *J. Ga. Entomol. Soc.* **3**, 23–24.

Hermann, H. R., Jr. (1968b). The hymenopterous poison apparatus. VII. *Simopelta oculata* (Hymenoptera:Formicidae:Ponerinae). *J. Ga. Entomol. Soc.* **3**, 163–166.

Hermann, H. R., Jr. (1969). The hymenopterous poison apparatus: Evolutionary trends in three closely related sub-families of ants (Hymenoptera: Formicidae). *J. Ga. Entomol. Soc.* **4**, 123–141.

Hermann, H. R., Jr., and Blum, M. S. (1967). The morphology and histology of the hymenopterous poison apparatus. III. *Eciton hamatum* (Formicidae). *Ann. Entomol. Soc. Am.* **60**, 1282–1291.

Holliday, M. (1904). A study of some ergatogynic ants. *Zool. Jahrb., Abt. Syst.* **19**, 293–328.

Hollingsworth, M. J. (1960). Studies on the polymorphic workers of the army ant *Dorylus (Anomma) nigricans* Illiger. *Insectes Soc.* **7**, 17–37.

Hung, A. C. F., and Vinson, S. B. (1975). Notes on the male reproductive system in ants (Hymenoptera:Formicidae). *J. N. Y. Entomol. Soc.* **83**, 192–197.

Huxley, J. S. (1927). Further work on heterogonic growth. *Biol. Zentralbl.* **47**, 151–163.

Jackson, W. B. (1957). Microclimatic patterns in the army ant bivouac. *Ecology* **38**, 276–285.

Johnson, R. A. (1954). The behavior of birds attending army ant raids on Barro Colorado Island, Panama Canal Zone. *Proc. Linn. Soc. N.Y.* **63–65**, 41–70.

Kannowski, P. B. (1969). Daily and seasonal periodicities in the nuptial flights of Neotropical ants. I. Dorylinae. *Proc. Int. Congr. Int. Union Study Soc. Insects, 6th, 1969* pp. 77–83.

Keast, A. (1972). Continental drift and the evolution of the biota on southern continents. *In* "Evolution, Mammals, and Southern Continents" (A. Keast, F. C. Erk, and B. Glass, eds.), pp. 23–87. State University of New York Press, Albany.

Kistner, D. H. (1972). A new genus of the staphylinid tribe Dorylomimini from Africa and its possible significance to ant phylogeny. *Entomol. News* **83**, 85–91.

Kistner, D. H. (1976). The natural history of the myrmecophilous tribe Pygostenini (Coleoptera: Staphylinidae). Section 3. Behavior and food habits of the Pygostenini. *Sociobiology* **2**, 171–188.

Kistner, D. H., and Gotwald, W. H., Jr. (1982). Biological observations on African army ants of the genus *Dorylus* (Hymenoptera:Formicidae). (In preparation.)

Kruuk, H., and Sands, W. A. (1972). The aardwolf (*Proteles cristatus* Sparrman) 1783 as predator of termites. *East Afr. Wildl. J.* **10**, 211–227.

Kutter, H. (1948). Beitrag zur Kennthis der Leptanillinae. Eine neue Ameisengattung aus Süd-Indien. *Mitt. Schweiz. Entomol. Ges.* **21**, 286–295.

Lamborn, W. A. (1913–1914). Observations on the driver ants *(Dorylus)* of southern Nigeria. *Proc. R. Entomol. Soc. London* **5**, 123–129.

LaMon, B., and Topoff, H. (1981). Avoiding predation by army ants: defensive behaviours of three ant species of the genus *Camponotus*. *Anim. Behav.* **29**, 1070–1081.

Lappano, E. R. (1958). A morphological study of larval development in polymorphic all-worker broods of the army ant *Eciton burchelli*. *Insectes Soc.* **5**, 31–66.

Lenko, K. (1969). An army ant attacking the "guaiá" crab in Brazil. *Entomol. News* **80**, 6.

Lepeletier deSaint-Fargeau, A. L. M. (1836). "Histoire naturelle des insectes. Suites à Buffon. Hyménopteres. 1." Roret, Paris.

Leroux J. M. (1975). Recherches sur les nids et l'activité prédatrice des dorylines *Anomma nigricans* Illiger (Hym. Formicidae). Memoire présenté a L'Université Pierre et Marie Curie, Paris VI, pour l'obtention du Diplome d'Etudes Supérieures de Sciences Naturelles.

Leroux, J. M. (1977a). Densité des colonies et observations sur les nids de dorylines *Anomma nigricans* Illiger (Hym. Formicidae) dans la region de Lamto (Cote d'Ivoire). *Bull. Soc. Zool. Fr.* **102**, 51–62.

Leroux, J. M. (1977b). Formation et déroulement des raids de chasse d'*Anomma nigricans* Illiger (Hym. Formicidae) dans une savane de Cote d'Ivoire. *Bull. Soc. Zool. Fr.* **102**, 445–458.

Leston, D. (1973). The ant mosaic—tropical tree crops and the limiting of pests and disease. *PANS* **19**, 311–341.

Leston, D. (1979). Dispersal by male doryline ants in West Africa. *Psyche* **86**, 63–77.

Lindquist, A. W. (1942). Ants as predators of *Cochliomyia americana* C. & P. *J. Econ. Entomol.* **35**, 850–852.

Linńe, Caroli (1764). "Insecta & Conchilla." Museum Ludoviciae Ulricae Reginae. Literis & Impensis Direct. Laur. Salvii. 111 pp.

Linné, Caroli (1767). "Systema Naturae," 12th ed., Vol. I, Pars II, Classis V. Insecta.

Loveridge, A. (1922). Account of an invasion of "Siafu" or red driver-ants—*Dorylus (Anomma) nigricans* Illig. *Proc. R. Entomol. Soc. London* **5**, 33–46.

Loveridge, A. (1949). "Many Happy Days I've Squandered." The Scientific Book Club, London.

Loveridge, A. (1953). "I Drank the Zambesi." Harper, New York.

Luederwaldt, H. (1926). Observações biologicas sobre formigas Brasileiras, especialment do Estado de São Paulo. *Rev. Mus. Paul.* **14**, 184–304.

Lund, M. (1831). Lettre sur les Habitudes de quelques Fourmis du Brésil, adressée à M. Audouin. *Ann. Sci. Nat.* [1] **23**, 113–138.

McGrew, W. C. (1974). Tool use by wild chimpanzees in feeding upon driver ants. *J. Hum. Evol.* **3**, 501–508.

Masner, L. (1976). Notes on the ecitophilous diapriid genus *Mimopria* Holmgren (Hymenoptera:Proctotrupoidea, Diapriidae). *Can. Entomol.* **108**, 123–126.

Masner, L. (1977). A new genus of ecitophilous diapriid wasps from Arizona (Hymenoptera: Proctotrupoidea:Diapriidae). *Can. Entomol.* **109**, 33–36.

Mayr, G. (1886). Ueber *Eciton-Labidus*. *Wien. Entomol. Z.* **5**, 33–36, 115–122.

Menozzi, C. (1927). Zur Kenntnis des Weibchens von *Dorylus (Anomma) nigricans* var. *molesta* Gerst. (Hymenoptera-Formicidae). *Zool. Anz.* **70**, 263–266.

Menozzi, C. (1936). Nuovi contributi alla conoscenza della fauna delle isole Italiane dell-egeo. *Boll. Lab. Zool. Gen. Agran. Portici* **29**, 262–311.

Mirenda, J. T., and Topoff, H. (1980). Nomadic behavior of army ants in a desert-grassland habitat. *Behav. Ecol. Sociobiol.* **7**, 129–135.

Mirenda, J. T., Eakins, D. G., Gravelle, K., and Topoff, H. (1980). Predatory behavior and prey selection by army ants in a desert-grassland habitat. *Behav. Ecol. Sociobiol.* **7**, 119–127.

Mukerjee, D. (1926). Digestive and reproductive systems of the male ant *Dorylus labiatus* Shuck. *J. Asiat. Soc. Bengal* **22**, 87–91.

Mukerjee, D. (1933). On the anatomy of the worker of the ant *Dorylus (Alaopone) orientalis* Westw. *Zool. Anz.* **105**, 97–105.

Müller, W. (1886). Beobachtungen an Wanderameisen (*Eciton hamatum* Fabr.). *Kosmos* **1**, 81–93.

Murray-Brown, J. (1972). "Kenyatta." Dutton, New York.

Normand, H. (1931). Nouveaux Coléoptères du nord de l'Afrique. *Bull. Soc. Entomol. Fr.* **36**, 124–126.

Onoyama, K. (1976). A preliminary study of the ant fauna of Okinawa-Ken, with taxonomic notes (Japan;Hymenoptera:Formicidae). *Ecol. Stud. Nat. Cons. Ryukyu Isl.* **2**, 121–141.

Patrizi, S. (1946). Stomach contents of a female ant-eater from Nairobi. *J. East Afr. Nat. Hist. Soc.* **19**, 67–68.

Patrizi, S. (1948). Contribuzioni alla conoscenza delle Formiche e dei mirmecofili dell'Africa Orientale. IV. Descrizione di un nuovo genere e di una nuova specie di stapfilinide dorilofilo dello scioa e relative note etologiche (Coleoptera Staphilinidae). *Boll. Ist. Entomol. Bologna* **17**, 158–167.

Paulian, R. (1948). Observations sur les coléoptères commensaux d'*Anomma nigricans* en Cote d'Ivoire. *Ann. Sci. Nat., Zool. Biol. Anim.* [11] **10**, 79–102.

Perkins, G. A. (1869). The drivers. *Am. Nat.* **3**, 360–364.

Petersen, B. (1968). Some novelties in presumed males of Leptanillinae (Hym., Formicidae). *Entomol. Medd.* **36**, 577–598.

Petralia, R. S., and Vinson, S. B. (1979). Comparative anatomy of the ventral region of ant larvae, and its relation to feeding behavior. *Psyche* **86**, 375–394.

Plsek, R. W., Kroll, J. C., and Watkins, J. F., II (1969). Observations of carabid beetles, *Helluomorphoides texanus,* in columns of army ants and laboratory experiments on their behavior. *J. Kans. Entomol. Soc.* **42**, 452–456.

Pullen, B. E. (1963). Termitophagy, myrmecophagy, and the evolution of the Dorylinae (Hymenoptera, Formicidae). *Stud. Entomol.* **6**, 405–414.

Raignier, A. (1959). Het ontstaan van kolonies en koninginner bij de Afrikaanse trekmieren. (Formicidae, Dorylinae, Dorylini). *Meded. K. Acad. Wet., Lett. Schone Kunsten Belg., Kl. Wet.* **21**, 3–24.

Raignier, A. (1972). Sur l'origine des nouvelles sociétés des fourmis voyageuses africaines. *Insectes Soc.* **19**, 153–170.

Raignier, A., and van Boven, J. K. A. (1952). Quelques aspects nouveaux de la taxonomie et de la biologie des doryles africains (Hymenoptères, Formicidae). *Ann. Sci. Nat., Zool. Biol. Anim.* [11] **14**, 397–403.

Raignier, A., and van Boven, J. (1955). Etude taxonomique, biologique et biométrique des *Dorylus* du sous-genre *Anomma* (Hymenoptera Formicidae). *Ann. Mus. R. Congo Belg.* **2**, 1–359.

Raigner, A., van Boven, J., and Ceusters, R. (1974). Der Polymorphismus der afrikanischen Wanderameisen unter biometrischen und biologischen Gesichtspunkten. *In* "Socialpolymorphismus bei Insekten" (G. H. Schmidt, ed.), pp. 668–693. Wiss. Verlagsges., Stuttgart.

Ray, T. S., and Andrews, C. C. (1980). Antbutterflies: butterflies that follow army ants to feed on antbird droppings. *Science* **210**, 1147–1148.

Reichensperger, A. (1934). Beitrag zur Kenntnis von *Eciton lucanoides* Em. *Zool. Anz.* **106**, 240–245.

Reid, J. A. (1941). The thorax of the wingless and short-winged Hymenoptera. *Trans. Entomol. Soc. London* **91**, 367–446.

Rettenmeyer, C. W. (1961). Observations on the biology and taxonomy of flies found over swarm raids of army ants (Diptera: Tachinidae, Conopidae). *Univ. Kans. Sci. Bull.* **42**, 993–1066.

Rettenmeyer, C. W. (1962a). The diversity of arthropods found with Neotropical army ants and observations on the behavior of representative species. *Proc. North Cent. Branch Entomol. Soc. Am.* pp. 14–15.

Rettenmeyer, C. W. (1962b). The behavior of millipeds found with Neotropical army ants. *J. Kans. Entomol. Soc.* **35**, 377–384.

Rettenmeyer, C. W. (1963b). The behavior of Thysanura found with army ants. *Ann. Entomol. Soc. Am.* **56**, 170–174.

Rettenmeyer, C. W. (1963b). Behavioral studies of army ants. *Univ. Kans. Sci. Bull.* **44**, 281–465.

Rettenmeyer, C. W. (1974). Description of the queen and male with some biological notes on the army ant, *Eciton rapax. Mem. Conn. Entomol. Soc.* pp. 291–302.

Rettenmeyer, C. W., and Akre, R. D. (1968). Ectosymbiosis between phorid flies and army ants. *Ann. Entomol. Soc. Am.* **61**, 1317–1326.

Rettenmeyer, C. W., and Watkins, J. F., II. (1978). Polygyny and monogyny in army ants (Hymenoptera:Formicidae). *J. Kans. Entomol. Soc.* **51**, 581–591.

Rettenmeyer, C. W., Topoff, H., and Mirenda, J. (1978). Queen retinues of army ants. *Ann. Entomol. Soc. Am.* **71**, 519–528.

Richards, O. W. (1968). Sphaerocerid flies associating with doryline ants, collected by Dr. D. H. Kistner. *Trans. R. Entomol. Soc. London* **120**, 183–198.

Roonwal, M. L. (1975). Plant pest status of root-eating ant, *Dorylus orientalis,* with notes on taxonomy, distribution and habits (Insecta:Hymenoptera). *J. Bombay Nat. Hist. Soc.* **72**, 305–313.

Santschi, F. (1931). La reine du *Dorylus fulvus* Westw. *Société d'Histoire Naturelle de l'Afrique du Nord. Bull. Algiers* **22**, 401–408.

Santschi, F. (1933). Contribution à l'étude des fourmis de l'Afrique tropicale. *Bull. Ann. Soc. R. Entomol. Belg.* **73**, 95–108.

Savage, T. S. (1847). On the habits of the "drivers" or visiting ants of West Africa. *Trans. Entomol. Soc. London* **5**, 1–15.

Savage, T. S. (1849). The driver ants of West Africa. *Proc. Acad. Nat. Sci. Philadelphia* **4**, 195–200.

Schneirla, T. C. (1933). Studies on army ants in Panama. *J. Comp. Psychol.* **15**, 267–299.

Schneirla, T. C. (1934). Raiding and other outstanding phenomena in the behavior of army ants. *Proc. Natl. Acad. Sci. U.S.A.* **20**, 316–321.

Schneirla, T. C. (1938). A theory of army-ant behavior based upon the analysis of activities in a representative species. *J. Comp. Psychol.* **25**, 51–90.

Schneirla, T. C. (1940). Further studies on the army ant behavior pattern. Mass organization in the swarm-raiders. *J. Comp. Psychol.* **29**, 401–460.

Schneirla, T. C. (1944). The reproductive functions of the army-ant queen as pace-makers of the group behavior pattern. *J. N.Y. Entomol. Soc.* **52**, 153–192.

Schneirla, T. C. (1945). The army-ant behavior pattern: nomad-statary relations in the swarmers and the problem of migration. *Biol. Bull. (Woods Hole, Mass.)* **88**, 166–193.

Schneirla, T. C. (1949). Army-ant life and behavior under dry-season conditions. 3. The course of reproduction and colony behavior. *Bull. Am. Mus. Nat. Hist.* **94**, 1–81.

Schneirla, T. C. (1953). The army-ant queen: Keystone in a social system. *Bull. Union Int. Etude Insectes Soc.* **1**, 29–41.

Schneirla, T. C. (1956a). A preliminary survey of colony division and related processes in two species of terrestrial army ants. *Insectes Soc.* **3**, 49–69.

Schneirla, T. C. (1956b). "Annual Report of the Board of Regents of the Smithsonian Institution," Publ. No. 4232, pp. 379–406. Smithson. Inst. Press, Washington, D.C.

Schneirla, T. C. (1957). A comparison of species and genera in the ant subfamily Dorylinae with respect to functional pattern. *Insectes Soc.* **4**, 259–298.

250 William H. Gotwald, Jr.

Schneirla, T. C. (1958). The behavior and biology of certain Nearctic army ants. Last part of the functional season, southeastern Arizona. *Insectes Soc.* **5**, 215–255.

Schneirla, T. C. (1961). The behavior and biology of certain Nearctic doryline ants. Sexual broods and colony division in *Neivamyrmex nigrescens. Z. Tierpsychol.* **18**, 1–32.

Schneirla, T. C. (1963). The behaviour and biology of certain Nearctic army ants: Springtime resurgence of cyclic function—southeastern Arizona. *Anim. Behav.* **11**, 583–595.

Schneirla, T. C. (1971). "Army Ants: A Study in Social Organization." Freeman, San Francisco, California.

Schneirla, T. C., and Brown, R. Z. (1952). Sexual broods and the production of young queens in two species of army ants. *Zoologica (N. Y.)* **37**, 5–32.

Schneirla, T. C., and Reyes, A. Y. (1966). Raiding and related behaviour in two surface-adapted species of the Old World doryline ant, *Aenictus. Anim. Behav.* **14**, 132–148.

Schneirla, T. C., and Reyes, A. Y. (1969). Emigrations and related behaviour in two surface-adapted species of the Old-World doryline ant, *Aenictus. Anim. Behav.* **17**, 87–103.

Schneirla, T. C., Brown, R. Z., and Brown, F. C. (1954). The bivouac or temporary nest as an adaptive factor in certain terrestrial species of army ants. *Ecol. Monogr.* **24**, 269–296.

Schneirla, T. C., Gianutsos, R. R., and Pasternack, B. S. (1968). Comparative allometry in the larval brood of three army-ant genera, and differential growth as related to colony behavior. *Am. Nat.* **102**, 533–554.

Seevers, C. H. (1965). The systematics, evolution and zoogeography of staphylinid beetles associated with army ants (Coleoptera, Staphylinidae). *Fieldiana, Zool.* **47**, 137–351.

Shuckard, W. E. (1840). Monograph of the Dorylidae, a family of the Hymenoptera Heterogyna. *Ann. Nat. Hist.* **30**, 188–201, 258–271, 315–328.

Shyamalanath, S., and Forbes, J. (1980). Digestive system and associated organs in the adult and pupal male doryline ant *Aenictus gracilis* Emery (Hymenoptera: Formicidae). *J. N. Y. Entomol. Soc.* **88**, 15–28.

Smith, F. (1858). Catalogue of hymenopterous insects in the collection of the British Museum. *Formicidae* **6**, 1–216.

Smith, F. (1863). Observations on ants of Equatorial Africa. *Trans. R. Entomol. Soc. London* **1**, 470–473.

Smith, K. G. V. (1967). The biology and taxonomy of the genus *Stylogaster* Macquart, 1835 (Diptera: Conopidae, Stylogasterinae) in the Ethiopian and Malagasy regions. *Trans. R. Entomol. Soc. London* **119**, 47–69.

Smith, K. G. V. (1969). Further data on the oviposition by the genus *Stylogaster* Macquart (Diptera: Conopidae, Stylogasterinae) upon adult calyptrate Diptera associated with ants and animal dung. *Proc. R. Entomol. Soc. London* **44**, 35–37.

Smith, M. R. (1927). A contribution to the biology and distribution of one of the legionary ants, *Eciton schmitti* Emery. *Ann. Entomol. Soc. Am.* **20**, 401–404.

Smith, M. R. (1942). The legionary ants of the United States belonging to *Eciton* subgenus *Neivamyrmex* Borgmeier. *Am. Midl. Nat.* **27**, 537–590.

Snelling, R. R. (1981). Systematics of social Hymenoptera. *In* "Social Insects" (H. R. Hermann, ed.), Vol. II, pp. 369–453. Academic Press, New York.

Strickland, A. H. (1951). The entomology of swollen shoot of cacao. I. The insect species involved, with notes on their biology, *Bull. Entomol. Res.* **41**, 725–748.

Sudd, J. H. (1959). A note on the behaviour of *Aenictus* (Hym., Formicidae). *Entomol. Mon. Mag.* **95**, 262.

Sumichrast, F. (1868). Notes on the habits of certain species of Mexican Hymenoptera presented to the American Entomological Society. *Trans. Am. Entomol. Soc.* **2**, 39–44.

Swynnerton, C. F. M. (1915). Experiments on some carnivorous insects, especially the driver ant *Dorylus;* and with butterflies' eggs as prey. *Trans. R. Entomol. Soc. London* Parts III, IV, pp. 317–350.

Tafuri, J. F. (1955). Growth and polymorphism in the larva of the army ant (*Eciton* (E.) *hamatum* Fabricius). *J. N. Y. Entomol. Soc.* **63**, 21–41.

Teles da Silva, M. (1977a). Behaviour of the army ant *Eciton burchelli* Westwood (Hymenoptera: Formicidae) in the Belém Region. 1. Nomadic-Statary Cycles. *Anim. Behav.* **25**, 910–923.

Teles da Silva, M. (1977b). Behavior of the army ant *Eciton burchelli* Westwood (Hymenoptera: Formicidae) in the Belém Region. 2. Bivouacs. *Bol. Zool.* **2**, 107–128.

Topoff, H. R. (1969). A unique predatory association between carabid beetles of the genus *Helluomorphoides* and colonies of the army ant *Neivamyrmex nigrescens*. *Psyche* **76**, 375–381.

Topoff, H. (1971). Polymorphism in army ants related to division of labor and colony cyclic behavior. *Am. Nat.* **105**, 529–548.

Topoff, H. (1972). Theoretical issues concerning the evolution and development of behavior in social insects. *Am. Zool.* **12**, 385–394.

Topoff, H. (1975). Behavioral changes in the army ant *Neivamyrmex nigrescens* during the nomadic and statary phases. *J. N. Y. Entomol. Soc.* **33**, 38–48.

Topoff, H., and Lawson, K. (1979). Orientation of the army ant *Neivamyrmex nigrescens:* integration of chemical and tactile information. *Anim. Behav.* **27**, 429–433.

Topoff, H., and Mirenda, J. (1975). Trail-following by the army ant, *Neivamyrmex nigrescens:* Responses by workers to volatile odors. *Ann. Entomol. Soc. Am.* **68**, 1044–1046.

Topoff, H., and Mirenda, J. (1978). Precocial behaviour of callow workers of the army ant *Neivamyrmex nigrescens:* Importance of stimulation by adults during mass recruitment. *Anim. Behav.* **26**, 698–706.

Topoff, H., and Mirenda, J. (1980a). Army ants on the move: relation between food supply and emigration frequency. *Science* **207**, 1099–1100.

Topoff, H., and Mirenda, J. (1980b). Army ants do not eat and run; influence of food supply on emigration behaviour in *Neivamyrmex nigrescens*. *Anim. Behav.* **28**, 1040–1045.

Topoff, H., Boshes, M., and Trakimas, W. (1972a). A comparison of trail following between callow and adult workers of the army ant *Neivamyrmex nigrescens* (Formicidae: Dorylinae). *Anim. Behav.* **20**, 361–366.

Topoff, H., Lawson, K., and Richards, P. (1972b). Trail following and its development in the Neotropical army ant genus *Eciton* (Hymenoptera:Formicidae:Dorylinae). *Psyche* **79**, 357–364.

Topoff, H., Lawson, K., and Richards, P. (1973). Trail following in two species of the army ant genus *Eciton:* Comparison between major and intermediate-sized workers. *Ann. Entomol. Soc. Am.* **66**, 109–111.

Topoff, H., Mirenda, J., Droual, R., and Herrick, S. (1980a). Onset of the nomadic phase in the army ant *Neivamyrmex nigrescens* (Cresson) (Hym. Form.): distinguishing between callow and larval excitation by brood substitution. *Insectes Soc.* **27**, 175–179.

Topoff, H., Mirenda, J., Droual, R., and Herrick, S. (1980b). Behavioural ecology of mass recruitment in the army ant *Neivamyrmex nigrescens*. *Anim. Behav.* **28**, 779–789.

Torgerson, R. L., and Akre, R. D. (1969). Reproductive morphology and behavior of a thysanuran, *Trichatelura manni,* associated with army ants. *Ann. Entomol. Soc. Am.* **62**, 1367–1374.

Torgerson, R. L., and Akre, R. D. (1970a). The persistence of army ant chemical trails and their significance in the ecitonine-ecitophile association (Formicidae:Ecitonini). *Melanderia* **5**, 1–28.

Torgerson, R. L., and Akre, R. D. (1970b). Interspecific responses to trail and alarm pheromones by New World army ants. *J. Kans. Entomol. Soc.* **43**, 395–404.

Trimen, R. (1880). On a supposed female *Dorylus helvolus* (Linn). *Trans. Entomol. Soc. London* pp. 24–25.

Tulloch, G. S. (1935). Morphological studies of the thorax of the ant. *Entomol. Am.* **15,** 93–131.

van Boven, J. K. A. (1958). Allometrische en biometrische beschouwingen over het polymorfisme bij enkele mierensoorten. *Verh. K. Vlaam. Acad. Wet., Lett. Schone Kunsten Belg., Kl. Wet.* **56,** 1–134.

van Boven, J. K. A. (1961). Le polymorphisme dans la caste d'ouvrieres de la fourmi voyageuse: *Dorylus (Anomma) wilverthi* Emery (Hymenoptera: Formicidae). *Publ. Natuurhist. Genoot Limburg* **12,** 36–45.

van Boven, J. K. A. (1967). La femelle de *Dorylus fimbriatus* et *termitarius* (Hymenoptera: Formicidae). *Overdruk Natuurhist. Maandblad* **56,** 55–60.

van Boven, J. K. A. (1968). La reine de *Dorylus (Anomma) kohli* Wasmann (Hymenoptera: Formicidae). *Nat. Can. (Que.)* **95,** 731–739.

van Boven, J. K. A. (1972). Description de deux reines d'*Anomma* (Hymenoptera: Formicidae). *Bull. Ann. Soc. R. Belge Entomol.* **108,** 133–146.

van Boven, J. (1975). Deux nouvelles reines du genre *Dorylus* Fabricius (Hymenoptera, Formicidae). *Ann. Zool. (Warsaw)* **33,** 189–199.

van Boven, J. K. A. and Levieux, J. (1968). Les Dorylinae de la savane de Lamto (Hymenoptera: Formicidae). *Ann. Univ. Abidjan, Ser. E* **1,** 351–358.

van Lawick-Goodall, J. (1968). The behaviour of free-living chimpanzees in the Gombe Stream Reserve. *Anim. Behav. Monogr.* **1,** 161–311.

von Ihering, H. (1912). Biologie und Verbreitung der brasilianischen Arten von *Eciton*. *Entomol. Mitt.* **1,** 226–235.

Vosseler, J. (1905). Die Ostafrikanische Treiberameise (Siafu). *Pflanzer* **1,** 289–302.

Vowles, D. M. (1955). The structure and connexions of the corpora pedunculata in bees and ants. *Q. J. Microsc. Sci.* [N.S.] **96,** 239–255.

Wang, Y. J., and Happ, G. M. (1974). Larval development during the nomadic phase of a nearctic army ant, *Neivamyrmex nigrescens* (Cresson) (Hymenoptera: Formicidae). *Int. J. Insect Morphol. Embryol.* **3,** 73–86.

Wasmann, E. (1904). Zur Kenntniss der Gäste der Treiberameisen und ihrer Wirthe am obern Congo, nach den Sammlungen und Beobachtungen von P. Herm. Kohl C. ss. C. bearbeitet. *Zool. Jahrb., Suppl.* **7,** 611–682.

Wasmann, E. (1917). Neue Anpassungstypen bei Dorylinengästen Afrikas (Col., Staphylinidae). *Z. Wiss. Zool.* **117,** 257–360.

Watkins, J. F., II (1964). Laboratory experiments on the trail following of army ants of the genus *Neivamyrmex* (Formicidae:Dorylinae). *J. Kans. Entomol. Soc.* **37,** 22–28.

Watkins, J. F., II (1976). "The Identification and Distribution of New World Army Ants (Dorylinae:Formicidae)." The Markham Press Fund of Baylor Univ. Press, Waco, Texas.

Watkins, J. F., II (1977). The species and subspecies of *Nomamyrmex* (Dorylinae:Formicidae). *J. Kans. Entomol. Soc.* **50,** 203–214.

Watkins, J. F., II, and Cole T. W. (1966). The attraction of army ant workers to secretions of their queens. *Tex. J. Sci.* **18,** 254–265.

Watkins, J. F., II, and Rettenmeyer, C. W. (1967). Effects of army ant queens on longevity of their workers (Formicidae:Dorylinae). *Psyche* **74,** 228–233.

Watkins, J. F., II, Cole, T. W., and Baldridge, R. S. (1967a). Laboratory studies on interspecies trail following and trail preference of army ants (Dorylinae). *J. Kans. Entomol. Soc.* **40,** 146–151.

Watkins, J. F., II, Gehlbach, F. R., and Baldridge, R. S. (1967b). Ability of the blind snake, *Leptotyphlops dulcis,* to follow pheromone trails of army ants, *Neivamyrmex nigrescens* and *N. opacithorax. Southwest. Nat.* **12,** 455–462.

Watkins, J. F., II, Gehlbach, F. R., and Kroll, J. C. (1969). Attractant-repellent secretions in blind snakes (*Leptotyphlops dulcis*) and army ants *(Neivamyrmex nigrescens). Ecology* **50,** 1098–1102.

Watkins, J. F., II, Gehlbach, F. R., and Pisek, R. W. (1972). Behavior of blind snakes *(Lep-*

totyphlops dulcis) in response to army ant *(Neivamyrmex nigrescens)* raiding columns. *Tex. J. Sci.* **23**, 556–557.

Weber, N. A. (1941). The rediscovery of the queen of *Eciton (Labidus) coecum* Latr. (Hym.: Formicidae). *Am. Midl. Nat.* **26**, 325–329.

Weber, N. A. (1943). The ants of the Imatong Mountains, Anglo-Egyptian Sudan. *Bull. Mus. Comp. Zool.* **93**, 263–289.

Wellman, F. C. (1908). Notes on some Angolan insects of economic or pathologic importance. *Entomol. News* **19**, 224–230.

Werringloer, A. (1932). Die Sehorgane und Sehzentren der Dorylinen nebst Untersuchungen über die Facetenaugen der Formiciden. *Z. Wiss. Zool.* **141**, 432–524.

Wheeler, G. C. (1938). Are ant larvae apodous? *Psyche* **45**, 139–145.

Wheeler, G. C. (1943). The larvae of the army ants. *Ann. Entomol. Soc. Am.* **36**, 319–332.

Wheeler, G. C., and Wheeler, J. (1964). The ant larvae of the subfamily Dorylinae: supplement. *Proc. Entomol. Soc. Wash.* **66**, 129–137.

Wheeler, G. C., and Wheeler, J. (1974). Ant larvae of the subfamily Dorylinae: second supplement (Hymenoptera:Formicidae) *J. Kans. Entomol. Soc.* **47**, 166–172.

Wheeler, G. C., and Wheeler, J. (1976). Ant larvae: Review and synthesis. *Mem. Entomol. Soc. Wash.* **7**, 1–108.

Wheeler, W. M. (1900). The female of *Eciton sumichrasti* Norton, with some notes on the habits of Texan *Ecitons*. *Am. Nat.* **34**, 563–574.

Wheeler, W. M. (1902). An American *Cerapachys,* with remarks on the affinites of the Cerapachyinae. *Biol. Bull. (Woods Hole, Mass.)* **3**, 181–191.

Wheeler, W. M. (1910). "Ants: Their Structure, Development and Behavior." Columbia Univ. Press, New York.

Wheeler, W. M. (1914). The ants of the Baltic amber. *Schr. Phys. öekon. Ges. Konigsberg* **55**, 1–142.

Wheeler, W. M. (1915). On the presence and absence of cocoons among ants, the nest-spinning habits of the larvae and the significance of the black cocoons among certain Australian species. *Ann. Entomol. Soc. Am.* **8**, 323–342.

Wheeler, W. M. (1920). The subfamilies of Formicidae and other taxonomic notes. *Psyche* **27**, 46–55.

Wheeler, W. M. (1921). Observations on army ants in British Guiana. *Proc. Am. Acad. Arts. Sci.* **56**, 291–328.

Wheeler, W. M. (1922). Ants of the American Museum Congo Expedition. A contribution to the myrmecology of Africa. *Bull. Am. Mus. Nat. Hist.* **45**, 1–1139.

Wheeler, W. M. (1925). The finding of the queen of the army ant *Eciton hamatum* Fabricius. *Biol. Bull. (Woods Hole, Mass.)* **49**, 139–149.

Wheeler, W. M. (1928). "The Social Insects: Their Origin and Evolution." Harcourt, New York.

Wheeler, W. M. (1930). Philippine ants of the genus *Aenictus* with descriptions of the females of species. *J. N. Y. Entomol. Soc.* **38**, 193–212.

Wheeler, W. M. (1933). "Colony-Founding Among Ants." Harvard Univ. Press, Cambridge, Massachusetts.

Wheeler, W. M. (1936). Ecological relations of ponerine and other ants to termites. *Proc. Am. Acad. Arts Sci.* **71**, 159–243.

Wheeler, W. M., and Bailey, I. W. (1925). The feeding habits of pseudomyrmine and other ants. *Trans. Am. Philos. Soc.* **22**, 235–279.

Wheeler, W. M., and Long, W. H. (1901). The males of some Texas *Ecitons*. *Am. Nat.* **35**, 157–173.

Whelden, R. M. (1963). Anatomy of adult queen and workers of army ants *Eciton burchelli* Westw. and *E. hamatum* Fabr. (Hymenoptera:Formicidae). *J. N. Y. Entomol. Soc.* **71**, 14–30, 90–115,158–178, 246–261.

Willis, E. (1960). A study of the foraging behavior of two species of ant-tanagers. *Auk* **77**, 150–170.

Willis, E. O. (1966). The role of migrant birds at swarms of army ants. *Living Bird* **5**, 187–231.

Willis, E. O. (1967). The behavior of bicolored antbirds. *Univ. Calif., Berkeley, Publ. Zool.* **79**, 1–132.

Willis, E. O., and Oniki, Y. (1978). Birds and army ants. *Annu. Rev. Ecol. Syst.* **9**, 243–263.

Wilson, E. O. (1953). The origin and evolution of polymorphism in ants. *Q. Rev. Biol.* **28**, 136–156.

Wilson, E. O. (1958a). The beginnings of nomadic and group-predatory behavior in the ponerine ants. *Evolution* **12**, 24–31.

Wilson, E. O. (1958b). Observations on the behavior of the cerapachyine ants. *Insectes Soc.* **5**, 129–140.

Wilson, E. O. (1964). The true army ants of the Indo-Australian area. *Pac. Insects* **6**, 427–483.

Wilson, E. O. (1971). "The Insect Societies." Belknap Press, Cambridge, Massachusetts.

Wilson, E. O., Carpenter, F. M., and Brown, W. L., Jr. (1967a) The first Mesozoic ants, with the description of a new subfamily. *Psyche* **74**, 1–19.

Wilson, E. O., Carpenter, F. M., and Brown, W. L., Jr. (1967b). The first Mesozoic ants. *Science* **157**, 1038–1040.

Wroughton, R. C. (1892). Our ants. *J. Bombay Nat. Hist. Soc.* **7**, 13–60, 175–203.

4

Fungus Ants

NEAL A. WEBER

I. THE ANTS

A. Introduction and Definitions

This chapter emphasizes the important advances of the 1970s since the publication of a monograph by Weber (1972a) on the subject of fungus ants. Its extensive bibliography of 331 titles is, for the most part, not repeated here.

SOCIAL INSECTS, VOL. IV
Copyright © 1982 by Academic Press, Inc.
All rights of reproduction in any form reserved.
ISBN 0-12-342204-3

Instead extensive tables of growth of colonies and their gardens that illustrate quantitatively their ecological impact are presented.

All ants are in the family Formicidae and the major subfamily, Myrmicinae, includes the tribe Attini, the fungus ants. Fungus ants are also known as leaf-cutting, fungus-growing, fungus-culturing, and gardening ants depending on which aspect of their activities is emphasized. The attine tribe is usually placed at the apex of the Myrmicinae and is distinctive in morphology, as well as in behavior. They have not clearly been derived from any other ants. All members of the tribe depend for food on a fungus that they grow in a fungus garden. With the exception of one species and its close relatives, the garden is also the site for the queen and brood.

Leaf cutting refers to the fact that the workers of the higher genera (*Trachymyrmex, Sericomyrmex, Acromyrmex,* and *Atta*) cut green leaves for their fungus garden. However, *Trachymyrmex* and *Sericomyrmex,* are not primarily leaf cutters. Fungus growing refers to the fact that the fungus is tended and grown by the ants rather than growing freely. Fungus culturing refers to the complex behavior by which ants gather substrate (what the fungus grows on), bring it to the nest, cut it to size, and add salivary and rectal excretions. The ants pick up adjacent pieces of the fungus and plant it on the substrate. Gardening refers to the fact that these species were probably the world's first gardeners in the human sense, providing the fungus garden with meticulous and continuous attention.

Ants grow their fungus on a substrate, which is characteristically plant material or insect droppings. The material cast out by the ants is termed exhausted substrate, since it is the portion remaining following metabolic activity of the fungus on the substrate; it is the residue (*basuras,* in Spanish) of the gardens. Dead ants are added to the exhausted substrate. Soil is cast out by the ants in the form of craters to the ant mound and it is not mixed with the exhausted substrate.

Perhaps fewer ants arouse more human interest and often hostility then this group. They form perhaps the largest ant nests known, with workers numbering in the millions, cutting huge quantities of leaves, often from plants needed by man. They are particularly disliked by people who grow roses. In nature they are of fundamental ecological importance.

B. Appearance of the Ants—Origin, Distribution, and Differentiation of the Tribe

While they have the general characteristics of all ants, the attines are distinctive in several anatomical respects. As myrmicines they have two small segments (petiole and postpetiole) separating the thorax or mesosoma from the gaster, the latter often being called the abdomen of which it is the major part (Fig. 1A–D). Workers and females have antennae of 11 segments; the

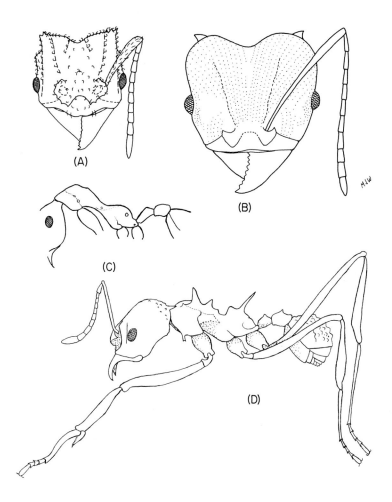

Fig. 1.   (A) Head of worker of *Trachymyrmex zeteki* Weber (Panama). (B) Head of worker of *Acromyrmex* (*Moellerius*) *striatus* (Roger) (Argentina). (C) Side view of worker of *Apterostigma peruvianum* Wheeler, showing nonspinous thorax typical of the genus (Peru). (D) Typical spiny worker attine in side view, *Acromyrmex* (*Moellerius*) *landolti* Forel (Guyana).

males have antennae of 13 segments. The mandibles are triangular with fine teeth. The worker eyes are moderately developed. Commonly, the head has one or more occipital spines or tubercles at each corner. The worker thorax in most genera is tubercular to spinous, often strikingly so, and usually there is a pair of spines, the epinotal, extending posteriorly, that may protect the petiole and postpetiole from being severed by an attacker. The gaster is roughly tuberculate in some genera and relatively smooth in others. The body has a covering of coarse hairs or a pilosity of finer hairs according to the genus. The surface of the body is variably punctate, being superficially smooth and shiny in some. The color is generally a reddish brown or ferruginous, varying from a dark chocolate brown in some species to a pale yellowish brown. The female has the general appearance of the worker but is larger, has a much larger thorax that bears wings, and has a triangle of ocelli on the frons posteriorly. The male is much darker in color, almost black. It has a small head with large compound eyes and its ocelli appear prominently on the constricted rear portion of the head (from feeble development of mandibular muscles that would be attached inside this area). The mandibles are more slender and weaker than in the worker or female. The male wings tend to be dusky brown with prominent darker spots; the female wings are pale and both have a stigma of feeble development (Fig. 2A–K).

The worker and female have a short, inconspicuous sting that may be hardly larger than terminal hairs (Hermann et al., 1970). The legs of workers, females, and males are long and slender, terminating in a pair of fine claws. The pecten or comb on the foreleg is prominent. Mouthparts are well developed and mobile; the glossa is often extended as a swab.

More detailed descriptions of the ants of the various genera are given in Section I, F.

The fungus ants are exclusively of American origin. In 1912, a Sumatran genus, Proatta, was erroneously attached to the Attini by Forel, but it has since been placed near Pheidole in Pheidolini. Santschi once identified a west African ant as a well-known American fungus grower, but it was clearly a case of incorrect locality.

The tribe has an exceptionally wide geographical range. From a probable original home in South America they have spread to 44° south latitude (Province of Chubut, Argentina) and 40° north latitude (New Jersey, formerly Long Island, New York, and Illinois). A recently unearthed tiny male of a primitive genus has been found at latitude 39°06′ in California, distinctly extending the northwestern range. No one genus is as widespread, covering 84° of the earth's surface; the genera Trachymyrmex and Acromyrmex extend from 39°S to 40°N and 44°S to 32°–33°N, respectively. One species, Trachymyrmex septentrionalis, was discovered on Long Island, New York (40°50′N) early in the 1900s and is presently found in the New Jersey Pine Barrens. It can be active in this area for only about 5 months (May–September) during most years and, rarely, 6 months in the warmest years. Although most species are found in lowland moist tropical forests, many of these areas are adjacent to forested mountains. Three species of Acromyrmex are recorded from Argentina at 3100 m, and one at 3500 m. The genus also extends to 2300–3000 m in Colombia, Ecuador, and Bolivia. The genus Atta does not

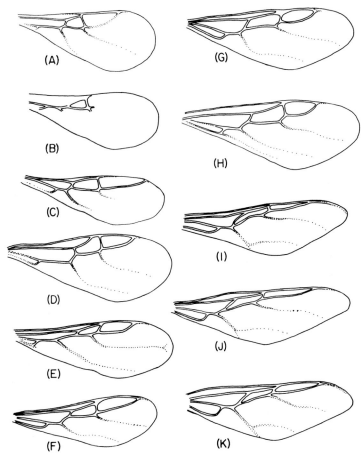

Fig. 2. Wings of attines. (A) *Cyphomyrmex rimosus fuscus* (curiapensis) (Venezuela). (B) *Cyphomyrmex costatus* Mann (Panama). (C) *Cyphomyrmex bigibbosus tumulus* Weber (Guyana). (D) *Myrmicocrypta occipitalis* Weber (Guyana). (E) *Apterostigma epinotale* Weber (Guyana). (F) *Trachymyrmex diversus* Mann (Guyana). (G) *Trachymyrmex ruthae* Weber (Trinidad). (H) *Trachymyrmex septentrionalis* (McCook) (New Jersey). (I) *Atta cephalotes* L. (Trinidad. (J) *Atta sexdens* L. (Guyana). (K) *Atta colombica tonsipes* Santschi (Panama).

extend as high, with the maximum elevation about 2000 m in Colombia and Mexico. Other genera are known from 1900 m elevation in Colombia.

Changes in the distribution of species in the twentieth century have been summarized by Weber (1972a). Local examples on the Isthmus of Panama were followed from 1938. Between 1938 and 1966 the common *Atta cephalotes isthmicola* and *Atta colombica tonsipes* of Barro Colorado Island, have fluctuated in their distribution at known locations along marked island paths. The latter species and *Atta sexdens* of the Pacific drainage have also fluctuated. *Atta sexdens* is a grassland species, while *A. tonsipes* is primarily found in the forest. At the time of the arrival of the Europeans, the Panamanian Indians had increased to perhaps one-million people and had reduced the forest by fires and agriculture. This must have permitted *A. sexdens* to occupy much of the Pacific slope of the Isthmus as the grasslands increased in size. With the rapid decrease in Indian population following the European conquest, the grasslands became partly overgrown with forest and *A. tonsipes* must have invaded the former *A. sexdens* territory. At the present time, grasslands are again becoming more extensive as human populations increase. In 1976, *Atta tonsipes* was still the common Canal Zone species but *A. isthmicola* as well as *A. tonsipes* occurred closely together at Ancon.

While ancient man was not a major factor in modifying the distribution of attine species, except in limited areas such as Panama, he has played an increasing role in the twentieth century. By cutting forests, he permitted the spread of grassland species and by creating small, local urbanized areas he created ecological niches for the same and other species. However, on a large scale the cutting of forests such as in São Paulo State, Brazil, has specifically fostered the spread of *Atta capiguara* and other *Atta* species. This alteration of plant habitats by man is rapidly accelerating and, in general will clearly reduce the number of rain forest species. The Amazon basin is highly vulnerable. Erosion that damages the soil surface over wide areas removes the essential nesting sites for most species. At the present time, in urbanized and cultivated lands, the ants are persevering, although precariously, along fence rows or other less disturbed islands of temporary refuge.

The areas occupied by other species of Attini will probably vary more from time to time than is generally thought to be the case. In addition to being influenced by humans, distribution will be affected by fluctuations in weather and interrelated factors such as changes in the survival of females or young colonies and changes in the substrate plants over a large area.

It is plausible that the ancestors of the attines collected seeds and other pieces of vegetation on which the adult and larval ants fed directly (Weber, 1958). Adventitious fungi growing on them or the middens may have been eaten indirectly and later directly. A shelter could then have been erected over the growths and certain fungi thrived and were nurtured by salivary and anal excretions applied by the ants. In the course of evolution the fungus lost the need to reproduce spores. The primitive fungus ants probably collected feces and pieces of vegetation before they acquired the habit of cutting living plants.

The hypothesis advanced (Weber, 1958) that attines could have evolved on the order of magnitude of 50 million years ago may or may not be true. However, the theory of plate tectonics suggests that if the attines are of some such comparable age their ancestors might be sought in Africa since they have no close relatives in North or South America. The progenitors of attines may have been present as the South American block drifted away from Africa and the tropical rain forests developed. Marshall *et al.* (1977) state that it is well established that South America was an island continent during most, if not all, of the Tertiary Period (65,000,000–2,000,000 years ago). As connections with Central America were established, waves of attines were permitted to migrate to North America.

The Pleistocene models of climatic and vegetational cycles (e.g., Meggus, 1975) suggest the means by which the grass-cutting attines and especially all attines of arid or subarid lands may have reached their present sites. Previously, many investigators considered that the entire Amazonian basin was a stable rain forest. It seemed to be the logical site for the evolution of major attines. During warm and arid periods of the Pleistocene, as now understood, there were scattered areas of forest in the Amazonian basin serving as refuges for animals. In late Pleistocene and post-Pleistocene periods the refuge for four families of woody plants were more widely scattered but were generally small. Zoologists generally agree that these localized islands existed. These refuges were probably more than sufficient to retain the present rain forest attine genera and species or their immediate antecedents. The extensive margins of these islands with presumed grasslands offered opportunities for the evolution of the grass-cutting species of the *Acromyrmex* subgenus *Moellerius* (on ecological grounds perhaps well considered to be a separate genus). Once adapted to using grass for substrate there were no barriers to their spread throughout South America and into Central America.

Parts of South America from sea level to 200 m elevation may well have been warm and humid enough to permit the ants and their fungi to survive when climatic changes were severe outside the tropics. The largest area of tropical South America which is at present no more than 200 m above sea level is the Amazonian basin, with considerable extensions north and south along the Atlantic coast. The Amazonian basin has connections with the Orinoco and Rio de la Plata basins below 500 m elevation. Extensive land connections extended into Argentina during both Lower and Upper Cretaceous.

When these ants migrated south into Argentina they developed greater resistance to lower temperatures than did those in North America. While moving south, the ants probably simultaneously adapted to the drier conditions of the grasslands, llanos, and pampas as well as to the colder winter temperatures of southern South America, reaching 44°S. This southern boundary has probably fluctuated markedly over the millenia and only small species which have adapted to semidesert conditions are currently found there. *Atta* does not now quite reach the Province of Buenos Aires, Argentina, but comes close at the present time at 33°S in neighboring areas, including Uruguay (Fig. 3), and may well be in the process of migrating south if present warm conditions continue in this hemisphere. From evidence based on their morphology and habits, the smaller species must have appeared much earlier and had time to make the necessary adjustments to temperate zone life. The ants have not been found anywhere in Chile. Evidently they could not invade the desert coast from the north nor could they cross the Andes from the east, where many species are found at the same latitude in Argentina. If introduced into Chile, there are habitats, especially gardens and plantations, which might be suitable for some attines.

From northern South America the ants now have routes which are at elevations 200 m or lower into and through Central America into Mexico. Through the millenia, the Isthmian connection has broken a number of times, and the resulting geographical isolation probably encouraged the development of the rich Central American fauna. The Peninsula of Yucatan is 200 m or lower and

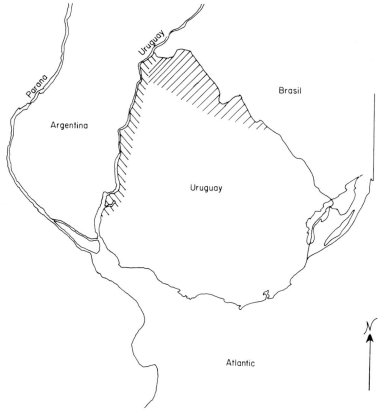

Fig. 3.   Outline of Uruguay showing distribution of *Atta vollenweideri* Forel (left) and *Atta sexdens piriventris* Santschi (right) (after Carbonell data).

it is connected to the United States along the Gulf of Mexico by a narrow zone of similar lowland. It may also have been the pathway to Cuba via a land bridge or by island hopping, where *Atta insularis* may have evolved from the ancestors of *A. texana* and *A. mexicana* (*sexdens?*). The west coast of Mexico is less favorable for migrations of tropical ants but does have a narrow belt of land under 200 m along the coast continuous with lower California. The ants invaded the United States presumably in several waves, but must have been exterminated along the upper East Coast during the Pleistocene.

C. General Ecological Role

In one sense, the fungus ants are agents of their fungi that feed on plant material as a major and vital group of saprophytes. Saprophytes feed on dead plant material. In another sense, the ants perform similarly to phytophagous insects in cutting and removing plant material, although they do not eat these plants directly, aside from imbibing some plant sap as they manipulate the cut leaf sections.

As leaf cutters, analogous to phytophagous insects, they are often condemned as destructive to the crops of man. This concept may be appropriate in the local view, but if carried to extremes may endanger the fundamental role the fungus ants have in natural ecosystems. Any species whose nests are as conspicuous and abundant as those of *Atta* must have evolved a basic role that is necessary to the natural turnover of plants.

The fundamental role of phytophagous insects as regulators of forest primary production has recently been reviewed by Mattson and Addy (1975). The principles may be applied to the attine ants that indirectly feed on the foliage. Mattson and Addy state, in part, that normal insect grazing (5–30% of annual foliage crops) usually does not impair annual plant (primary) production but rather may serve to accelerate growth. After an outbreak has subsided they point out that there is evidence that the residual vegetation is more productive than the vegetation that was growing immediately before the outbreak.

> For almost a century, research on phytophagous insects has focused primarily on aspects of their population biology and dynamics and their short-term impact on host plant growth and survival. Only recently has attention been directed at understanding and elucidating their long-term interactions with such fundamental ecosystem processes as primary production and nutrient cycling. (Mattson and Addy, 1975, p. 515)

Mattson and Addy emphasize in their summary that insect grazers function in a manner similar to cybernetic regulators of primary production in natural ecosystems; i.e., they ensure consistent and optimal output of plant production over the long term for a particular site. Their actions or activities appear to vary inversely with the vigor and productivity of the system. This inverse relation is probably a consequence of the long history of coevolution between plant systems and their usual consumers.

A similar conclusion was reached long ago by this author (Weber, 1976b) and still holds.

> Dicotyledonous plants of great variety may be attacked and, when the ants move into a cultivated area, the plants may be temporarily damaged. In some cases the ants may be considered to be one of several pruners to which the plants respond by putting out more foliage or flowers. The process is one of accelerating this phase of a fundamental ecological cycle, reduction of green plants to a form or forms used by other organisms. This ant phase markedly accelerates the green plant to the fungus saprophyte stage whereas plant-feeding insects interpolate one or more other stages. The specialized grass-cutting *Atta* and *Acromyrmex* species cause the same response that domestic cattle and other grass-feeders cause, a quick response from the leaf bases and fibrous root system to growing more grass. (Weber, 1976b, p. 828)

The tables which follow list the substrate, usually fresh green leaves (in grams), correlated with the corresonding growth of the fungus garden on a monthly and yearly basis for a number of laboratory colonies, duplicating the situation in nature. For example, a 10-year-old colony used 51 kg of substrate during its lifetime. Commonly a young colony of leaf cutters will use 3–6 kg of fresh leaves in its third year. The great numbers of incipient colonies started each year may use 1 kg of substrate in nature in the first year. Only a small fraction survive the first year, but in the meantime they have taken this organic material underground and left much of it there.

1. Strategies for Obtaining Substrate and Survival

Based on the geographic spread of attines and the variety of habitats invaded, it appears that attine strategies for obtaining substrate and for survival have developed over many thousands if not millions of years.

Basically attine strategy is to obtain and maintain the optimum environment for their fungus garden. The ants themselves have flexibility in meeting environmental change but their fungus does not. A brief exposure to dry air is sufficient to wither the hyphae, at least temporarily, and a quick method of killing an entire garden is to dry it.

An example of how the attine maintain ideal conditions for the fungus is given by the fecundated female who, before leaving the nest, takes fragments of the hyphae of the garden into her infrabuccal pocket. This normally is the only time in the life of the ant fungus that it leaves the safety of the humid nest. The pellet is not ejected until she has reached a safe, secluded, and humid spot, usually the cell in the damp earth that she excavates. The act of manuring it with her fecal droplets causes an early and vigorous growth for which constant moisture is vital. This behavioral pattern may induce the female to manure sand grains if she does not have this pellet (Weber, 1937).

Invasion of desert and semidesert areas is possible by timing the nuptial flight to night, when relative humidity is the highest, or immediately following rains. This strategy for survival thus involves critical timing both for reasons of humidity and for reducing the period of predation on the comparatively slow flying and nutritive female.

The maturation of the sexual brood occurs usually during the dry season. Males and females are then ready to leave the nest when correct external conditions have developed, such as a rise in temperature and humidity, following a rain. However, there is not a complete emptying of the nest of all mature sexual forms at one time. Frequently, this occurs over a period of days or even weeks. This has adaptive value since an entire brood may be destroyed by a local catastrophe. For example, a large group of *Acromyrmex lundi* flew over the broad Rio de la Plata, apparently from Argentina, and many were found the next day drowned or drowning on the opposite shore in Uruguay.

There are now a number of examples in the literature which indicate that the attine fecundated females return to the surface of the soil to forage for substrate. In *Acromyrmex lundi* they were noted to remove a plug of soil grains at the nest entrance, forage and return, and then stop up the entrance from below. This is a primitive trait among ants and illustrates the combination of primitive with specialized characteristics that contribute to their widespread success.

The first broods also show variability in the time each is active on the surface of the soil. The hours of opening and closing the nest entrance vary. Again, this flexibility in behavior has survival value in reducing the hazards to an entire local population of young colonies.

The choice of raw materials for the garden contributes greatly to the success of the attines. The small and primitive species choose the dried insect feces in the form of woody pellets of Coleoptera or the leafy pellets of Lepidoptera and Orthoptera. These are widespread on the soil surface and are easily picked up and carried to the garden.

However, the outstanding success of the attines is based on the exploitation by the larger attines of the most widely available and abundant organic material, green plants. By selecting this material and carrying it underground they provide an inexhaustible food source for the fungus combined with safety in the largely unused subsurface soil. In the natural tropical or subtropical forest this exploitation is not ultimately harmful to the plants. Were it not for animals, these plants would otherwise be choked by their own exuberant growth. The subsurface tropical soil is penetrated by relatively few animals and the roots of few large trees.

Table I compares minor attines with respect to colony size, nesting characteristics, fungus, substrate, and distribution. These species are representative of little known genera that have seldom been reported in the literature.

TABLE I

Comparison of Some Minor Attines

Species	Workers in colony	Nest surface	Nest	Fungus and garden	Substrate	Distribution	Reference
Cyphomyrmex rimosus	to 300	preexisting aperture in soil or wood	brood separate from garden in preexisting cavity	masses of yeast cells on garden	insect feces	United States to Argentina	Weber, 1955c
Cyphomyrmex costatus	ca. 100	soil crevices	soil cavity	typical mycelium	plant parts, insect feces, and carcasses	Panama, Colombia	Weber, 1957a
Mycetophylax conformis	several hundred	circular crater of sand grains	sand cavities	scanty mycelium	loose grass sections	South America	Weber, 1945, 1972a
Mycocepurus smithi	ca. 500	crater of fine soil grains	soil cavities	typical mycelium	plant parts	Caribbean, Panama	Weber, 1945
Myrmicocrypta buenzlii	to 1700	irregular crater of fine sand grains	soil cavities	typical mycelium	insect feces, plant parts	Trinidad	Weber, 1945, 1972a
Apterostigma mayri	ca. 100	preexisting aperture	soil or wood cavities	scanty mycelium	plant parts, insect feces	Central and South America	Weber, 1941, 1945
Apterostigma urichi	ca. 100	preexisting aperture	soil or humus cavity	scanty mycelium with veil	insect feces	Trinidad, Venezuela	Weber, 1941, 1945
Apterostigma dentigerum	100–200	preexisting aperture	soil cavities	scanty mycelium with veil	insect feces, plant parts	Central America	Weber, 1941
Sericomyrmex urichi	ca. 1800	circular to semicircular crater	soil cavities	mycelium	plant parts	Trinidad	Weber, 1972a
Trachymyrmex ruthae	ca. 300	collar of plant stems	humus cavities	mycelium	plant parts	Trinidad	Weber, 1945
Trachymyrmex septentrionalis	ca. 300	semicircular crater	soil cavities	mycelium	plant parts, insect feces	United States	Weber, 1966a,b,c
Trachymyrmex urichi	ca. 1000	turret and crater	soil cavities	mycelium	plant parts	South America (N), Panama	Weber, 1966b

However, they represent important stages in the evolution of attines and their fungus and have revealed significant information (cf. Weber, 1972a). A colony of the *Sericomyrmex urichi* was kept for nearly 11 years, during which 56,127 ants were produced (37,216 workers, 2585 females, and 16,321 males). A colony of *Myrmicocrypta buenzlii* had the unexpectedly high total of 1700 workers, which is exceptional for minor attines. Its cultured fungus produced sporophores identified as *Lepiota*.

2. Impact on Energy Flow in Tropical Forest and Grasslands

Over large areas of tropical and subtropical America, the attines are locally among the dominant animals, or may be codominant together with termites. The activities of large mammals except domestic species are not as important as they are in Africa, although peccary, deer, and armadillos may be widespread. These may be indirectly associated with *Atta* nests as shown in the section that follows.

The attine impact on energy flow may be divided into their effects on the soil and those that bear directly on the organic world.

a. Effects on the Soil. Both *Acromyrmex* and *Atta* overturn the soil in excavating their chambers. Soil is always brought up from below and cast out on the surface. On a slope in open or rocky areas this soil may be washed away; in level places and in tropical forests it may accumulate. The mound-building *Acromyrmex,* with variance according to the species, may use some soil from underground chambers to build up a soil base for the thatch of the mound or the soil may filter down the thatch to produce a mixed thatch–soil mound.

In *Atta vollenweideri* of the largely level Chaco of Paraguay and Argentina, the nest is distinctly paler than the surrounding soil (see Fig. 4). The surface soil has a higher pH and carbonate content than the deeper part. This deeper layer is pale, calcareous soil that weathers to near white and has a high salt content. The recent studies of Jonkman (1976, 1977), earlier evidence by Bucher and Zuccardi (1967), and personal communications by students of vertebrates (e.g., Brandt and Mayer, 1977) indicate that the nest surface concentrates salts. The ants tolerate this salt concentration and, in addition, the salts are attractive to armadillos, peccaries, deer, and possibly other vertebrates.

In all cases the soil of *Atta* nests is loosened by the ant activities; all types of nests absorb more rainfall than the surrounding soil, thereby decreasing the run-off of water (see Fig. 5). This adds to the water table and benefits surrounding plants. It would appear to be particularly desirable in mesophytic and xerophytic areas where most of a sudden shower is retained, instead of flowing away in creeks and rivers. It is doubtless a factor in the plant succession on *Atta vollenweideri* mounds.

Fig. 4. Mound nest of numerous small conic craters created by *Atta vollenweideri* (Argentina).

The aeration of the soil by nest building is also an ecological factor. In the heavy clay of rain forests, the *Atta* tunnels are one of the few passages (comparable to decaying tree roots) that admit air which promotes activity of other plants and animals.

Fig. 5. Characteristic site of a nest of *Atta laevigata* under bushes. The chambers are under the fused craters created by the ants (Venezuela).

b. Direct Effects on the Organic World.   Through the formation of their fungus gardens, the attines are perhaps the animal species chiefly responsible for introducing organic matter into the soil in tropical America. The tunnels and chambers that *Atta* constructs in the soil are numerous and extend deeper than those of vertebrate animals, with few if any exceptions. Only on the pampas of temperate Argentina and Uruguay does the tucotuco (*Ctenomys*) mammal rival *Atta* and *Acromyrmex* in the extent of its burrows, and these do not have the concentration of organic matter underground that is contained in a mature *Atta* nest. The armadillos may form larger tunnels in the soil or in *Atta* nests.

The underground ant nests are somewhat comparable in bulk to the large termite mounds above ground in Africa (*Bellicositermes* or *Macrotermes*) and may have a comparable role with respect to organic matter. Termites (*Amitermes*) that construct mounds as high as 3 m coexist with the leaf-cutting ants in grasslands of the Americas, and together form the chief animal component of the ecology of such regions. The excavations have the marked physical effect of aerating and draining the soil, and a host of other animals follow the ants and termites for shelter or for food.

Where the ants cast their refuse to the surface there is an increased superficial root development (Haines, 1973). This refuse has a higher nitrogenous component than plant detritus. When the refuse is processed through an ecological chain, the result is an increase in root structures of various plants. When the refuse is underground, the improvement in plant development is less obvious, but in the case of *Atta vollenweideri,* this may be a major cause of the rapid development of cacti, palms, and dicots at the periphery of the dying nests. However, Jonkman (1976, 1977) found that there was also a physical factor: the center of the old nest collapses, thereby creating a pool in wet seasons. He found that these nests in some areas were responsible for the change from grasslands through the creation of these "woods-nuclei." As the woody vegetation spreads from the old nest sites, the former grasslands are altered to another stage in the formation of woodlands.

3. Impact on Energy Flow in Grasslands

The impact on energy flow in grasslands has led to exaggerated claims of damage by attines because they appear to be in competition with domesticated grazing mammals. The specialized grass-cutting *Acromyrmex* and *Atta* species do in fact use the same species of grass as the mammals, but the fundamental character of grass is often overlooked. The grass responds to the increased fibrous root systems and pruning of leaves with an increase in growth in these mild to tropical climates. In a small fenced-in enclosure with comparatively large numbers of mammals, grazing of all species would, of course, be destructive. However, in South America the cattle ranges tend to be large, and ample grass is available for both mammals and ants.

The studies of Jonkman (1976, 1977) in Paraguay on the grass-cutting *Atta vollenweideri* show that previous studies of this species elsewhere have led to overestimation of the quantities of grass taken. Statistical methods of analysis were shown to be in error.

All the species of *Acromyrmex* and *Atta* found in grasslands areas are not necessarily exclusive grass cutters. Grasslands have parklike areas of patches of trees and bushes, especially along watercourses. Coexisting with grass is a wide variety of herbs. A listing of plants in a typical Venezuelan grassland (Weber, 1947) by the ecologist J. G. Myers, (in the company of this author), was as follows (see tabulation below):

Grasses
    Trachypogon plumosus (wooly variant)
    (*Cymbopogon rufus?*) (edge of road)
    Andropogon condensatus
    Axonopus aureus
    Aristida sp.
    Paspalum anceps (shade of *Curatella*)
Other herbs
    Polygala sp.
    Paepalanthus capillaceus
Woody plants
    Curatella americana
    Bowdichia virgilioides
    B. coccolobaefolia
    Byrsonima crassifolia
    Casearia carpinifolia (in shade of *Curatella*)

The dominant attine here was *Acromyrmex* (*Moellerius*) *landolti*. This widespread ant occupies the interior grasslands of much of South America north of Argentina and is currently adapting to the spread by human agency of western Venezuelan grassland (J. R. Labrador and J. Martinez, unpublished). Because of the need for more grassland for cattle raising, the forests in the 1970s have been cut down and an imported Guinea grass (*Panicum maximum*) has been planted. The *Ac. landolti* invaded this new area in great numbers in the 1970s, forming an enormous number of colonies in 4 or 5 years. There were few competitors for the new grass except for cattle. The lushness of the grass may also make it easier for the nuptial flights to be successful; the females find ready shelter in the dense grass where predators would be few and where the soil remains moist.

4. Some Special Atta Situations

The species of *Atta*, being prominent wherever they occur, have been the subject of special studies treated in this section. The paradigms, *A. cephalotes* and *A. sexdens*, are treated in Section I,E.

5. Growth Rate of Atta vollenweideri Nests

The Jonkman (1976, 1977) studies of individual nests of *Atta vollenweideri* in Paraguay showed that the growth of young nests during 3 years followed a normal-shaped curve. Growth slows in the winter and accelerates in the summer. Most of the nest openings are closed in the winter.

The mature nests went deeper than Argentina investigators had revealed. The general structure was a dome or hemispheric area containing hundreds of chambers with fungus gardens, below which were much larger chambers used for refuse storage (exhausted substrate). At the periphery were large, deep, empty chambers. Above the dome was the upper surface area of the mound, penetrated by numerous tunnels leading to the active chambers. The refuse chambers appeared to be constructed long before they were to be used.

The deepest nests were those in low water table areas. Flooding is prevented by the position of the garden chambers. The tunnels do not extend to the water table. In this species, the maturity phase, when the sexual brood is first developed, is in the 5- to 6-year-old nest, compared with 3-year-old nests in *sexdens* (Autuori). The sexuals leave in December–January.

The nests of 3.5–4 years of age had only males, while older nests contained females. Nuptial flights were not observed but may take place in November–December after a rain.

The youngest nests observed were estimated to be about 6 months old. They were surmounted by discs 30–50 cm in diameter and about 10 cm high. One nest was observed for 2.5 years. Based on this nest scale it was determined that nests of this age and older could be detected in aerial photographs from a height of 3280 m (10,000 ft).

In western São Paulo State, Brazil, the rolling grass to sparse brush or tree-covered region or cerrado has three species in abundance, *Atta laevigata, Atta capiguara,* and *Atta sexdens rubropilosa*. The former two live on the open grassland, the latter under trees, including planted *Eucalyptus* groves. The *A. capiguara* specializes on grass, the *A. laevigata* cuts some grass but also more miscellaneous herbs and bushes, and *A. sexdens rubropilosa* harvests leaves of trees and is considered to be a major defoliator of the *Eucalyptus*. Orange or lemon trees planted by man on the grasslands close to *A. capiguara* and *A. laevigata* nests are ignored by these ants, but *A. rubropilosa* ants will form trails from some distance to defoliate them.

6. The History of Atta capiguara

The history of *Atta capiguara* is an excellent story of changing distribution with respect to man (Gonçalves, 1973 *in litt.*). It was described by Gonçalves in 1944. In 1941, *Atta* occupied a limited savanna area south of the city of São Paulo. These areas were separated by continuous forests from the north and west of this state. Later, the forests were cut down and crops were planted, particularly coffee. When the coffee ceased to be sufficiently productive, it was removed and other crops were planted; some land was converted into pastures for grazing cattle. *Atta* appeared as a pasture problem in 1963. Gonçalves felt that the almost continuous new pasture areas permitted the ant to spread and is quite sure that *Atta* was not a pest before 1954. Some time between 1941 and 1964 it appeared in the south of the state at latitude 22°55′S, longitude 48°25′W where it was common in 1973. It has been the subject of intensive surveys by Amante (1967a,b,c, 1968). Personal observations by this author with Amante in 1973 demonstrated that this species was keeping pace with the deforestation and resulting grasslands, scattered tree stumps still remaining to show the original forested nature. It has since been found to be abundant in a small area of Paraguay (as noted further on).

This species may harvest large quantities of grass that may be dumped along trails or at the entrances. Tropical species may also perform in this manner, e.g., when they are caught in a shower while carrying leaves.

7. Three Atta Species in Panama

Three *Atta* species are found in Panama and have been noted since 1938 in Panama City and in the Panama Canal Zone (Weber, 1969b). One, *Atta colombica tonsipes,* is found across the Canal Zone from the Atlantic (Caribbean) to the Pacific coasts. It and *Atta cephalotes isthmicola* are found on Barro Colorado Island in Gatun Lake, the latter tending to occupy the hill or mountain tops in dense forest here and in adjacent areas. The *A. colombica tonsipes* is a general lowland species that adapts well to man and was first described in 1929 from grassland as well as forest in Panama. The third species, *A. sexdens,* came up from the grasslands of South America to the Pacific slopes of Panama and also occupies grasslands here. It was dominant in the main plaza (Plaza de Lesseps) of Panama City in 1938, and continued to be abundant there through the mid-1950s, when a major building program extinguished most of its nesting and foraging area. By 1971 it had been replaced by a struggling colony of *A. colombica tonsipes* and in 1973 only a small colony of *A. sexdens* was found; *A. sexdens,* however, remained an urbanized species in parts of the city in 1976. Immediately across the Republic of Panama–Canal Zone boundary at the foot of Ancon Hill and at the base of the former Tivoli Hotel a colony of *A. cephalotes isthmicola* existed in 1973 and within 1 km in 1976 a younger colony was observed. Clearly all three species have withstood the encroachment of man for 450 years in this urbanized area. A fourth attine, *Acromyrmex octospinosus,* has had equal or better success here, its colonies being smaller and less conspicuous.

8. Ecological Niches and Condensed Cycles

The microhabitats or ecological niches of the minor attines vary according to the species and even the genera.

Within the tropical rain forest, where most species live, a species may be found nesting in rotted wood lying on the ground or on top of a fallen tree, in the soil or humus beneath, or in accumulations of debris above ground between the roots or leaves of ephiphytes. A flourishing colony of *Acromyrmex octospinosus* was found nesting in the crown of a palm tree in the Botanical Gardens in Trinidad (Weber, 1945).

Each species and each colony of the same species is an independent unit that is not in competition with its neighbor, in most cases, because the substrate for the fungus is always in abundant supply and there is ample space for innumerable small colonies. Three species of *Apterostigma* (*urichi, dorotheae,* and *epinotale*) were taken in the same rotten stump within 45 cm of each other in Guyana (Weber, 1946). A dense concentration of small nests of *Mycocepurus smithi, Myrmicocrypta buenzlii, Sericomyrmex urichi,* and *Trachymyrmex urichi* occurred under a row of trees in Trinidad (Weber, 1972a). All were in the clay soil and under the shade of three trees. In an area of 81 m² there were 51 colonies or 1 nest per 1.6 m². An adjacent area of 20.25 m² had 36 nests or 1 nest per 0.56 m². In these cases the density was such as to suggest that many of the colonies were young and would not reach maturity. When colonies were young there was little or no competition for substrate, but as the colonies expanded, space might be in short supply.

Other species may occupy special niches, such as a snail shell for *Cyphomyrmex rimosus* in Cuba, or among bromeliad roots attached to the stems of a cactus forest (*Cereus*) on Patos Island, Venezuela, or in the crevices of limestone rocks at Key West, Florida. This species was taken at a height of 30 m (92 ft) in a *Cassia* tree that was 64 m (195 ft) high in Guyana (Weber, 1946).

In general, the tropical forest with its diverse epiphytic growths furnishes a great variety of niches for small species. In mesophytic or grasslands areas there are far fewer sites and the subsurface soil is usually the only location available for most species. In all locations, suitable sites may be under partly buried stones or in the crevices between rocks.

The general ecological role (see Scheme I) of the attines may be summarized in two cycles, one emphasizing the botanical side and the other including, as well, the animals involved.

D. Historical Development of Knowledge

The Indians of tropical America had a sharp awareness of attines as attested by Guatemalan and other tales. That they are keen observers to the present day was illustrated by an Arawak Indian assistant in British Guiana (now Guyana) in 1935 who identified a number of species and applied the name "kuyamaru" (phonetic) to *Acromyrmex*. The same name was applied by Indians of the Brazilian border of Guyana in 1970 to *Atta* as well and probably signifies carrying ants (N. A. Weber, unpublished). The aforementioned

Ecological cycle of attines emphasizing plants

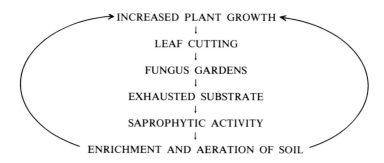

INCREASED PLANT GROWTH
↓
LEAF CUTTING
↓
FUNGUS GARDENS
↓
EXHAUSTED SUBSTRATE
↓
SAPROPHYTIC ACTIVITY
↓
ENRICHMENT AND AERATION OF SOIL

Ecological cycle of attines including plants and animals

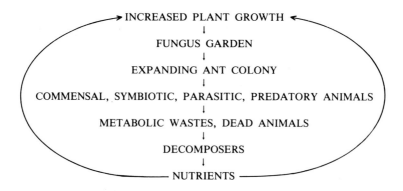

INCREASED PLANT GROWTH
↓
FUNGUS GARDEN
↓
EXPANDING ANT COLONY
↓
COMMENSAL, SYMBIOTIC, PARASITIC, PREDATORY ANIMALS
↓
METABOLIC WASTES, DEAD ANIMALS
↓
DECOMPOSERS
↓
NUTRIENTS

Increased plant growth leads to more leaf-cutting and more fungus garden and the cycle is thus repeated.

Scheme I

Arawak assistant named numerous other species of ants and one genus, *Sericomyrmex harekulli* n. sp. (Weber, 1937, 1946), records the Arawak term. A long list of vernacular names in the different Latin American countries (Weber, 1937, 1946, 1972a) for *Atta* and *Acromyrmex* shows that the Indian names were adopted by the subsequent Europeans.

Almost as soon as the Spanish arrived in the New World they made note of the depredations of ants, which probably were *Acromyrmex* and *Atta*. In 1559, Bartolome de las Casas described the failure of the Spaniards in Hispaniola to grow cassava and citrus trees because of ants whose nests, at the bases of the trees, were "white as snow" (probably the fungus gardens and brood).

Depredations of ants, including the attines were mentioned in seventeenth century accounts. P. Bernabe Cobe, in his *Historia de Neuvo Mundo* (1655), described effects of ants in Santo Domingo, giving the name of one ant, Iczan, strangely like the common Isau name for *Atta* queens in southern South America. However, *Atta* is not known to occur in Santo Domingo. Other accounts of the sixteenth and seventeenth centuries clearly refer to leaf cutters on citrus, introduced into the New World by Columbus, and on other plants. Specimens of *Atta* were brought back to Europe in the eighteenth century and listed by Linnaeus in 1758.

Article 19 of the cedula proclaimed by the King of Spain on November 20, 1783, for opening Trinidad to immigration states that "the Government was to take the utmost care to prevent the introduction of ants into Trinidad," recognizing their impact on agriculture.

The nineteenth century opened with other observations. Felix de Azara in *Viajes por America Meridional* (1809) wrote of the use of *Atta* females as food in Argentina.

Credit for the first more than casual account of the habits of these ants in British Guiana (Guyana) should probably go to Richard Schomburgk who, through the assistance of Alexander von Humboldt, secured the support of King Frederick William IV of Prussia in investigating the natural resources of the colony in the years 1840–1844. His brother, Robert Schomburgk, had explored this area in 1838–1839 under the sponsorship of Queen Victoria of England and the Royal Geographical Society of London. Earlier travelers, especially of ecclesiastical orders, traversed portions of this area intermittently following the discovery of the New World and must have observed these conspicuous ants. W. E. Roth translated the two-volume account of Richard Schomburgk and quotations have been published (Weber, 1946), especially on *Atta cephalotes* observed in an area draining to the Orinoco Delta.

The middle and late nineteenth century was marked by widespread observations, including those of Buckley (1860a,b) and McCook (1880) on *Atta texana* and *Trachymyrmex septentrionalis*. H. W. Bates (1863) on the Amazon added more regarding *Atta*.

Two outstanding students of these ants came in this period. One was Thomas Belt, in Nicaragua, and the other, A. Möller, was especially interested in ant fungi in South Brazil. Belt's 1874 study was the first to show correctly the role of the ant fungus. He had gone to Central America as a mining engineer and became fascinated by the habits of *Atta cephalotes*. He excavated a large nest uncovered by a mine shaft and came to the conclusion that the ants formed gardens by chewing up the leaves and permitting a fungus to grow on them, thus being actually fungus growers. This discovery was widely disseminated in Europe and had a marked effect on other naturalists, especially Möller. The late nineteenth century was marked by many observations by Brazilians, including Von Ihering, Sampaio and Huber, and Urich of Trinidad among others.

The foundations of attine systematics were laid by Auguste Forel of Switzerland and Carlo Emery of Italy in the latter part of the nineteenth century. Forel and Felix Santschi made a journey to Colombia in 1896, and the latter also made many contributions to systematics. These three identified the ants that were studied by South American investigators for many years.

One of the most important contributions to the investigations of ants was made by W. M. Wheeler. He was a master of the literature and of the organization of knowledge of all ants. His 1907 study, *The Fungus-Growing Ants of North America,* gave an excellent survey of the attines to that date. His own studies of the four major United States species [*Cyphomyrmex rimosus, Trachymyrmex septentrionalis, Acromyrmex (Moellerius) versicolor,* and *Atta texana*] were aided by three of his students (C. G. Hartmann, A. L. Melander, and C. T. Brues) and for decades were considered an important contribution. He also reviewed Möller's ant fungus work. Much of this information was included in his famous ant book (1st ed., 1910) that had a large distribution. A 1925 study dealt with Panamanian species. Oddly enough he never did study the biology of the tropical *Atta* to any great extent, perhaps because he did not have assistants comparable to the early youthful Hartmann, Melander, and Brues, who became authorities in other fields, nor had he the necessary time to spend in the field. Perhaps for this reason he and Thomas Barbour sponsored this author's research for the summer of 1933 in Cuba and a year in Trinidad (1934–35) during and following graduate work at Harvard University as Wheeler's last student. The latter study led to Wheeler's final book (1937) that was originally intended as a joint work with this author. The book, *Mosaics and Other Anomalies among Ants,* dealt principally with an unprecedented anomalous colony of *Acromyrmex octospinosus* that this author discovered in Trinidad. The manuscript occupied much of Wheeler's time and effort until the day he died (April 19, 1937). A good account of his influence on myrmecology appears in Evans and Evans (1970), *William Morton Wheeler, Biologist.*

E. The Paradigms, *Atta cephalotes* and *Atta sexdens*

The paradigms, *Atta cephalotes* and *A. sexdens,* were the first attines to be clearly recognized because of the unusual soldier. Bates (1863), *Naturalist on the River Amazon* contains interesting notes on both *A. cephalotes* and *A. sexdens* and it was *A. cephalotes* that was the subject of Belt's perceptive observations after he arrived in Nicaragua on February 15, 1865.

Later in the nineteenth century, Brazilian investigators initiated the modern studies, as noted in historical accounts. They dealt especially with the common subspecies of *A. sexdens* (*rubropilosa* Forel) that was found in the more southern parts of the country. Von Ihering in 1882–1888 published important observations on both species and on colony founding (von Ihering, 1882). The work of Sampaio De Azevedo (1894) and Goeldi (1905a) followed. The results of these studies were fully verified by subsequent investigators. Möller's (1893) pioneer work on ant fungi dealt with *Acromyrmex* and lesser genera rather than with *Atta*, as the generic term has been used since his day. Brent (1886) described *A. cephalotes* in Trinidad. Wheeler (1907) gave wide dissemination to those accounts published by 1907 and the United States Smithsonian Institution translated Huber's work (1905) on *Atta sexdens* colony founding into English in 1907.

The first half of the twentieth century was marked by the studies of other investigators in South America, such as those by Eidmann (1935) and Stahel and Geijskes (1939). Especially important are the works on *A. sexdens rubropilosa* of Autuori (1941–1950). His work at the Instituto Biologico, São Paulo, Brazil was carried on by Elpidio Amante, who dealt with this and other *Atta* species, especially *A. capiguara*. The decades of the 1960s and 1970s contain special studies on *A. cephalotes* by various students. Those by Weber span the period 1937–1977, based on field work in various countries since 1934 and dealt with the ant fungi as well as with other aspects of ant biology. Those by Martin and associates introduced the vital biochemical aspects.

Other species of *Atta*, such as *A. texana* (Moser), *A. capiguara* (Amante), and *A. vollenweideri* (Jonkman) are considered elsewhere. In this section the accounts generally refer to *A. cephalotes* and *A. sexdens* unless a specific species is indicated.

1. A Comparison of the Paradigms

Both leaf cutters have an especially broad distribution in South America and both range into Central America but *A. cephalotes* ranges into Mexico and *A. sexdens* extends only to Costa Rica. Both are widespread in Brazil and their distribution has been mapped by Gonçalves (1960) and Amante (1972). The subspecies *A. sexdens rubropilosa* is found in the more southern parts. This author's first experiences with *A. cephalotes* were during 1934–1935 in Trinidad and with both species in Venezuela and Guyana in 1935. In the 1960s and 1970s many younger investigators specifically dealt with *A. cephalotes*. Their studies in different countries created a need for a careful taxonomic study of the possible taxa. Two described subspecies, *A. cephalotes opaca* Forel of Colombia and *A. cephalotes isthmicola* Weber, appear to be morphologically and geographically distinctive. The latter from Panama has sometimes been kept alive in the laboratory side by side with typical *A.*

cephalotes of Trinidad and the two are easily separable. They are as antagonistic to one another as are two full species. The 1950 taxonomic review of Borgmeier that synomomized the two was based apparently on limited museum material then available to him and thus the entire genus should be reexamined. There appear to be other undescribed related taxa in Central America. The *A. cephalotes lutea* Forel 1893 of Barbados is an extinct form of this or *A. colombica*.

The two occupy different ecological niches. That of *A. cephalotes* is characteristically the interior of tropical forests while that of *A. sexdens* is the forest–grasslands ecotone or out on the grasslands.

Where *A. cephalotes* and *A. sexdens* occur closely together, as in Venezuela, Panama, and the Guianas, colony size and activities have been shown to be comparable (e.g., Stahel and Geijskes, 1939; Weber, 1946, 1947, 1969). Along the upper Courantyne basin between Surinam and the former British Guiana only *A. sexdens* was found in 1936, but *A. cephalotes* occurred at the mouth. At the junction of the Mazaruni and Cuyuni Rivers in Guyana, about 300 km from the Courantyne, the two species occur closely together as they do at the head of the Orinoco Delta in Venezuela. This general region of Guyana may be the area from which the Linnean specimens came over 200 years ago.

At the Cuyuni–Mazaruni junction, *A. cephalotes* nested in clay soil while a few meters away *A. sexdens* occurred in sand. This is not always a rigorous choice in nesting sites. The white sand ridges of forested Guyana have *A. sexdens* and clay elsewhere is the common soil of forests. In Panama there is a similar distribution of the two species, sandy soil often being the soil of the Pacific drainage, where *A. sexdens* is confined. The *A. cephalotes* distribution in Trinidad and adjacent countries includes swamps, the ants adapting by having especially shallow gardens. Young colonies of *A. sexdens* have been described by Bazire-Benazet (1974), among others.

2. *Atta cephalotes*

Colonies of *A. cephalotes* are started independently by a single female. Before the nuptial flight the winged females acquire in their infrabuccal pocket a pellet from the parental nest containing hyphae from the fungus garden.

Following the flight the fecundated female descends to the ground at a point probably determined by the strength of her wings, air currents, rain, natural barriers, or other factors largely of chance. She may also be attracted to forested areas resembling the one she left. She then loses her wings at a predetermined suture and digs into the soil. The odds, of course, are greatly against her and many ants fall prey to other animals or fail to land on suitable soil.

Virgin females taken from their nests and kept with parts of their colonies retain their wings indefinitely. When such a single female is taken from the nest, and placed in a small container with soil (after her wings are first cut off with scissors), she immediately starts making a cell. If she is provided with only a shallow layer of moist and coarse sand she uses this to wall up the sides and ceiling of the container, thus cutting out the light as well as providing a chamber where the humidity would be uniform throughout. If more suitable and deeper soil is provided she digs a tunnel. In one instance, two such artificially dealated virgin females were placed in the same container. No animosity was shown to exist. By the next morning they were found to have excavated a narrow tunnel several centimeters deep and were busily engaged in deepening it. There being room in the tunnel enough for but one at a time, they were found taking turns, one waiting for the other to come out with a load of sand and then immediately darting in before the other could deposit the load at the periphery of the chamber and return. They actually jostled one another in their "eagerness" to excavate. For 8 days they occupied the tunnel together but on the morning of day 9, one was found, dead but undamaged, firmly packed in sand at the end of the tunnel (Weber, 1937).

Several times virgin females, taken from the parental nest before their normal time for emergence and without a hyphal pellet, attempted to develop a fungus garden by gathering a cluster of sand grains and manuring them.

The depth of the initial chamber in the soil is a few centimeters. The results of measurements are presented in Table II. Time is a vital parameter here since the first requirement is to get out of sight quickly; the next is to excavate only as deeply as the first chamber will protect the first garden. Moist soil is imperative, so that during a brief dry spell her garden will not dry out. The first

TABLE II
Depth of the Original *Atta* Chamber Made by the Queen

Species	Soil	Locality	Depth (mm)	Average (No.)	Reference
A. cephalotes	Clay	Trinidad	65–128	105 (8)	Weber, 1937
A. cephalotes	Clay	Trinidad	100	100 (1)	Weber, 1945
A. cephalotes	Clay	Trinidad	80–100	90 (2)	Weber, unpublished, 1965
A. cephalotes	Clay	Guyana	65–75	70 (3)	Weber, 1946
A. cephalotes	Sand	Guyana	127	1 (1)	Weber, 1946
A. colombica tonsipes	Clay	Panama	35–100	58 (13)	Weber, 1977a
A. sexdens	Clay	Guyana	60	60 (1)	Weber, 1946
A. sexdens	Sand	Guyana	21–67	30 (7)	Weber, unpublished, 1970
A. sexdens	Clay	Panama	120–190	133 (3)	Weber, unpublished, 1966
A. sexdens rubropilosa	Clay	Brazil	85–150	113 (11)	Autuori, 1942

chamber will have to suffice for her first broods that are too small to dig coarse sand or wet clay. If the chamber is too deep the efficiency of the first broods will be much impaired by the time they need to go up and down when bringing in substrate or later excavating. The first chamber suffices for several months. The next chamber is usually deeper and much larger. The site of this initial chamber can often be identified the following year.

After completing the tunnel, with its enlarged end in the form of a small cell, the ant ejects the infrabuccal pellet and manures it with her own liquid feces. The hyphae grow and a miniature fungus garden is soon developed on which the ant lays her eggs.

She eats some of the eggs that she lays and all are laid directly on the incipient garden, becoming embedded in the mycelium (Fig. 6). Eggs hatch into minute larvae and these grow as they are fed crushed eggs or strands of the fungus. They mature into naked pupae that gradually mature into semipupae, then callow workers. These are called the minima. The workers then bring in small particles of vegetal debris, including sections of dead leaves.

Some of these very young colonies develop a chimney or turret entrance to the nest from the clay soil carried up in their excavating; others deposit the soil, if it is loose, in the form of a low disc or crater. A turret started on a steep

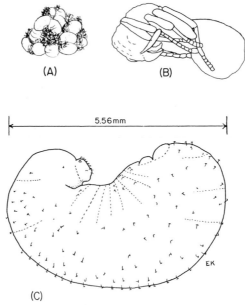

(A)   (B)

5.56mm

(C)

Fig. 6. (A) Cluster of eggs embedded in mycelium of *Trachymyrmex septentrionalis.* (B) Female pupa of *Trachymyrmex septentrionalis.* (C) Larva 5.56 mm long of *Acromyrmex* (*Moellerius*) *heyeri* Forel.

bank turns upward from the base and attains a vertical position. A common turret size is 50 mm high by 60 mm in diameter.

Of eight colonies in this early stage, the average distance of the ceiling of the chamber from the surface was 105 mm. The maximum depth was 128 mm, while the minimum was 65 mm. There was no relation between the downward depth and the diameter of the chamber. The maximum diameter of the chamber varied from 45 to 150 mm and this variation probably merely reflected differences in the ages of the colonies. The maximum diameter was sometimes in the vertical plane and sometimes in the horizontal plane. The table contains data on two other species that show them to be similar.

At this time ants are diurnal and remain close to the entrance of the nest for the night by 1700 or 1800 h, pulling parts of leaves into the nest opening. Increasing daylight in the morning, perceived through the loose leaf plug, causes them to remove the pieces and start foraging.

The flourishing young colony eventually outgrows the initial chamber and builds a second chamber deeper in the soil. The fungus garden is moved here and the first chamber becomes a temporary refuse chamber for exhausted substrate and dead ants. In one nest the second chamber was only 5 cm below the first, but the shallowness is readily explained by the location of the nest in poorly drained soil in a swamp. In the majority of nests the vigorous young colony builds the second chamber as deeply as the drainage and other conditions permit.

Much of the difficulty in controlling the leaf-cutting ants arises from the ease with which new colonies are initiated and become established. After the female has dug into the soil, made her initial chamber, and started the garden, there is a period of about 2 months before the first broods mature. The first broods are workers, 2 or 3 mm long. They remain underground for several weeks, following which they break out to the surface.

Such young colonies often go unnoticed in the garden or field for perhaps 6 months or more, especially if the area is weedy. The trails increase in size but are irregular and often overhung with vegetation. The ants by now number in the thousands, but they scatter as they cut the vegetation, and their damage may not be apparent. By this time, there may be several gardens in the soil and a crater entrance that is fairly conspicuous if it is out in the open. Colonies 2 or 3 months old may be extremely numerous in an area that offers a favorable site during the nuptial flight at a time when the weather is propitious.

3. The Nest after the First Year

Following the young stage of the colony (first and second chambers), the intermediate stages consist of colony growth and rapid excavation of the third and successive chambers up to about 40 in a 2-year-old nest. These are smaller chambers. By the end of the first year, the colony may have the original chamber and two large gardens; one contains the queen and brood, and the other contains full-sized soldiers; both have workers.

The nest expands rapidly in the third year. Development of the first sexual brood may occur by the end of this period. Hereafter, in the mature phase, a sexual brood is produced annually toward the end of the long dry season. The chambers increase exponentially and eventually number in the hundreds. Most of these will contain gardens, some will have gardens with brood, and others of large size may be used for the storage of exhausted substrate and dead ants.

The form of the mature *A. cephalotes* nest is highly variable and represents adaptations to edaphic conditions. Furthermore, the nest never reaches a static phase but is continually altered by the ants as the colony grows and as any important part of the environment changes.

A particular nest at the end of its first year had three craters with diameters of 20, 40, and 50 cm, respectively, while a sister nest nearby had a single cone 15 cm high and 50 cm in diameter. These nests demonstrate the differences in growth rates of individual colonies that have been shown equally well in the laboratory. Both had soldiers of full size.

The faster growing nest at the end of the second year was examined during a particularly long dry season. Little above ground activity was taking place and no new gardens were being constructed. A shaft 1 m² was excavated in the nest center and 12 garden chambers were exposed at various depths and with cell dimensions (in cm) as shown in Table III. It was a sister colony to that noted by Weber (1976b). Four of the gardens were weighed immediately after they were placed in the standard 2.25 liter plastic countainer. All four collapsed; however, a rough estimate of the equivalent in volume was made. The

TABLE III

Garden Chambers under 1 m² of a 2-Year-Old Nest of *Atta cephalotes*

Chamber	Depth (cm)	Lateral diameter (cm)	Height (cm)	Garden weight (gm)	Garden volume (ml)
1	8	20	15	—	—
2	9	24	16	929	3700
3	9	15	14	—	—
4	10	20	11	338	1050
5	10	15	17	—	—
6	11	12	9	209	1500
7	11	9	12	—	—
8	11	18	12	—	—
9	12	11	11	91	700
10	12	14	12	—	—
11	14	9	5	—	—
12	20	10	8	—	—

929- and 338-gm garden volume estimations gave ratios of 1:4 and 1:3, respectively, clearly too low an estimate of volume. The two garden weights of 209- and 91-gm garden volume estimations of ratios of 1:7 and 1:8 ratios were more nearly correct.

It was estimated that there was a total of 40 gardens. During the third year many of the small chambers would have been enlarged as soon as the rains brought new plant growth. There would probably be several hundred gardens by the end of the third year and, if the species behaved as *A. sexdens,* a sexual brood would have developed. However, this particular nest, being disturbed by excavation and by other human interference, moved in less than 2 months to a site several meters away. It was then poisoned by local inhabitants.

4. *Mature Nest of Atta cephalotes*

Excavation of a mature nest of *A. cephalotes* is a major undertaking that few have had the manpower to complete. This has been done with government support and manpower by Stahel and Geijskes (1939) in Surinam.

One nest of a surface area of 16 × 16 m had 283 chambers of the following diameters: 4 or 1% were 30 cm or broader, 129 or 46% were 20–29 cm, 146 or 52% were 10–19 cm, and 4 or 1% were 9 cm.

Of 75 fungus gardens the weights (in gm) were: 1–30, 16–50 to 99, 32–106 to 190, 12–200 to 299, 12–300 to 399, 1–425, and 1–550. The average was 181 gm. This corresponds well with data obtained from the laboratory (Weber, 1976b). Another nest had 592 chambers of which 347 had fungus gardens and 245 were empty.

Several nests of *A. cephalotes* were excavated by Weyrauch (1942) in the Andes of Peru. One was influenced by a shallow water table at a depth of about 1.4 m and the ants had built a mound above the soil level but all chambers were at depths of 0.3–1.3 m below the general surface. The extreme of this type was a nest in the Orinoco Delta (Weber, 1947) where the tide covered the land twice a day and the gardens were entirely in the mound above the surface created by the ants. Weyrauch measured 82 chambers with gardens in one nest and found a variation in size similar to Stahel and Geijskes, the largest being 30 × 50 cm and 24 × 22 cm and a common size being 12 × 20 cm.

Photographs of an undisturbed nest of 11.5 × 13.8 m in external dimensions in Guyana and one of the subspecies *A. cephalotes isthmicola* of 6 × 10.5 m in Panama have appeared (Weber, 1937, 1938, 1941, respectively). Both show the alteration of the plant environment caused by the ant activities.

5. *Trails*

Trails made by the ants of a mature nest have been repeatedly studied and their traffic analyzed. The antecedents of these trails are not as well known.

Trail formation, as noted, is initiated within days of the emergence to the soil surface of the first workers in an incipient colony. These ants are the minima and their first foraging is diurnal, a brief and random venturing out from the nest entrance to pick up vegetal debris in the vicinity. Because they tend to follow one another by the trail pheromone emitted (assuming they follow the general ant pattern), they tend to move over a few radiating paths. These are invisible to the human eye (cf. the path taken by a *Trachymyrmex isthmicus* colony over fallen dead leaves in Weber, 1941) but become visible as they pick up debris in these paths.

Within 1 or 2 weeks, the narrow, tortuous trails of *A. cephalotes* may be perceived. These still radiate in a spidery fashion. Within weeks fewer and more clearly outlined trails appear that lead to particular sources of vegetal debris, then to suitable leaves close to the soil. Several hours of daylight foraging suffice for their needs. These trails become less tortuous as obstacles in them are removed.

A colony that emerged July 2 had on July 16 one main file that was 320 cm long. It proceeded straight for 60 cm, then angled with three or four curves for 260 cm to the area of leaf cutting as determined by P. Weber (unpublished).

There was nothing in the path taken that was a physical trail visible to the human eye and it was presumably a chemical trail. The path was not physically altered, but the ants followed one another exactly as though it was a pheromonal trail.

In 35 min, Weber counted 199 ants coming back to the entrance and 235 going out from 0815 to 0850 h. During this period the traffic was as shown in the tabulation below:

Number of ants (5 min intervals)						
Incoming	16	30	30	32	27	25	39
Outgoing	18	39	43	40	32	30	33

This is indirect evidence for the population of this colony (a sister one to Weber, 1976b) that originated probably from a female fecundated the preceding May. Some of these ants may have made several trips in this period, but clearly the total population was several hundred.

By the following May this colony, now 1 year old, was foraging some 15 m from the nest according to Williams. The population was indicated by a count of 250 ants on the trail passing a point in 14.33 min.

When it was 2 years old, there were three major trails that extended to maximum east–west distance of 76 m and north–south 27 m. The ants foraged over most of the 2052 m^2 area.

The trails of a mature colony may be strikingly broad, clear, and long (cf. figures in Weber 1958, 1966a,b, 1972a). A trail 30 cm wide in places was kept clear, in part, by heavy rains that tended to wash away fallen loose material.

Those of half that width or less and passing through grassy areas were cleared primarily by the ants. Those in heavy forest and traversing a steep slope diagonally were deepened by the action of rain in combination with the ant traffic. There were six principal trails and three lesser ones in the colony of some 2 years age (perhaps nearly three) used for study by Cherrett (1968).

6. *Foraging by the Adult Colony*

Year-long foraging patterns of three colonies each of *A. cephalotes* and of *A. colombica* in a Costa Rican seasonal evergreen forest (Rockwood, 1975) showed similar reactions to the seasons by both species. However, that of *A. cephalotes* preceded the season change by 10 days or more. Foraging in both reached a low point in March, when there was little or no rainfall. The high point for *A. cephalotes* came in May and June as generally found elsewhere; largest quantities of leaves were harvested at the beginning of the dry and rainy seasons when the greatest number of plant species were producing new growth of leaves and flowers. Mature leaves were available all year but few plant species were used as sources for them. Mostly the ants switched from one group of plants to another as new growth became available and in only one case was a tree completely defoliated. This study was later used (Rockwood, 1976) in part in a search for a suitable mechanism that would allow workers to transmit biochemical information about leaves and the state of the fungus garden to each other.

The daily foraging pattern has been examined in an attempt to reconcile the conflicting findings of numerous earlier investigators. There was little doubt that sometimes the species was nocturnal, sometimes diurnal. Investigators often failed to take into account the previous immediate history of the colony under study, whether it was young or old and what the internal requirements of the fungus gardens or brood were. The latter factors are difficult to examine in nature. Trail activity and numbers of workers of two *Acromyrmex octospinosus* colonies on them during continuous 24-hr periods are shown in Weber (1966b). In the laboratory (Weber, 1976b) the ants foraged day and night indiscriminately, provided the gardens were ready for new substrate. A recent examination (Lewis *et al.,* 1974a) concluded that neither photoperiodism, nor lunar cycles, nor any other widespread component in the physical enviornment adequately accounted for the long-term or diel rhythms observed. A companion article (Lewis *et al.,* 1974b) suggested that "by collecting fragments in daylight when sugar, amino acid and aromatic content of leaves are greater than in darkness, the ants may supply larvae directly with trace nutrients or provide the fungus with readily assimilable nitrogen." However, no evidence has been presented to show that the leaf sap is directly and in quantity supplied to the larvae and alternative explanations are available for the fungal requirements.

The literature contains a long list of plants whose leaves are used by *Atta cephalotes* in nature or in the laboratory (e.g., Cherrett, 1968; Rockwood, 1975; Weber, 1945, 1972a, 1976b). In general, it may be assumed that in nature they will take the young leaves of almost any tree in their area at some time. Möller (1893) maintained that other leaf cutters take the leaves of any plant sooner or later in the course of a year in south Brazil. One of the few plants that semed to be repellent for *A. mexicana* (quoted in Weber, 1972a) in Mexico was *Senecio ehrenbergianus* Klatt. In the laboratory *A. cephalotes* generally avoids *Quercus* (oak) leaves and many succulent flowers such as those of rhododendron and azalea.

7. Leaf Cutting

Leaf cutting has been an outstanding feature of popular interest, of which there are many accounts. The ants attack a leaf relentlessly and may cut much of it up before moving on. The ant works from the margin, taking a firm stance by attaching the claws of the legs to the leaf (Fig. 7B,C). It faces inward and proceeds to cut an arc, using the body as the radius, and turning. The base of the gaster or hind legs tend to be largely fixed points. The size of the section therefore tends to vary with the size of the ant. The mandibles are then applied to opposite sides and they start a scissorlike action, alternating the closing of one mandible, with the other.

The speed of cutting depends on the nature of the leaf and the size of the mandibles. Ants of a Trinidad colony on the average cut sections in 2–3 min (Lewis *et al.*, 1974a). Flowers (modified leaves) are cut fastest of all, but the ants also do well with rhododendron.

Commonly the ants leave the midrib of the leaf if it is large. Every part of fresh rose flowers, however, it taken down to the last bit of the fibrous receptacle. They will also cut sections of ash leaf petioles of 2 or 3 cm length and place these on the garden.

Factors in the laboratory affecting the site and pattern of leaf cutting activity have been described in detail (Barrer and Cherrett, 1972). Fresh ant cuts increased the probability of further ant cuts occurring on the same leaf. It was concluded that this was due to the increased availability of sensory information concerning the chemical composition of the leaf and partly to the physical geometry of the cut.

Photographs by Weber (1972a,b,c) show leaf cutting in the Swarthmore laboratory colony that was maintained for a 10-year period. The colonies with their gardens were in clear plastic chambers connected by tubes. The normal build-up of the gardens was measured. Only in this way was it possible to determine that a given weight or volume of leaves was actually incorporated into gardens and the subsequent increase in volume of gardens determined.

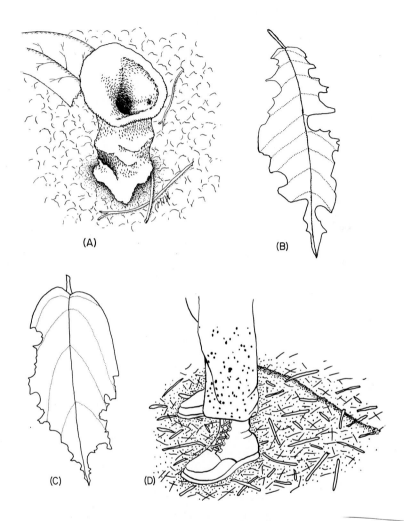

Fig. 7. (A) Turret built with 40 × 37 mm flared orifice, 19 mm above the ground at one point. Nest of *Trachymyrmex phaleratus Wheeler* (Guyana). (B) Leaf cut by *Atta cephalotes* in Costa Rica. (C) Leaf cut by *Atta cephalotes* in Guyana. (D) Khakai trouser leg spotted by anal or rectal excretions of *Acromyrmex ambiguus* Emery (Argentina).

8. *Rates of Leaf Input to the Nest*

Many studies have produced rates for the transport of leaf sections on the trails toward or into the nest. It is difficult to observe the incoming files without speculating on the total quantities of leaves that must be cut. Studies in this species and in other leaf cutters indicated a general rate of travel of the ants bearing leaf sections to be about 1 m/min at a variety of temperatures.

Ingenious and accurate methods of counting the ants per unit time and of measuring the weight and volume have been devised. The investigations of leaf input generally suffer from a lack of knowledge of the internal size of the nest to which the ants bring leaves and the current total volume of the gardens. Other conditions being equal, the larger the total garden volume the greater the ant population and the greater the input will be. These other conditions include the cycle that the colony is in at the moment, the cycle of plant growth, whether this is the beginning or end of a major period of activity such as the end of a dry season, or the maturation of a sexual brood. The 10-year laboratory record of an *Atta cephalotes* colony, including volumes of gardens compared with monthly input of leaves of known weight, may be compared with these field data.

The rates of leaf input to the *Atta* nest have been the subject of recent research in the field and have been summarized in part by Lugo *et al.* (1973). The field data listed in tables by Lugo apply principally to *Atta cephalotes* and to *Atta colombica tonsipes* (referred to as *A. colombica* in Lugo's tables). For *A. cephalotes* the leaf input varied for a 24-hr period from 13.1 gm/hr to 64.0 gm/hr for entire nests (Table I). For nests of *A. colombica* (Table VII) other parameters of total impact on the forest were listed.

Absent from the literature is a natural correspondence between the leaf input to the nest and growth of the gardens. General observations show that there is a marked growth of the gardens as indicated by the rapid excavations taking place at times following considerable leaf-cutting acvtivity. The correspondence in the laboratory between specific quantities of leaves and the subsequent growth of gardens is described in the Section I,E,9 and in Weber (1972a,b).

Individual ants on the trails were marked with paint and followed into and out of a Trinidad nest (Lewis *et al.,* 1974b). At times of peak foraging the ants carrying leaf sections into the nest stayed 10–30 min before returning for another section. Of ants going into the nest, 13% were unladen. Individual ants spent 3–5 hr out of the nest on a foraging mission. The maximum direct distance they traversed was 250 m.

9. *Limits to the Nest and Colony Size*

No *Atta* nest grows indefinitely. With the one queen known from any nest there is a practical limit to the transport of her eggs and resulting brood to adjacent gardens. These adjacent gardens act as incubators and the resulting workers fan out to establish peripheral gardens. As leaves are harvested, trails radiate for scores of meters. The limit to nest and colony size depends on the balance between energy expended in growing and that expended in maintenance.

A primary limit is the years that a queen can live and the years that her ovarioles can produce viable eggs, which are not exactly the same. Once she dies, the colony disintegrates approximately in the year her last brood lives. Disintegrate means that the colony organization falls apart because it lacks the queen's integrating role. Her massive pheromone production, shown by the numerous ants constantly attending and licking her, ceases. The workers in all attine colonies gradually cease their active substrate gathering and garden construction. A garden will seldom flourish for more than a few months without the queen.

In the laboratory, not all queens from similar incipient colonies or the same nuptial flight survive the same length of time. Results in *Atta colombica tonsipes* (Weber, 1972c, 1977a) show a variability in the queen's live weight and the differences in years of surviving under similar conditions. This is equally true of *A. cephalotes* and *A. sexdens* in the same laboratory. Apparently all queens do not leave the parental nest with identical food reserves and vitality. The detailed history of their colony formation will differ.

Queens of *A. cephalotes* are known to have lived 10 and 12 years. It is probable that they could live 15 years in nature or in the laboratory under normal conditions and 20 years may be possible.

The survival of a nest may differ from that of the original queen (plus 1 year for her last workers) if a new and young colony could develop on the periphery. Also possible is the adoption of a new queen by the colony soon after the old has died.

10. Colony Population Size

Although complete counts of *A. cephalotes* colonies have not been done, counts of samples of a large colony of *A. colombica tonsipes* have been conducted (Martin *et al.,* 1967). The samples, used for bio-chemical analysis, weighted 6.35 kg in total and were estimated to have been 65–85% of the nest. Random samples of workers weighed 4–8 mg and on that basis the nest was estimated to have contained 1.0 to 2.5 million ants. However, this estimate of this nest was doubled (Weber, 1972a) based on the fact that the method of collection tended not to include the minima since they clasped the garden particles (this author was present during part of the collecting). The garden particles were not retained. The minima comprise 60–65% of the ants in a garden and therefore the population was more probably 2–5 million. This colony was comparable in size to a mature *A. cephalotes* colony and the proportions of worker sizes identical.

The counts of workers in *A. cephalotes* and *A. sexdens* isolated gardens of known volumes permit an estimation of total population in large colonies. A 1700 ml garden of *A. cephalotes* had 8762 ants (1:5.15 ratio), a 1500 ml garden of *A. sexdens* had 3825 ants (1:2.55 ratio), and also a 794 brood; a 1300 ml garden of *A. sexdens* had 4307 ants (1:3.31 ratio). The above 4500 ml of gardens thus contained 16,894 ants. The average number of ants to a milliliter of garden was 3.75.

Average gardens in nature are often 1500 ml and a mature nest may have 500 gardens. This ratio of 3.75 ants to 1 ml of garden would make a total population of 2,812,500 ants, comparable to the estimation of 2.5–5 million ants in the aforementioned large colony of *A. colombica tonsipes*.

11. Brood Production and Appearance of Worker Sizes

As earlier noted, the first brood of the young colony consists of the minima workers which are 1.5–3 mm long, then workers of up to 4 mm length. The maximum size of the worker increases gradually during the first year. Of five sister colonies in Trinidad (Weber, 1976b) originating from a May 1965 nuptial flight the first soldier was produced on September 17, 1965 (11 mm long).

Under comparatively good conditions for viewing (a rare event in a colony) a female of *A. cephalotes* was observed for 1 hr. She was laying eggs under normal, undisturbed conditions. A media worker with mouthparts and forelegs appressed to the queen stood against the left side of the gaster at the suture between the first tergite and the first sternite. At first, there were 5–10 smaller workers on or beside the female with antennae constantly exploring her integument. Several additional minima stood with mandibles open and faced the apex of the gaster. The gaster, at the beginning of observations, rested on the garden. Her left rear leg moved up and down slowly. Eleven minutes later she moved her body for the first time, raising the gaster above the garden by less than 1 mm. During the next 5 min she raised the gaster momentarily three times off the garden. Three minutes later a minima licked the exposed apex of the gaster and six more minimas were on or at the gaster. Then the apex was brought down briefly while the right leg went up and down. Three minutes later she raised the gaster and several ants attended the apex. An egg appeared 30 sec later, which was then hidden by gastric movements. During all this time the media ant maintained its position. Shortly thereafter the gaster was again raised and a small media and a minima licked the area.

A second egg appeared 10.5 min after the first but its progress could not be followed. Six minutes later another appeared and after 2 min two more eggs emerged. One minute later there was a string of four eggs extruded. Immediately a minima removed the eggs as the female momentarily raised the gaster 1 or 2 mm. After 55 min of being stationary the media ant finally changed its position when the female moved slightly. Another egg appeared 6 min after the string of eggs was removed.

In summary, during 1 hr of continuous observation 10 eggs were seen within a space of 26 min during the hour and more could have been overlooked.

The eggs are often in clusters and are always embedded in the mycelium

(Fig. 6A). When an egg matures the larva inside rotates so that its head is exposed after emergence. There is little or no concrete evidence that the future larval head is appressed to the mycelium for feeding, contrary to that stated in the literature.

a. Larvae and Pupae.   The external anatomy of the larvae of all attines is characterized as attoid by Wheeler and Wheeler (1974), a form that in lateral view is plump, with the head below the anterior part and directed backward (Fig. 6C). The almost naked body has few hairs and these are mostly restricted to the ventral surface. The spinulose mandibles and short ventral hairs about the mouthparts form a basket in which the workers place the fungus. The head of the larva is always exposed and the fungus is brought to the mouthparts by the workers.

Rearing larvae experimentally (Weber, 1966b, 1972a) was continued by Schreiber (1974). Earlier it was found that isolated larvae would accept staphylae on the point of an insect pin or crushed eggs of their species. Isolated larvae and pupae would mature but pupae could not emerge to the callow stage without the assistance of the workers (Weber, 1966b). Schreiber verified these findings with *Atta cephalotes, A. sexdens,* and *A. colombica tonsipes* and determined that contamination by alien fungi or bacteria prevented normal pigmentation of the pupae.

The mature larva becomes a semipupa in which the head region bulges beyond the mouthparts, that are ventrally located. These are in front of the anal hillock.

The larval envelope is removed by adult workers at the time of pupation, leaving the pupa naked (rather than in a cocoon) and the appendages free but appressed to the body (Fig. 6B). The head, mandibles, and antennae are directed down and posteriorly, against the underside of the thorax. The posterior legs in these long-legged insects extend beyond the body. When the soldier is ready to pupate there is a bizarre stage when the pupal envelope slips down over the bulging head, constricting it tightly as a tight stocking. The worker pupa viewed from the side shows the bent legs extending as far as the occipital spines of the head.

Worker and soldier pupae alike are white at first, then the eyes darken first, greatly contrasting with the pale body. The edges of the mandibles are next to darken, then the entire integument turns from pale brown to bright reddish brown; apices of spines and ridges darken first. These color stages could be useful for establishing ages of the brood (as in Weber, 1966c, for *Trachymyrmex).*

At the conclusion of the pupal stage there are slight tremors of the body and probably hormonal stages that indicate to the nursemaids that the pupae have attained maturity. In the meantime the adults have removed the pupae to the periphery of the garden. Here the ants lick the pupae closely, thus removing the thin pupal envelope, and exposing the callow stage, The callow is much paler than the adult and stands unsteadily on its feet. It moves its legs slowly, then moves jerkely about. The callow stage is of variable age but is a few days for the workers, and 1 week or so for the soldier.

b. Life Span of Worker.   The life span of the first worker broods appear to be between 6 months and 1 year. Toward the end of the first year (June 27, 1966) in a sister colony of the 10-year *A. cephalotes* colony 156 dead workers were cast out. Earlier (March 15–21) another sister colony lost 1800 workers in a moat; some of these may have accidentally drowned.

12. Polymorphism and Castes

Polymorphous literally (Figs. 8 and 9) means "having, assuming, or passing through many or various forms, stages or the like" (Random House Dictionary, 1971). Polymorphic means multiform. This has been the usual sense in myrmecology. It has been graphically portrayed in drawings and photographs in *Atta cephalotes* (compare the living soldier and its workers in Weber 1972c, Fig. 3 with these same specimens in Wilson, 1975, Fig. 14-1, from Weber's laboratory table and photographed by C. W. Rettenmeyer). Polymorphic emphasizes morphology but is commonly associated with behavioral traits.

The discussions of Wilson (e.g., 1953, 1971, 1975) have introduced more precision in myrmecological usage. As used by him (e.g., Wilson, 1975) a caste is limited to one or more roles and in advanced polymorphism the intermediate workers in the ant colonies have dropped out and the two remaining size classes are strikingly different, the larger class serving as soldiers. Polymorphism, summarized by Wilson (1971), "can be interpreted as a function of two variable characters of the adult females of any species; allometric growth series and intracolonial size variation." Allometric growth, termed by Huxley as heterogonic growth, may be defined by equations. Investigators commonly refer to it as a differential or disproportionate growth of parts of the body. These parts tend to grow exponentially, such as the soldier head of *Atta* where the inflated head houses enlarged adductor muscles for the powerful mandibles.

Wilson differs from most writers in excluding the male as a caste but undoubtedly the male will continue, at least in ants, for some time to be referred to as the third caste, together with the queen and the workers. Any writer emphasizing polymorphism of castes and division of labor needs to refer to Wilson's discussion.

Since the above was written, Wilson (1980a,b) has uniquely explored in great detail caste and division of labor in *Atta sexdens,* using a colony which

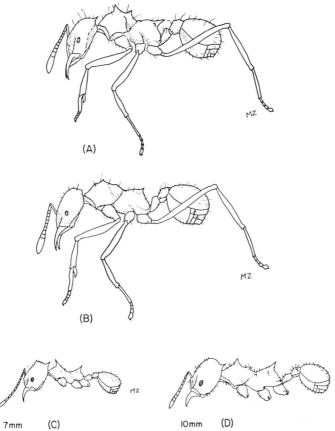

Fig. 8. (A) Minima worker 2 mm long of *Atta cephalotes* (Trinidad). (B) Minima worker 2 mm long of *Atta sexdens* (Guyana). (C) Small maxima worker 7 mm long of *Atta sexdens* (Guyana). (D) Soldier 10 mm long of *Atta sexdens* (Guyana).

this author gave him in 1974. This colony was a sister colony to others the author took in Timehri, Guyana in 1970, one of which lived 10 years. Emphasis is elsewhere in this chapter. The present usage is to consider the minima, media, and maxima worker together with the soldier as castes in the Wilsonian sense, but it should be pointed out that the functioning of these morphological castes shows flexibility. For example, minima are functioning mostly within the garden as nursemaids to the brood and caretakers to the fungus but also some go out on trails and ride the sections back, swabbing the leaf surfaces. The queen fulfills all roles of the female, including those of the workers (Weber, 1960) while commonly performing worker functions only to the rearing of her first broods, and then becoming primarily an egg layer. She can revert to worker roles if deprived of the garden and furnished with

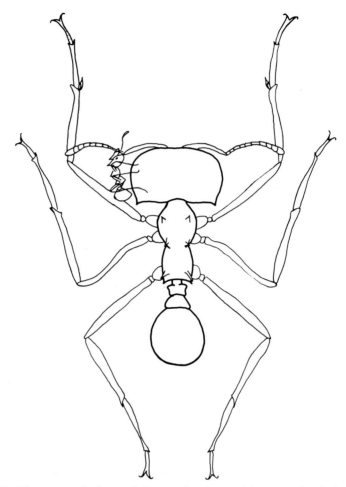

Fig. 9. The extremes in *Atta cephalotes* worker size, a minima grooming the head of the soldier (Trinidad).

another fragment. Attine females may also forage for fresh substrate while the first brood is developing.

A generally overlooked anatomical feature in the largest worker caste of *Acromyrmex* and the soldier of *Atta* is the occurrence of a more or less completely double median ocellus on the frons (Weber, 1947a). Three ocelli commonly occur in many insects, in addition to the usual lateral compound eyes, but the doubling of the median ocellus is rare and is not known to occur in the other sizes of workers of these two genera or in any other genera. Future research on polymorphism and on the evolution of sense organs might be directed to this subject.

Tables IV–VI present measurements (to the nearest mm) of entire worker

TABLE IV

Census of Worker Sizes[a] in Entire Laboratory *Atta* Gardens

Species	Worker size (mm)												Total
	1.5	2	3	4	5	6	7	8	9	10	11	12	
A. cephalotes (Garden 1700 ml)	173	3493	1718	1737	1135	367	93	28		17		1	8762
A. sexdens (Garden 1300 ml)	2	2280	604	490	427	263	121	32	33	36			4288
A. sexdens (Garden 1500 ml)[b]		2401	605	426	252	91	22	8	20[c]				3825

[a] Measured to nearest millimeter, except smallest workers.
[b] Also with 226 larvae and 568 pupae.
[c] 9 and over.

TABLE V

Census of Five Worker Sizes[a] in *Atta cephalotes* Gardens in Nature

Garden	Worker size (mm)					Total
	2–3	4–5	6–7	8–9	10 plus	
A (1/5)	1436	634	134	5	14	2223
B (1/5)	518	132	52	14	11	727
C (1/7)	1304	671	212	24	21	2232
D (1/4)	684	191	44	3	1	923
Total 4 samples	3942	1628	442	46	47	6105
Percent of total	65	27	7	1	1	

[a] Measured to nearest millimeter.

populations in *Atta cephalotes* and *A. sexdens* gardens and censuses of these sizes. These tables show that there are gradations in sizes from 1.23 to 12 mm; however, the categories of minima, media, maxima, and soldiers are convenient to use when correlated with the behavior of these sizes as described in Section I, E, 13. It is also apparent that those of 2–3 mm size comprise 60–67% of the entire population. These are the sizes not commonly seen outside the nest except for the few minima that are found on the trails.

Table IV presents the census of worker sizes (to nearest mm) from entire laboratory gardens of both *Atta cephalotes* and *A sexdens*. A census of the larvae and pupae in one of the *A. sexdens* gardens is included. The others lacked

TABLE VI

Length and Weights of Living Workers in *Atta cephalotes*[a]

Length (n) (mm)	Thorax length (mm)	Average thorax length (mm)	Weight (mg)
1.23 (1)	0.9		0.51
2 (10)	0.87–1.24	0.97	0.42
3 (10)	1.49–1.85	1.71	1.24
4 (10)	1.93–2.40	2.17	5.29
5 (10)	2.56–2.90	2.78	7 est.
6 (10)	3.07–3.28	3.18	9.26
7 (10)		3.38	12 est.
8 (10)		3.90	14.51
9 (10)			20.02
10 (2)			34.35–38.39
11 (5)	4.3–4.4		39.2–55.6
12 (4)	4.7–5.5		54.3–60.7

[a] (n) = No. specimens.

brood and are average gardens, in that respect, in any medium-age to mature nest.

Table V presents the census of five worker sizes (to nearest mm) from the 2-year-old *Atta cephalotes* colony of Table III. The four gardens sampled are those whose weights are given in Table III, and the proportion of garden sample is roughly estimated as 1/5, 1/5, 1/7, and 1/4, respectively. These fractions were isolated quickly after the gardens were weighed and are deemed representative. The disturbances of excavation may have caused some of the largest ants to leave but all of the smaller ones would have remained.

Table VI gives the length (to nearest mm) and weight (in mg) of living *Atta cephalotes* workers including soldiers in the laboratory. A 1.23-mm worker is unusually small and most of the minima are about 2 mm–3 mm long. The thorax length (in the sense of Weber, 1938b, Fig. 17, and generally used since then) is a more precise measurement and does not vary with the contortions of the ant, live or dead. The total length in Tables IV–VI is that in the walking position with head advanced at a slight angle and is easily measured through clear plastic sides of containers. Ants were isolated in tubes directly from the nest to the Mettler semimicro balance for accurate weighing. The sensitivity of the balance is great enough to reflect the comparative fullness of the digestive tube and glands. Any ant in this species greater than 9 mm long is a soldier, having a disproportionately large head. Soldiers as heavy as 92 mg were not raised in the laboratory but were taken in the field. The smallest soldier weighed 38.39 mm.

The largest soldiers weighed in the subspecies *A. isthmicola* weighed 103 mg, that of *A. sexdens* 44.66 mg, *A. colombica tonsipes* 38.9 mg, and *A. laevigata* 65.73 mg, the latter directly from a garden in nature. Larger soldiers in all were known from the field but not weighed.

The smallest workers in the aforementioned species were smaller in size and weight than those of *A. cephalotes*. Usually they weighed 0.34–0.7 mg.

13. Specialized Behavior

Two different interpretations of the function of the minima workers appearing on the *Atta* trails and riding back on the leaf sections preceded an understanding of the actual function. Belt (1874) realized that their primary function was inside the nest as nursemaids and garden tenders. He has

> seen them running out along the paths with the others; but instead of helping to carry in the burdens, they climb on the top of the pieces which are being carried along by the middle-sized workers, and so get a ride home again. It is very probable that they take a run out merely for air and exercise (Belt, 1874, p. 82).

This explanation sufficed for nearly a century for many people until I. Eibl-Eibesfeldt (1967) and Eibl-Eibesfeldt and Eibl-Eibesfeldt (1967) alleged that the minima functioned in this way to fend off parasitic flies. They observed phorid flies hovering over the files, saw the minima rise up on the leaf sections, and concluded that the minima drove off the parasites. However, longer observations in the field or in the laboratory (e.g., Weber, 1972a,b, 1976b) would have shown that these ants are swabbing the leaf sections with their mouthparts, starting the process of preparing the leaf sections for incorporation into the garden. The hovering of phorid flies does alert the *Atta* soldier (Weber, 1945); however, the ants of all sizes are ineffective in driving the flies away.

The general function of the minima workers, as noted, is to work over the garden minutely, going over every particle of substrate that is commonly one-half the length of the ant. They lick the hyphae and the finely ridged glossa enable them to rasp the cell wall and imbibe the protoplasm. This minute and close attention is also given to the exhausted substrate, particularly if it is moist. Possible functions here are to ingest and thus inactivate toxins or products of bacterial and alien fungal action. They may be recycling nutrients from dead ants cast out with the substrate. It is the minima that function as nursemaids to the youngest larvae and care for and also lick the brood. Along with other classes, they have a general function of caring for the queen, licking her integument and presumably ingesting the female pheromones. These are passed to other members of the colony by trophallaxis.

The media (4–6 mm) are more specialized for construction of new gardens, excavation of soil and cutting of leaf sections to small enough sizes for incorporation into the garden. The larger media go out on the trails to cut leaves. They imbibe water or plant sap, may attack intruders, and carry the brood about. While grooming is a general behavioral pattern it is most prominent in the media.

The maxima workers (7–9 mm), below the soldier in size and weight, are aggressive and effective in protecting the colony. They cut and carry leaves, excavate soil, and clear trails. When media and maxima attack a person they may leave pungent droplets of amber fecal droplets on the clothing; this also occurs in *Acromyrmex* (see Fig. 7D).

The soldiers (over 9 mm) rarely perform the aforementioned duties of the smaller sizes. The massiveness of their heads and mandibles make feeding themselves difficult, if not impossible, and they receive liquids by trophallaxis. They usually patrol near the entrances of the nest and rush out at the slightest disturbance. Their mandibles cut human flesh with ease, commonly making a bleeding cut of 5 mm length. The mandibles cut through clothing and may take halfmoon sections out of leather shoes. In the laboratory they will stand motionless on open surfaces or entrances to the chamber for many minutes at a time. They stalk out on the trails and may occasionally tug at a piece of debris. A photograph of one young soldier (Weber, 1945) shows it standing and facing a female pupa about to mature as a callow. It maintained this position for many days.

Several types of queen behavior have been described and in addition the *Atta* queen probably retains vestiges of generalized worker functions as found in *Trachymyrmex* (Weber, 1960) and *Acromyrmex* (Weber, 1972a, 1977b). These traits would enable her to revert to worker behavior in exigent situations. However, she differs from all other attine queens in that she is so massive and clumsy that she is primarily an egg layer and originator of integrative pheromones.

The male has a passive role in the nest and is fed and groomed by the workers.

14. The Grooming

The peculiar importance of grooming in attine ants has been emphasized (e.g., Weber, 1958). The details are common to all species.

In grooming itself an ant uses the pecten at the end of the tibia of each foreleg, drawing the legs of the opposite side between the pecten and the curved, adjacent first tarsal segment. The near surface of the latter bears a row of dense, short bristles which together with the pecten make an effective comb. The funicular segments of the antennae are kept constantly cleaned in the same manner. The legs and antennae are also drawn between the partly opened mandibles and the other mouthparts, thus cleaning them more thoroughly and moistening them with saliva. Detritus from this cleaning tends to collect in the infrabuccal pocket off the pharynx. The body, in general, is scraped with the short dense bristles of the ventral surfaces of all tarsal segments and the total effect is to keep the ant externally clean.

The ants spend much time cleaning one another. In cleaning one another, the one being cleaned remains passive while one or more members of the colony examine it thoroughly with their submandibular mouthparts. The mandibles may be opened or closed and are not used. The glossa of the groomer is appressed to the integument and is applied thoroughly to irregular crevices as well as to flat surfaces. It is in the form of a minutely ridged swab that is moist with salivary secretions.

The sexual forms, and especially the queen, are thoroughly cleaned by the workers and this care may extend to the wings. The queen remains in one place for long periods. In *Atta* she is so bulky and large that she is incapable of cleaning all parts of her own integument. The smallest workers go down into her narrowest crevices, such as between the head and the thorax, and lick them carefully.

The amount of time spent in grooming depends on the "dirtiness" of the environment to which the ants are exposed but this is always a major activity.

What is accomplished by this? First, the antennal funiculi appear to be the primary sense organs and must be immaculate to be used efficiently. The terminal segments are continually plied over the substrate and the fungus and actually touch them or come extremely close. They obviously perceive substrate as being suitable for incorporation into the garden and can perceive alien fungi from a variable distance of several millimeters, depending presumably on their pungency. Second, foraging ants come into contact with a host of organisms such as alien fungi, bacteria, protozoa, and parasites that would be harmful if taken into the nest and these must be removed. Third, the ant integument may at the least be a passive if not nutritive site for organisms that might proliferate in the constantly humid air in and about the garden. Living closely together, as these social animals do, perhaps requires more constant care of the integument than is common in solitary animals. To this has been added their dependence on a fungus susceptible to contamination.

15. Salivary and Rectal Excretions

Salivary and rectal excretions, which are always applied by the ants to the gardens, have long been believed to be a vital factor in this symbiotic or mutualistic relationship (Fig. 7D). Early statements regarding this include "these substrate particles are manured with the ant's liquid excrement, the latter probably contributing an important part of the proteins and other foods necessary to the metabolism" (Weber, 1941) and "that the chemical environment created by the malaxated (chewed) substrate and the ant fecal droplets favors the particular fungus grown by the ants" (Weber, 1947). Later statements (such as Weber, 1954) that "in the production of pure cultures (of fungi) the salivary and anal excretions of the ants may play an important role" led to the significant work of Martin and associates (1969 ff) that have gone far in analyzing the biochemical aspects. These are summarized in Section II,E since they do not apply specifically to *A. cephalotes* but are fully applicable to this and the other attine species.

16. Communications and Pheromones

The ants communicate with each other by chemical (pheromonal), tactile, auditory, and visual means, of which the chemical communication is pervasive. The chief chemical sense appears to be most precisely located in the apical or terminal antennal segments. The antennae are continually used by the ant for exploration of the environment, fellow ants, and the garden. When an ant approaches a fellow member of the colony, it touches the other ant with its antennae. The touching may involve either the antennae of the other ant or the other's head and mouthparts. When an ant approaches an alien ant of another colony of the same species or some other kind, it waves its antennae as the ant comes nearer. Recognition of the alien ant is as instantaneous as it is for alien fungi.

Tactile methods are correlated with antennal use. The ants groom one another frequently, licking each other with the mouthparts below the mandibles extruded and the finely ridged glossa (tongue) swabbing the skin. This attention is given to all parts of the body. When one ant receives food through regurgitation (trophallaxis), pheromonal and tactile means are employed. Auditory means involve stridulation and visual means appear to involve the sexes during the nuptial flight. The exploration of pheromones and their use has been performed with other ants and is more fully considered in Section I,F.

17. Stridulation

Stridulation, the production of sound (in ants) by rubbing one abdominal segment against another, has not been thoroughly explored from functional standpoints. The anatomy, physical characteristics, and one possible function have been described in detail by Markl (1965, 1970) for *Atta cephalotes*. Markl found that the ant produces trains of short, transient clicks by pulling a file at the end of the first gastric tergite against a scraper on the first postpetiolar tergite in front of it. The pattern is determined by such external factors as temperature and mechanical stimuli and the nervous system localized in the petiolar ganglion. It has been suggested that if an ant were buried in soil, then stridulating would enable it to be found and rescued by nest mates.

However, burial in soil is not a particular hazard in attines, but there are other possible functions of stridulation. In the laboratory, if an ant is accidentally held by the head, when a cover is closed against it, it stridulates long and loudly. The loudest of the stridulations are those observed in *Acromyrmex octospinosus* and one can always demonstrate this to students by holding an ant by its head and bringing it to the person's ear. Tape recordings of the stridulation of a number of species, including *A. cephalotes,* have been produced accurately by R. Hewitt with the cooperation of C. P. Haskins. When these were played back to the ants in the laboratory the results were inconclusive. However, each species had its own pattern fully distinguishable to the human ear. It may be that ants on the trails or while leaf cutting may become temporarily lost, especially if it drops from a tree and is separated from its fellows. By stridulating the ants may find each other. It would appear to be more useful for stridulation to be perceived through the air than through the soil. Tropical clay in which *A. cephalotes* live is not likely to collapse from ant activities, but stridulation perceived through the tunnels and chambers could have a useful communicative function. For example, it could alert ants in one chamber or tunnel to a disturbance in another.

Males and females use stridulation on their nuptial flight and this again is audible to the human ear. Amante (in Weber, 1972a) found that the large swarms of mating *Atta* in Brazil create a large volume of sound high in the air. An unrelated ant, having a long series of nuptial flights along a large area of semiarid Andean slopes, was found to be stridulating at the entrances to the nests in the soil (Weber, 1963).

18. Pheromones

Pheromones, as noted earlier, are vital chemical agents used for communication. The literature regarding attines contains vast numbers of references to behavior attributable to pheromones, whether the term pheromone is used or not. The trail-making pheromone was one of the first to be chemically analyzed (Moser and Blum, 1963; Blum *et al.*, 1964; Moser and Silverstein, 1967; Moser *et al.*, 1968; Tumlinson *et al.*, 1971; Riley and Silverstein, 1973). They are presently being explored as a method of biological control of the leaf cutters (Cherrett, 1972, Cherrett *et al.*, 1968; Robinson *et al.*, 1974; Seaforth *et al.*, 1974, Littledyke *et al.*, 1976). Alarm pheromones have been considered by Crewe *et al.* (1972). Those pheromones adaptive to the sender have been termed allomones and those adaptive to the receiver called kairomones.

Wilson (1975, pp. 231–235) has placed them in the general context of animal behavior and speculated, with J. B. S. Haldane much earlier, that they are the lineal ancestors of hormones.

The recent report of the Proceedings of the XV International Congress of Entomology contains articles by Pasteels (1977), Shorey (1977), and Duffey (1977) which have extensive bibliographies on pheromones and derivatives, such as allomones, that update this greatly expanding subject. The principles apply to attines, and attines were used specifically in some cases.

19. The Atta sexdens Paradigm

Early accounts of *Atta* by Europeans dealt with both *A. cephalotes* and *A. sexdens,* sometimes without discrimination.

Fritz Möller referred in a letter directed to Charles Darwin (1874) (quoted by W. M. Wheeler, 1907) to what must have been *A. sexdens* ssp. *rubropilosa* in Rio Grande do Sul, Brazil. Shortly thereafter, investigators in Brazil were making invaluable and pioneering observations, especially von Ihering (1882) and Sampaio De Azevedo (1894) and later Goeldi (1905a,b) and Huber (1905) as noted. While von Ihering discovered the method of transfer of the fungus from one ant generation to the next, it was Huber who described the process in great detail. The most extensive accounts of all phases of *A. sexdens rubropilosa* biology were published by Autuori in a series of articles (1941, 1950, 1956).

Where *A. sexdens* and *A. cephalotes* occur, their general habits appear to be significantly similar. One would have difficulty in determining any significant differences in nest architecture. In northern South America the individual craters of *A. sexdens* in sand or loam may be higher (e.g., 15 cm) and the rims sharper (Figs. in Weber, 1972a). Stahel and Geijskes (1939) pictured and mapped both nests in Surinam. In Panama *A. sexdens* nested just above high tide on the Pacific side in pure sand and a nest of *A. cephalotes* in Trinidad was noted to be similar under coconuts on the Middle Atlantic sandy shore (Fig. in Weber, 1937). In Trinidad, the species usually nested well back from the shore in clay or swamp loam. In Guyana *A. sexdens* nested in local patches of savannah or on sand ridges and *A. cephalotes* tended to be within the forest in clay, as was noted (Weber, 1946).

The Brazilian *A. sexdens* ssp. *rubropilosa* formed somewhat similar nests to the aforementioned, but in open areas or under isolated trees and in Rio Grande do Sul State it harvested the leaves of young *Eucalyptus* that other species did not touch.

A comparative study in an area where *A. sexdens* and *A. cephalotes* both occurred would probably reveal differences in plants attacked. However, the gardens and the mycelium are similar under the microscope.

Colony foundation of *A. sexdens* ssp. *rubropilosa* has been followed carefully by Autuori and is described in Section I,E,20. Much general information has been published since then by Amante and others. In general, however, *A. sexdens* as a paradigm is much like *A. cephalotes* and is not described as completely here. Reference to the literature citations permits a more detailed comparison to be made.

20. Growth Rates of Atta sexdens Colonies

Unpublished rates of growth of *Atta sexdens* colonies may be compared with those of *Atta cephalotes*. Colonies of *A. sexdens* were taken in Panama in 1966, and in Guyana in 1970, and the colonies survived for periods up to 10 years with their original queens. Table VII shows the growth of three colonies from the Pacific coast of Panama correlated with the intake of fresh substrate. They were given the same plants that were furnished to *A. cephalotes* in the same time span. They were of comparable size when taken, about 2–3 months of age. Under these laboratory conditions they did not attain the maximum size of *A. cephalotes* and the maximum was 7700 ml reached by one in 1968 when it also took in the maximum monthly weight of substrate, 720 gm. Probably the ants would have expanded more with a different selection of plants. The number of days in each month and grams of substrate provided for representative years of 1969 and 1972 are given for Colony C in Table VIII.

Table XVI compares six attine species with respect to rate of cutting

TABLE VII
Monthly Use of Fresh Substrate and Growth of the Gardens in Three
Atta sexdens **Colonies**

Year	Colony	Substrate (gm)		Yearly total	Garden size (ml)	
		Minimum	Maximum		Minimum	Maximum
1966	A	10	30	79	30	350
	B	5	56	115	30	600
	C	5	56	122	50	450
1967	A	43	440	3016	500	5400
	B	42	360	2515	520	3920
	C	48	460	3001	520	5400
1968	A	217	720	5512	2200	7700
	B	180	540	3962	1350	4200
	C	140	452	2996	100	3100
1969	A	80	500	3409	1600	4700
	B	130	245	2141	800	3400
	C	100	530	3414	1900	5600
1970	A	205	470	4187	2600	5700
	B	75	460	3096	1400	6200
	C	170	505	4176	2800	6900
1971	A	125	425	3301	2600	5900
	B	225	475	4110	2600	6200
	C	280	395	3899	3700	5800
1972	A	175	410	3226	2000	4800
	B	95	426	2818	1200	4600
	C	190	425	3866	3500	5600
1973	A	10	167	544	30	1400
	B	48	440	2136	1000	2800
	C	125	770	3458	1900	5300
1974	A[a]	—	—	—	—	—
	B	8	58	1590	300	900
	C	49	300	813	900	1400

[a] Colony died.

grapefruit pith, a generally acceptable attine substrate. The *A. sexdens* colonies varied in the rate of cutting, but were generally as fast or faster than those of *A. cephalotes,* considering the volumes of the colony gardens. In general, the larger the garden is, the larger the colony population is.

Table XVII compares garden volume with weight among five species. The *A. sexdens* garden was slightly heavier than the others, probably reflecting a later stage in garden life history, when it tends to be heavier.

F. The Ant Genera

The ant genera are briefly reviewed here. The wings of genera and representative workers appear in Weber (1966a) and the general treatment in Weber

TABLE VIII

Monthly Use of Substrate and Number of Days in the Month on Which It Was Given to a Colony of *Atta sexdens* L.

Year		Jan	Feb	Mar	Apr	May	June	July	Aug	Sept	Oct	Nov	Dec
1969	Substrate (gm)	118	100	160	236	200	280	410	490	460	530	385	245
	No. of days	16	16	20	22	16	14	17	19	19	17	16	13
1972	Substrate (gm)	425	340	340	391	305	320	330	410	190	350	260	185
	No. of days	13	11	12	12	10	8	9	8	5	9	7	6

(1972a) contains numerous references and figures and a key to the workers of the genera. The wings of species other than those discussed previously are shown here since they are of value in identification when a sexual brood is present (Fig. 2).

There are some 200 species of attines (187 in Kempf, 1972) divided among 12 genera. Kempf of Brazil at the time of his death in 1976 was engaged in extensive studies in neotropical ant systematics, including the attines. He had indicated generic and specific changes that it is hoped will be published posthumously. However, the main outlines are clear and the present sequence is one primarily of convenience. It does not represent all evolutionary concepts. A complication in attine systematics is the intimate relations that the ants have with their fungi that are not considered by the systematist. The systematics of the ant fungi may be relevant to ant systematics, assuming coevolution. The particular emphasis here is on the nature of the genus *Cyphomyrmex* that includes a widespread species which cultures yeast and others which culture a conventional mycelium. Attached to the species are numerous subspecies or varieties (86 in Kempf) that at the least show variation that may be leading to more speciation.

Kusnezov (1963) divided the Attini into the Palaeattini (*Apterostigma, Myrmicocrypta,* and *Mycocepurus*) and the *Neottini* which include the remainder of the genera (including *Cyphomyrmex*). His last four genera (omitting *Pseudotta* as a derivative of *Acromyrmex*) are the ones also in this sequence (*Sericomyrmex–Trachymyrmex, Acromyrmex–Atta*) as generally used in this author's studies (e.g., Weber, 1937, 1972a, 1976a) for biological reasons. These four are the higher genera, as noted earlier. The genus *Cyphomyrmex* seems better considered with the Palaeattini and is here considered as the primitive genus.

1. Distribution and Appearance of the Genera

The genus *Cyphomyrmex* is large, heterogeneous, and widely distributed. One species, *C. rimosus* and its subspecies, is unique in that it cultures a fungus in the form of yeast. This ant is also the most widely distributed attine, ranging from Florida and Texas to Argentina; several subspecies have been described. In this genus the workers have unusually large and broadly separated frontal lobes and complete antennal scrobes. The thorax generally has rounded tubercles instead of spines.

The genus *Mycetosoritis* is represented by a species in Texas and others in Brazil. The genus appears to relate some *Cyphomyrmex* species with *Trachymyrmex*. It has not been well studied.

The genus *Mycetophylax* lives primarily in the semi-desert or grasslands of interior South America as far south as the state of São Paulo, Brazil. One species, *M. conformis,* on Trinidad and other islands in the Caribbean, is adapted for living in the sand just above tide level and uses grass for its gardens. The genus has not been found in Central America. The workers lack spines except for the epinotal pair.

The genus *Mycocepurus* is South American with the exception of *M. smithi* which is found in the Caribbean and Panama. The ants have small eyes and are unusually spiny.

The genus *Mycetarotes* has two Brazilian species, one of which has been taken from north Argentina. The biology is not well known, but the fungus gardens of this and the two preceding genera appear to be similar in some important respects.

The species of *Myrmicocrypta* are small, spindly ants with elongate heads that are rounded behind. The eyes are small and strongly convex. Spines or tubercles are conspicuous on the thorax and bear squamate hairs which are characteristic of the genus. Males have huge hemispherical eyes, and the head is strongly sculptured behind. The ants appear to be ants of the forest, but the biology of only two species has been studied.

The species of *Apterostigma* are the only completely nonspinous fungus growers except for *A. epinotale* which has distinct but short epinotal spines. All species have small frontal lobes that conceal the antennal insertions and have heads evenly rounded at the occiput. Workers and females are pilose. The males have small heads with disproportionately long antennae. The wings of both sexes are dark brown. *Apterostigma* is characteristically found in humid tropical areas.

Species of *Sericomyrmex* have the appearance of unusually smooth and silky *Trachymyrmex*. However, their heads lack occipital spines and are cordate in outline. The pronotum is high and convex anteriorly and surmounted by a pair of spines. The males have unusually small heads with almost hemispherical black eyes. The wings of both sexes are dark brown. The genus has recently been shown to exceed *Trachymyrmex* in colony population and size of fungus gardens so that it is probably not primitive as was formerly stated and may be closely related to *Trachymyrmex*. The genus *Sericomyrmex* is restricted to the continental tropical mainland.

The ants of the genus *Trachymyrmex* are generally spiny. The head always has a number of spines or sharp tubercles posteriorly. The thorax is spiny, in some species similar to *Acromyrmex*, and these spines often themselves bear tuberculate hairs at the base. The gaster is more or less densely tuberculate. The female is not much larger than the worker, being distinguished most frequently by a thicker thorax. The males are much smaller and darker. The ants resemble the comparably sized small workers of *Acromyrmex*, but the mandibles of the latter tend to be more massive and species of *Trachymyrmex* are largely monomorphic. From comparable sizes of *Atta* the ants are distinguished by the tuberculate head and gaster. There are more species of

Trachymyrmex than there are in any other genus of attines, and they are distributed from the United States to Argentina.

The genus *Acromyrmex* is broadly distributed and comparable to *Atta* in its biological and economical importance. Emery (1905), Santschi (1925), and Gonçalves (1960) have published taxonomic studies on the South American species. The genus includes species in highly diverse habitats ranging from lowland rain forest and grassland to deserts and high Andean mountains. The typical subgenus, *Acromyrmex*, of Central and South America, comprises most of the species, including the important economic species, *A. lundi* of southern South America, and *A. octospinosus*. The latter species ranges throughout most of South America and has been found on Trinidad, Tobago, Curacao, Guadeloupe, and Cuba. The other subgenus, *Moellerius*, absent from the Caribbean, extends from California to Argentina in semidesert or grasslands areas. The subgenus *Moellerius* differs from *Acromyrmex* (*Acromyrmex*) by having short, stout mandibles, not sinuous on their outer border, and by a lack of lateral spines above the eyes of the worker. One species, *Ac. landolti*, of which *A. balzani* appears to be a synonym or a subspecies, ranges throughout central South America from Colombia, the south half of Brazil and Paraguay to Argentina. The most common grassland and semidesert fungus-cultivating ants are of the two subgenera of *Acromyrmex*. Although the habitats appear to be ecologically similar, the species vary in the extent to which they collect grass. Some, such as *Ac.* (*Moellerius*) *versicolor* in arid parts of southwestern United States, collect pieces of leaves from bushes and herbs rather than from the grasses, which may be absent.

Species of *Acromyrmex* are highly polymorphic in the size of the body, spines, and tubercles. While generally resembling some *Trachymyrmex*, the polymorphism of the worker and the much larger nests and gardens make separation easy. The females and males are also much larger than those of *Trachymyrmex*.

The only species of *Pseudoatta, Ps. argentina* (Gallardo, 1916), was redescribed by Bruch (1928). It appears to be a social parasite on *Acromyrmex lundi* and is known only as males, females, and brood from Argentina.

The most recent publications on the taxonomy and distribution of *Atta* were by Borgmeier (1950, 1959), Gonçalves (1944, 1960), Weber (1968, 1969b, 1972a) and Kempf (1972). The northernmost species are *A. insularis* (Cuba), *A. texana* (United States), and *A. mexicana* (Mexico). Several of the other species are primarily Brazilian, and distribution maps for them have been published by Gonçalves (1960). *Atta* is the one genus with soldiers. It has the largest colonies and does the most conspicuous damage to vegetation. The paradigms, *A. cephalotes* and *A. sexdens*, are typical of the species and data on other important species such as *A. colombica tonsipes* and *A. vollenweideri* are scattered through this chapter.

Maps showing the ranges of primary *Atta* species have been published in recent years (Weber, 1966a, 1969b, 1972a). Brazil has nine species of which four have been recorded only from this country and have not been re-evaluated in recent years. Bolivia and Paraguay have five each and Colombia and Venezuela four each. The species occurring in the most countries is *A. cephalotes,* which has been found in 17 countries. *Atta sexdens* has been found in 14 countries, but has not been found as far north in Central America as *A. cephalotes.*

Table IX lists the names and distribution of the species of *Acromyrmex* and *Atta.* There are many more species of *Acromyrmex;* they are more widely distibuted and show greater variability than the species of *Atta.* The variability is reflected by the subspecific names, some of which may be untenable.

TABLE IX

The Species of *Acromyrmex* and *Atta* and Their Distribution[a]

Acromyrmex Mayr		
Names	Investigator	Distribution
ambiguus	Emery	Argentina
aspersus	Fr. Smith	Brazil, Peru, Argentina
aspersus fuhrmanni	Forel	Colombia
coronatus	Fabricius	Bolivia, Peru
coronatus andicola	Emery	Ecuador, Peru, Bolivia
coronatus augustatus	Forel	Costa Rica
coronatus globoculis	Forel	Guianas
coronatus importunus	Santschi	Costa Rica
coronatus panamensis	Forel	Panama, Costa Rica, Peru
coronatus rectispinus	Forel	Costa Rica
crassipinus	Forel	Paraguay, Argentina, Brazil
disciger	Mayr	Brazil
gallardoi	Santschi	Argentina
hispidus	Santschi	Bolivia
hispidus fallax	Santschi	Argentina, Brazil
hispidus formosus	Santschi	Argentina, Brazil
hystrix	Latreille	Guianas, Brazil, Peru
hystrix ajax	Forel	
laticeps	Emery	Brazil, Uruguay, Paraguay, Bolivia, Argentina
laticeps nigrosetosus	Forel	Brazil, Bolivia
lobicornis	Emery	Brazil, Uruguay, Argentina, Paraguay, Bolivia
lobicornis cochlearis	Santschi	Argentina
lobicornis ferrugineus	Emery	Argentina, Paraguay
lobicornis pencosensis	Forel	Argentina
lobicornis pruinosior	Santschi	Argentina

(cont.)

TABLE IX (*Continued*)

Acromyrmex Mayr		
Names	Investigator	Distribution
lundi	Guerin	Argentina, Brazil, Uruguay, Paraguay
lundi parallelus	Santschi	Argentina
lundi boliviensis	Emery	Bolivia
lundi corallinus	Santschi	Argentina
lundi nigripes	Santschi	Argentina
lundi carli	Santschi	Brazil
lundi decolor	Emery	Argentina
lundi pubescens	Emery	Paraguay, Brazil, Argentina, Uruguay
lundi chacoensis	Santschi	Argentina
niger	Fr. Smith	Brazil
nobilis	Santschi	Brazil
octospinosus	Reich; Forel	Brazil
octospinosus echinatior	Wheeler	Cuba
octospinosus echinatior	Forel	Ecuador, Colombia, Panama, Costa Rica, Guatemala, Mexico
octospinosus ekchuah	Wheeler	Mexico
octospinosus inti	Wheeler	Peru
octospinosus vulcanus	Wheeler	Costa Rica
rugosus	Fr. Smith	Brazil, Uruguay, Paraguay, Argentina, Bolivia, Colombia
rugosus santschii	Forel	Colombia
rugosus rochai	Forel	Brazil
subterraneus	Forel	Brazil, Paraguay, Argentina, Bolivia, Peru
subterraneus ogloblini	Santschi	Argentina
subterraneus brunneus	Forel	Brazil
subterraneus molestans	Santschi	Brazil
subterraneus peruvianus	Borgmeier	Peru

Acromyrmex (Moellerius)		
heyeri	Forel	Brazil, Uruguay, Argentina
heyeri gaudens	Santschi	Argentina
heyeri lillensis	Santschi	Argentina
landolti	Forel	Colombia, Venezuela, Brazil
landolti balzani	Emery	Paraguay, Brazil, Argentina
landolti miltituber	Santschi	Bolivia
landolti cloosae	Forel	Colombia
landolti fracticornis	Forel	Brazil, Paraguay, Argentina
landolti myersi	Weber	Guyana
landolti pampanus	Weber	Bolivia

(*cont.*)

TABLE IX (*Continued*)

Acromyrmex (Moellerius)		
landolti planorum	Weber	Venezuela
mesopotamicus	Gallardo	Argentina
pulvereus	Santschi	Argentina
silvestrii	Emery	Argentina, Uruguay
silvestrii bruchi	Forel	Argentina
striatus	Roger	Brazil, Uruguay, Bolivia, Argentina
striatus laeviventris	Santschi	Argentina
versicolor	Pergande	United States, Mexico
versicolor chisoensis	Wheeler	United States

Atta Fabricius		
bisphaerica	Forel	Brazil
capiguara	Gonçalves	Brazil, Paraguay
cephalotes	Linnaeus	Brazil to Mexico
cephalotes isthmicola	Weber	Central America, Colombia
cephalotes lutea	Forel	Barbados (extinct)
colombica	Guerin	Colombia
colombica tonsipes	Santschi	Central America
goiana	Gonçalves	Brazil
insularis	Guerin	Cuba
laevigata	Fr. Smith	Brazil to Colombia
mexicana	Fr. Smith	United States, Mexico, Central America
opaciceps	Borgmeier	Brazil
robusta	Borgmeier	Brazil
saltensis	Forel	Argentina, Paraguay, Bolivia
sexdens	Linnaeus	Paraguay to Costa Rica
sexdens rubropilosa	Forel	Paraguay, Brazil, Argentina
sexdens piriventris	Santschi	Brazil, Uruguay, Argentina
texana	Buckley	United States
vollenweideri	Forel	Argentina, Uruguay, Paraguay, Bolivia, Brazil

[a] From Kempf, 1972; Weber, 1972a.

G. Availability and Utility of *Attine* Species for Study

Most of the species are obscure and biologically unknown. Others have some special advantages for study in their ranges and a sampling is given here for the United States, tropical America, and temperate South America (Fig. 7A). A comparative study is always valuable since no one species can represent all important features. The ones most commonly studied in nature because of their ecological or economic importance are not necessarily the ones most convenient for detailed study in the laboratory. Frequently the entire colonies of the small, slowly moving species will show, under the binocular stereoscopic microscope, aspects of their behavior that are not readily perceived in samples of the volatile *Atta* workers. Dozens of species scattered through the nine principal genera have been under study in this author's laboratories in Trinidad, Guyana, Panama, home in Argentina, and the United States (Pennsylvania, Wisconsin, and Florida). They have been personally collected in eight South American countries, two of Central America, six West Indian islands and most of the American states that have them (Fig. 10). Distribution of the genera is represented in maps in Weber (1972a).

1. Laboratory Temperatures

Ordinary laboratory temperatures were found to be satisfactory in research conducted during 1933 in Cuba and during 1934–1935 in Trinidad, and later during years in the tropics (Fig. 11) . Similarly Möller in 1893 in south Brazil did not alter the temperature where he worked. Autuori (1942) found that the ordinary laboratory temperature of 20° C or more in São Paulo, Brazil, was satisfactory. In a few short periods in the winter the temperature went below 20° and he used an electric heater to bring it to 22°C. During the summer, when the temperature went above 33°C, distilled water was provided to *Atta sexdens* two or three times daily.

When a laboratory was created in 1954 for the study of tropical species removed to Swarthmore, Pennsylvania, the chief difficulty was keeping the temperature low enough during the summer and warm enough during the winter. Temperature records were maintained (Weber, 1956) and a temperature of close to 25°C was considered desirable. This was verified as ideal for tropical species by taking a series of measurements in the Panama Canal Zone (Weber, 1959) where subsurface soil temperatures in the shade of *Atta* nest sites were found to be close to 25°C at all times of the year.

In July, 1963, colonies of *Trachymyrmex septentrionalis* were kept in two laboratories, one with temperatures that fluctuate from 27° to 33°C with most hours within 1° or 2°C of 30°C. Under these conditions the colonies required more water and development was accelerated. The other laboratory was uniformly 23° ± 0.5°C.

An air-conditioned laboratory was available at the University of Wisconsin during the sabbatical year of 1955–1956. This was kept at about 25°C and was entirely satisfactory for Panamanian colonies (see also Weber, 1979b).

In 1964, a Swarthmore College laboratory kept at 25°–27°C was used for tropical colonies and was maintained until July 2, 1974. Occasional minor fluctuations in temperature took place and both relative humidity and maximum–minimum temperatures were recorded daily.

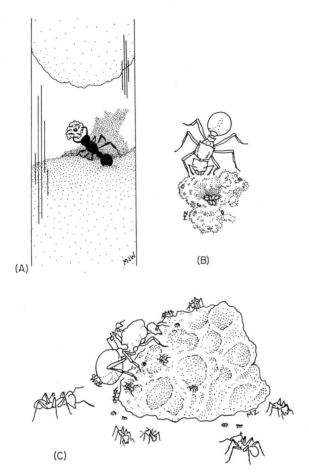

Fig. 10. (A) Incipient garden created by a female of *Acromyrmex octospinosus* after isolation in a vial (Trinidad). (B) Incipient garden with eggs created by a female of *Acromyrmex coronatus* (Venezuela). (C) Growing garden with queen and workers of *Atta cephalotes* (Tobago Island).

2. Species of the United States

The most useful species for study is *Trachymyrmex septentrionalis* because of its wide distribution and ease of study (Weber, 1956, 1960, 1966c). As far north as New Jersey along the Atlantic Coast, its colonies are active May–September. The species is easily recognized by its semicircular crater in open, sandy pine-oak. The sexual brood can be observed June–July.

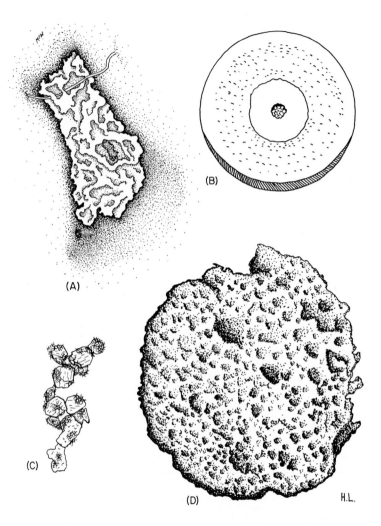

Fig. 11. (A) Garden of *Trachymyrmex septentrionalis* in sand (New Jersey). (B) Small garden of *Trachymyrmex septentrionalis* in the center of an originally sterile nutrient agar petri dish. The ants have removed all agar from the vicinity of the garden and islets of contamination dot the remaining agar. (C) Particles of substrate in a garden of *Trachymyrmex septentrionalis* showing islets of growth of the mycelium planted by the ants. These will develop into gongylidia in a few days. (D) Underside of a 15-cm garden of *Atta cephalotes* from a mature nest in Trinidad. This basal and yellowed part has small cells and thick septa.

In the Southwest, *Acromyrmex* (*Moellerius*) *versicolor* is available in the southernmost areas in deserts and semidesert. The ants frequently form craters of a smaller size than those of *Atta*. Information on the habits starting with Wheeler (1907) and most recently by Gamboa (1975a,b) show marked similarities within the genus as it occurs through the tropics but it is not a thatch-mound builder as is the subgenus in Argentina.

The northernmost *Atta texana,* of Texas and Louisiana, was the subject of the early studies of Wheeler (1907); it was much more common then than it is presently. More recently Moser (1963, 1967) has used it to good advantage for showing its place in *Atta* biology. This species differs from most *Atta* in that it uses pine needles for substrate, something not available to tropical *Atta*. In the early part of the 1900s it nested commonly in more or less open fields. It has now been confined partly to pine woods.

An obscure attine, *Cyphomyrmex rimosus,* is known to culture yeast and is of unique significance in mutualism studies. That and other habits in Florida have been described (Weber, 1955c). It occurs along the Gulf of Mexico coast and in Texas. It is about 2 mm long, dark brown, and slow moving. Perhaps the best field characteristic is its habit of slowly carrying insect feces, such as barrel-shaped caterpillar pellets, in its mandibles to its nest. Some nests may be between layers of loose pine bark close to the ground, others may be under pieces of rotted wood or loose pine bark close to the ground, while others may be under pieces of rotted wood or leaves on the ground, but they are always in moist areas. In the central and southern part of Florida it may be active on all but the coolest days. Its small size makes it possible to have an entire colony confined to a petri dish and placed on the stage of a stereoscopic binocular microscope. Here, under 6–20 power magnification, much may be learned of its behavior toward the brood and the yeast (*bromatia*). Part of a colony of a common West Indian ant, *Wasmannia auropunctata,* nesting in similar situations, may be included in the observation dish and presents an unusual opportunity for examining reactions with a nonattine ant of even smaller size.

In addition to these four species there are others that offer a challenge to the investigator who could readily unearth completely new information of value. Such would be *Cyphomyrmex wheeleri* of Texas to California, *Trachymyrmex jamaicensis* of the West Indies found once on the Florida east coast (for habits, see Weber, 1967a), *T. turrifex, T. desertorum,* and *T. arizonensis, Mycetosoritis hartmanni* (see Wheeler, 1907), and *Atta mexicana,* found just within the border of Arizona. Several more recently described *Trachymyrmex* are unknown biologically.

3. Tropical America

Outstanding for study are *Atta cephalotes* and *A. sexdens* because of their ecological and economic importance. Prominent, however, for what is known of the general biochemical significance of the ant excretions and the fungus is a species of limited distribution, *Atta colombica tonsipes,* because of the work of Martin and associates (*q. v.*). The species was chosen because of its availability in Panama where it is the common *Atta* species and the one that is distributed across the Isthmus (Weber, 1977a).

The other *Atta* species, extending south to the Temperate Zone are *A. capiguara* and *A. vollenweideri* studied by Amante and Jonkman, respectively, and discussed elsewhere in this chapter. In addition any species of *Atta* would well repay field study in the region where it occurs, especially the widespread *A. laevigata* (Da Silva, 1975) of the interior South American grasslands.

Of the leaf cutters of the genus *Acromyrmex,* it is *A. octospinosus* that is being most studied from the aspect of control by Cherrett and his associates (Cherrett, 1972; Cherrett *et al.,* 1968; Robinson *et al.,* 1974; Seaforth *et al.,* 1974; Littledyke *et al.,* 1976). It has also been the subject of life history and fungus garden studies (Weber, 1967b, 1969c, 1972a,b,c, 1977b). Other widespread species of the forest such as *A. coronatus* and *A. hystrix* or, in fact, any species of the genus would well repay study. Techniques available are listed in the aforementioned citations for *A. octospinosus.*

The grass-cutting subgenus *Moellerius* is ecologically important, especially the widespread *M. landolti,* studied by Labrador *et al.* (1972) and considered in this chapter.

The species of the genus *Trachymyrmex* have the same advantages for study as *T. septentrionalis* of the United States. A colony of *T. zeteki* from Panama was kept in the laboratory with the same queen for 5 years, producing 660 workers and 44 males (Weber, 1964). Fragmentary biological studies of many species are found in Weber (1941, 1945, 1946). Queens of *T. septentrionalis* lived 5 years and queens of *T. urichi* lined more than 3 years. Species recognition in the field may depend on unusual characters (Fig. 7A).

The genus *Sericomyrmex* is as useful for study as *Trachymyrmex* although the colonies are larger. The 10-year colony record of *S. urichi* (Weber, 1966b, 1969a, 1976a) is unique for longevity and biological data. If this is representative of the possibilities for research, it may be considered to be the ideal compromise between the small size colonies of *Trachymyrmex* and the colonies of *Acromyrmex* and *Atta* that are much more difficult to keep in the laboratory. The biology of other species such as *S. amabalis* (Wheeler, 1925; Weber, 1941) shows the value of such species for study of unusual guest ants living with them.

Unexpected rewards came from studying colonies of two minor genera in the form of sporophores of their fungus appearing in artificial cultures (Weber, 1966a). One, a colony of *Cyphomyrmex costatus* from Panama, was the source for the first sporophores of any ant fungus to be produced in culture (Weber, 1957a). The significance lay in the fact that only sporophores can be positively identified to species or even genera. The other colony was one from Trinidad of *Myrmicocrypta buenzlii,* whose queen was kept for over 3 years. The ants themselves and their gardens were representative of the numerous species of small and rarely recognized genera whose study would probably be rewarding from many aspects. They are difficult to see because of their 2–3 mm size, inconspicuous color, and slow movement. They become immobile at the slightest disturbance and when that occurs the ants are impossible to discern. They are best recognized when they are carrying substrate or by their nest superstructure (Weber, 1972a). The colonies of *Mycocepurus* are similar in these respects and those of *Mycetosoritis* and *Mycetarotes* are also rarely seen.

The colonies of *Apterostigma* culture a fungus sometimes showing the clamp connections typical of Basidiomycetes and the hyphae lack the distinct gongylidia of most attine fungi. Several species have gardens that are surrounded by the thinnest possible veil of the elongate hyphae (Weber, 1941, 1945). This was thought to be caused by the ants until the fungus was cultured on sterile rose mash in flasks in the absence of ants. The hyphae grew straight out from the irregular protuberances of the medium, then became interwoven millimeters away to form a veil with much clear space below, as from substrate in nature.

Generic distinctions between some of these genera was a subject of study by Kempf up to his death in 1976 and biological studies are needed to answer some of the questions that he raised. This author took specimens of a species in São Paulo State, Brazil, that do not fit the usual generic distinctions and Kempf had other material of this type.

In Kempf's last letter to this author, August 8, 1976, the month of his death, he wrote that "the biggest problem (from the point of view of systematics) is the generic distinction of the group (*Trachymyrmex*), especially from *Acromyrmex, Mycetosoritis,* and perhaps also *Cyphomyrmex* and *Sericomyrmex.*"

Currently, renewed interest is expressed in the fungus cultured by some of the small, obscure genera such as *Myrmicocrypta* and *Cyphomyrmex.* The correct generic and specific names of the host ants are important for this and because of the interest in evolutionary biology of the entire tribe of attine ants. The only existing collection of fungus cultures from a broad spectrum of attine ants (as the Weber collection of the New York Botanical Garden) has been used for fungus systematic, antibiotic, and other unusual qualities. Questions of degree and quality of mutualistic and symbiotic associations will be aided in solution by accurate systematic identification of both the ants and the fungi. A complication is the fact that most of the ants are found in primitive Tropical American areas that are often being modified to some degree by man.

In general, then, many aspects of attine biology remain to be studied in tropical America.

4. Temperate South America

The fauna of temperate South America is a direct and ancient extension of that of tropical America and speciation may well have developed other than by geographic isolation in some instances. Clearly species evolved in the temperate climates both in the lowland pampas of Argentina and Uruguay and in the Patagonian steppes of semiarid and cool climates. A particular feature is the evolution of species of *Acromyrmex* that form mounds of thatch consisting of stems of grass and herbs (Bruch, 1917a,b; Weber, 1972a). In these mounds is either one large and irregular garden or a series of smaller ones (Bonetto, 1959; Gonçalves, 1961; Bucher, 1974; Montenegro, 1973). These nests are significant ecological features of the pampas. An unusual feature is that they are the site for snakes and lizards which may lay their eggs in them (Vaz-Ferreira *et al.,* 1970, 1973; Zolessi and de Abenante, 1973). Gallardo (1951, 1964) regularly found amphibians as well as reptiles in these nests.

The smallest ants in this southern area to adapt to semiarid conditions are those of the genus *Myceptophylax.* They have simple crater entrances to the nest in bare soil and their adaptations to the climate and vegetation are largely unknown. One species, *M. conformis,* is found on the Trinidad and Grenada, West Indies beaches many hundreds of kilometers away from Patagonia and uses grass for substrate to make an unusually loosely structured garden deep in beach sand.

The Argentine fauna is the southernmost attine fauna. It would be rewarding study to determine the factors preventing the spread of *Atta* farther south than 33–34°S.Lat. (35° inland). The common leaf cutter, *Acromyrmex lundi,* is highly urbanized and is a major pest of gardens in the cities.

A comparison of the northernmost species of *Trachymyrmex, septentrionalis* in the United States, and the southernmost, *T. pruinosus* in Argentina, should reveal significant convergent evolution to these perimeters; it is already known that the northern species is normally active no more than 5 months of the year.

There is an *Apterostigma* (*A. steigeri*) of Buenos Aires that also must have made some adaptations to leaving tropical rain forest.

For ease of examination and for general importance *Acromyrmex lundi* is recommended for study in Argentina. The nuptial flights in October–December and their general habits have been described (Weber, 1972a). The newly fertilized queens are easy to capture in metropolitan Buenos Aires and may be isolated and the resulting incipient colonies reared. No special temperature or humidity are required. The species is a leaf cutter of universal habits and it is easier to rear than *Atta* colonies. On the pampas the mound-building *Acromyrmex* species are easy to find and rear. North of Buenos Aires four species of *Acromyrmex* may be found nesting in the same region (Bucher and Montenegro, 1974; Bucher, 1974; Montenegro, 1973) and are significant in several types of ecosystems. Similar species are available for study in Uruguay (Carbonell, 1943; Vaz-Ferreira *et al.*, 1970).

The vast size and vast fauna of Brazil have the range of species outlined under tropical America (Section I,G,3). The southern states such as Rio Grande do Sul have a fauna that includes the species noted previously for Argentina and Uruguay. However, because of their ecological and agricultural importance the *Atta* species are recommended for field studies (e.g., Amante, 1967a,b, 1968, 1972; Mariconi, 1974; Da Silva, 1975). Colonies of the numerous *Acromyrmex* species are also freely available and easy to keep in the laboratory.

A unique study (Jonkman, 1976, 1977) makes use of aerial photographs of the same areas over a period of years to correlate with present *Atta vollenweideri* nest distribution in Paraguay. This is made possible by the fact that these nests are large enough to be detected from heights of 3000 m and contrast as white dots against the surrounding darker soil and vegetation. Two photographs of the same area in 1944 and 1965 combined with 1971–1974 ground studies show successional changes in vegetation. There are many more nests on the predominately grasslands of 1944 than in 1965, but the exact sites of former nests were clearly replaced by "wood-nuclei" or islets of woody vegetation by 1965. These successional changes from grassland to woods were shown in 1971–1974 to be attributable to the sprouting of woody vegetation when the nests were abandoned.

A study of the distribution of eight species of attines and their proximity to one another in an area 18 × 77 m over a 5 year period in Trinidad (Weber 1966b, 1972a) shows an exceptional density of colonies. This is clearly of biological significance.

5. *Latitudinal Limits of Attines*

Kusnezov (1949, 1963) and Weber (1970, also unpublished) have determined the southern (Argentine) and northern limits of *Attini* (United States) in degrees of latitude as shown in the following tabulation:

Genus	Southern	Northern
Cyphomyrmex	38°03′ Province of Buenos Aires	39°06′ California
Mycetophylax	44° Territory of Chubut	not present
Trachymyrmex	38° Province of Buenos Aires	40°22′ Illinois (formerly 40°50′ New York)
Acromyrmex	44° Territory of Chubut	33°10′ California
Atta	35° Province of San Luis	32°40′ Arizona

In a posthumously published article, Kusnezov (1963) considered the zoogeography of ants of South America with particular reference to Argentina, where he had been living. His modified family tree of ant genera (disregarding one that is not accepted by others) is as follows:

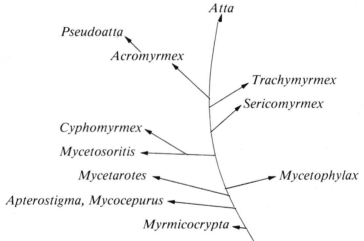

As noted earlier, this author considers *Cyphomyrmex* to be at the bottom of this list, below *Myrmicocrypta*.

H. Ten-Year Colonies in the Laboratory

Colonies of three genera of attines that were kept in the laboratory for 10 years have produced data regarding known rates of substrate intake that correlated with the corresponding size of the gardens (Weber, 1976a,b, 1977b). Some of the data for *Atta cephalotes* were given in Section I,E. Populations of gardens have been determined that may be extrapolated to the situation in nature. Colonies of *Atta cephalotes, Acromyrmex octospinosus,* and *Sericomyrmex urichi* were originated in 1965 from single fecundated females, the normal pattern in attines, and were kept in the laboratory within weeks, or in one case within the hour, of her fecundation. All were taken in Trinidad, less than 1 km from each other.

These records illustrate some factors in research. For example, none was established as a 10-year project and these data were collected from the survivors of a great number of colonies collected over the years. What made the records possible were the daily systematic taking of notes, weighing of substrate, and the monthly (or more often) estimation of resulting volumes of gardens. Aside from the problems of arranging the trips to the tropics and finding the colonies in the first place, all colonies survived many human induced hazards, including difficulties in creating and maintaining a laboratory, absences of the investigator and necessity of transporting the colonies to other states.

The records are among a larger number kept under similar conditions in the same laboratory including one published for *Atta colombica tonsipes* (Weber, 1977a). A 1970–1980 colony of *Atta sexdens* remains to be described. All expansion rates of the gardens and ant colonies show a sigmoidal curve of growth. The three colonies of 1965, and a number of others that dated from July, 1965, expanded slowly at first, then grew rapidly by the end of the year. This expansion rate accelerated markedly in the years 1966 and 1967. The peak was reached in two cases in 1968. In *Atta cephalotes* the maximum amount of substrate was used in 1968, but the gardens expanded to their maximum volumes in the next 3 years. By 1971, the rate of expansion was slowing in all colonies. Until 1974, it was fairly steady and the decline in mid-1974 and later was attributable to external causes.

The rate of substrate use and the volumes of gardens accurately reflect the populations of the ants (Weber, 1972a). The relationship between the volume of the garden and ants is direct. Only a large population of ants can cut, incorporate, manure, and constantly tend a large volume of fungal mycelium. Wherever there is an accidental loss of a large number of ants, the gardens immediately show deterioration, unless a large brood of ants is maturing at the same time.

1. The Sericomyrmex Colony

This account of the colony of *Sericomyrmex urichi* Forel (Weber, 1969a, 1976a) for the first time established the intermediate position of the genus in the evolution of attines and its closeness to *Acromyrmex* and *Atta* in leaf cutting and fungus culture. Field evidence had shown that a colony had 1691 counted workers and this was higher than known for any *Trachymyrmex* species. Species of the genus itself had been occasionally mistaken for *Trachymyrmex* because of the general worker similarities in feeble polymorphism and habitus. No record of any *Trachymyrmex* colony to date has approached the present one in size of colony or garden.

Species of *Sericomyrmex* are known only from the tropical mainland (considering Trinidad and Tobago as essentially mainland because of their relatively recent separation from Venezuela). The present species is common in Trinidad. Specimens of the colony had been furnished to other investigators for pheromonal studies (e.g., Blum and Portocarrero, 1966). They did not follow trails of other attines.

The ants are shown in Weber (1966a) and have small queens, proportionate to those of *Trachymyrmex* in size, although distinctly larger than the worker. The living weight of a female was 7.62 mg, that of a male 2.85 mg, and the average weight of 10 workers was 1.97 mm. There was the queen, about 20 workers, and about 70 ml of garden when the incipient colony was found in clay soil at a depth of 20 cm above a young *Atta cephalotes* colony. The colony was collected on July 15, 1965, and had probably originated from the female being fecundated in the preceding late May. It was transferred to the United States laboratory and kept at 25°–27°C with relative humidity in the laboratory ranging from 10–35% in January–April, 27–52% in May–August, and 9–36% September–December over the years. For the remainder of 1965 the garden was generally smaller than when originally collected.

In late 1965, the small plastic box in which the original collecting vial was placed was itself placed in the standard 2250-ml plastic box used in this laboratory for colonies of many species. This box remained the site of the garden until the colony died in April, 1976. The capacity included the space under the domed cover and the ants attached the uppermost cells to this ceiling (Weber, 1969a, Fig. 1, shows an intermediate condition in 1967).

This single container was filled with the garden in early 1968, and remained essentially filled through 1974, 7 years in all. During this period there were brief periods in the summer, when humidity was higher and more flowers were available, when a second and similar container connected with the original had a small garden up to a maximum of 400 ml. This garden was removed by the ants in the fall and was not essential to their normal development. In nature such additional gardens would be formed and seven gardens were present in one colony. No garden in nature of 2250 ml volume is known.

Substrate and Its Monthly Use. The substrate furnished the ants included most of those acceptable to *Acromyrmex* and *Atta* colonies but was used in smaller quantities. Granulated cassava meal, the pith (albedo) of grapefruit and orange, entire rose flowers, and young rose leaves were often given. All flowers and leaves were weighed while fresh. The annual totals of substrate used by the ants are represented in Table X. About 1 kg of substrate annually was used to maintain the garden of approximately 2250 ml. During

TABLE X

**Annual Use of Substrate and Annual Maximum
Growth of Gardens by the Ten-Year Colony of
*Sericomyrmex urichi***

Year	Substrate used (gm)	Maximum garden size (ml)
1965	10 (est.)	70
1966	188	1100
1967	828	2150
1968	1093	2550
1969	1221	2150
1970	1052	2250
1971	981	2250
1972	1069	2250
1973	1146	2620
1974	561[a]	1800

[a] To July 1.

1966–1974, 8139 gm were taken in the 9.5-year period. During the first 6 months the intake was probably less than 100 gm making a total of approximately 8200 gm in 10 years compared with 51 kg used by the much larger *Atta cephalotes* 10-year colony.

The estimated volume 1 year later (mid-July, 1966) and on subsequent anniversaries (unpublished) was as follows:

Year	Volume (ml)	Year	Volume (ml)
1965 (original)	15	1970	2250
1966	60	1971	2250
1967	1900	1972	2250
1968	2550	1973	2620
1969	2150	1974	2250
		1975	1800

The colony died in April 1976 and had an earlier garden volume of 1000 ml.

The first spurt in garden volume took place in the second half of July, 1966, when it increased from 60 ml to 100 ml. During this month it used 10 gm of substrate, mostly rose flowers. As the result of increased brood maturing, the garden expanded to 1100 ml by the end of December. On January 2, 1967, 106 ants were counted at one time while cutting orange pith for substrate and 100 additional ants were visible. A total population of 500 to 1000 was estimated. On January 15, 1968, there were 371 ants at one time cutting grapefruit pith and others visible brought the total to 667.

By November 16, 1967, the garden had increased to 2150 ml, close to the capacity of the chamber (2250 ml) and weighed 270 gm with ants. The high point in workers visible at one time totaled 1309 on December 31, of which those on grapefruit pith were 278. On February 8, 1968, there were 1055 ants visible at one time, including 409 cutting pith. In all cases as many as three to five times as many ants were estimated to be in the garden cells at work as would be outside it at any time.

Throughout the life of the colony, after 1967, all dead were collected monthly, separated into workers, females, and males and counted. Tables in Weber (1972a, 1976a) show their monthly totals through 1975. Yearly totals may be combined as shown in the following tabulation:

Year	Workers	Females	Males
1967	407[b]	0	0
1968	2505	372	666
1969	5878	1	3888
1970	5121	85	3649
1971	3816	235	2405
1972	7663	366	1047
1973	4603	603	3014
1974	4866	849	1651
1975	1156	42	6
1976 to May	1207	32	0
Total 1965–1976[a]	37,222	2585	16,326

[a] From 1965–1966 no deaths of females or males occurred.
[b] Includes worker deaths in 1966.

In nearly 11 years of colony life a total of 56,127 ants were produced by one queen. The ratio of males to females actually collected was 6.32:1. It had been 9.4:1 until May, 1971, a difference attributable to either an actual difference in ratio in different years, as is possible, or a failure to count some dismembered males in one period.

2. The Acromyrmex Colony

The colony of *Acromyrmex octopinosus* was precisely dated from the day of capture in Trinidad of the newly fertilized queen. She was part of a nuptial flight seen at 0645 hr on July 1, 1965, and was taken with others by 0730 hr when the relative humidity approximated 100%, the air temperature was 30.6°C in the sun and it was 27.0°C in the shade. It was then calm; 41 mm rain fell the previous afternoon and rain fell later in the morning. Each female was isolated in her own container and was not carrying anything between her mandibles (Fig. 10). Six of the females ejected the contents of her infrabuccal chamber and initiated incipient fungus gardens. Three queens produced their own colonies independently and survived for years in the aforementioned *Sericomyrmex* laboratory and under the same conditions. First they were given hibiscus leaves. One of the three had a garden approximately 2 × 2 mm on which four eggs were found July 5. This garden was 2 × 3 mm the next day and 7 × 6 mm July 15, and had fully formed staphylae with gongylidia (see below under Section II,A, Fungus). In an early account of the first year's growth (Weber, 1967b), these three colonies were labeled colonies B, C, and D; colony A was older and captive for only 17 days. Colony C survived for over 10 years. The living weight of Trinidad workers of this species were 0.65–0.85 mg for 10 minima, 2.89–8.85 mg for 10 media, and 12.62–15.59 mg for 10 maxima. Others of intermediate sizes were not weighed in this series but were found at other times to weigh 0.40–8.85 mg. An alate female weighed 36.44 mg and an egg-laying female weighed 26.65 mg, the difference probably being largely attributable to metabolism of wing muscles.

The queen of the 10-year-old colony was originally placed in a shell vial and the garden did not thrive at first. On August 8, she was given a fragment of a 1 or 2 ml mature garden from another colony. By September 3, she had not raised a brood, primarily because of neglect from the absence of the investigator. The shell vial was placed in the standard 2.25 liter plastic box of 2250-ml capacity on January 6, 1966, by which time she had reared the first broods. The next day she cut a small part of a leaf of *Bryophyllum calcynum,* having previously used rose and other leaves. On March 7, there were an estimated 12 ml of garden, a few media, and mostly minima workers. The garden deteriorated in the next month, then expanded to 45 ml by June 13. On July 7, there were approximately 80-ml garden and an estimated 150 workers of which about 33% were minima and 66% were media.

a. Substrate.   Beginning with hibiscus leaves in Trinidad, the next substrate furnished on removal at the United States laboratory was rose leaves, followed by minor amounts of other plants. The chief plants used in the 10-year study included: leaves of ash (*Fraxinus americana*), *Bryophyllum calcynum,* camelia, *Chamaecereus Silvestrii, Crassula arborescens,* flowering cherry, forsythia flowers, *Graptopetalum paraguayense,* grapefruit pith, *Kalanchoe pinnata,* orange pith, pecan, rhododendron, *Sambucus canadensis, Sedum Adolphii,* and *Stachys floridana.*

In 8.5 years, 19,408 gm of fresh substrate was used by the ants (Table XI). The maximum amount was used in 1968, when the maximum garden size was reached. In the next 4 years the garden maximum size declined as did the substrate use. One of the colonies declined and, by October 9, 1970, had a garden of 80 ml. It gradually revived and, in 1971, had an estimated population of 500–600 workers when the garden was 300 ml. The garden size was the same in March 1973, and the colony died, in July 1974, from mechanical shock and

TABLE XI

Annual Use of Substrate and Annual Maximum
Growth of Gardens by the Ten-Year Colony of
Acromyrmex octospinosus

Year	Substrate used (gm)	Maximum garden size (ml)
1966	500	480
1967	2214	3900
1968	4722	6100
1969	3929	5300
1970	2689	5000
1971	1496	3100
1972	1001	2700
1973	2137	2900
1974	720[a]	2500
1975	—[b]	1700[a]

[a] To July 1
[b] Not weighed.

temperature changes when it was moved to another state. The second colony died, in March 1973, from neglect during the investigator's absence.

The populations of the three July 1, 1965, colonies were estimated to be 10,000 workers in two and 8000–15,000 in the 10-year-old colony, on January 16, 1968.

b. Sexual Broods. Female sexual broods appeared in the three July 1, 1965, colonies as follows (referring to the 10-year-old colony as C, the others as B and D):

1967. March 2. Two callows (Colony D), the first to appear in any colony. There were eight here March 31, 13 on May 16, and 20–25 on September 28.

1968. February 25. New female brood in Colony D. Colony C had its first maturing female on April 15. The rest of the 1967 female brood of Colony D died during the first 3 months while the new brood was maturing in February. This colony was not healthy, perhaps from using an excess of azalea leaves, cast out some worker pupae May 20 and on the May 28, cast out 10 dead females that appeared to be unusually small. The colony had been using too many flowers for substrate. The alate females of the two colonies fed on the staphylae of the gardens as did workers. Some lost their wings in August. Several of these dealate females cut and carried rhododendron leaves to the gardens in November; they also worked over the exhausted substrate as did workers, ingesting or inactivating something unknown. Both colonies had mature female broods the remainder of 1968.

1969. Colony C had a female brood of larvae and pupae on February 7. On the February 14, all three colonies had female broods for the first time. Some remained alive for the remainder of 1969. A fourth and unrelated Trinidad colony taken in 1964 also had a female brood by the end of 1969.

1970. Colonies B, C, and D retained their female broods into June, few remaining alive in September.
1971. Colonies B and D had a few females during the year, only D retained any at the end.
1972. Colonies B and D had females during the year, a few surviving to the end in Colony B.
1973. No females seen in Colonies C and D in January, many in Colony B of which 10 at one time were cutting and carrying pith to their garden. The queen of Colony B was found dead on March 19, immaculate and entire. She was 8 years old. Females of the last brood were smaller than she was and, when dead, were carried by workers to a refuse pile and left there. The colony came to an end during the absence of the investigator.
1974. All ant colonies had to be moved to another state on July 2, and the queen and workers of Colony D died from mechanical shock and temperature changes affecting the garden. Colony C survived a second move to Florida. The queen died early 1976 and the last workers died in June.

c. Other Sexual Broods of Trinidad Colonies. A colony of *A. octospinosus* taken in July 1964, by Peter Weber, when it was then fully 1 year old, was more vigorous than the above colonies for some years. The queen was 11 years old when she died as a result of moving the colonies to another state. The colony produced several female broods. Another 1965 colony of unknown age produced both male and female broods by July 13, 1966. Lewis (1975) has studied this species in the field in Trinidad and has furnished data (Table III) on laboratory colony growth.

3. The Atta cephalotes Colony

The general account of the species is given above under Section I,E. The following account applies specifically to the 10-year-old colony (Weber, 1976b) as do Tables XII and XIII.

The first brood of workers emerged to the soil surface on July 2, 1965. The queen had probably been fertilized the preceding May. She was one of five from the same nuptial flight that descended to the ground under trees within a few square meters of each other. Two of the resulting sister colonies were also maintained in the same laboratory, affording comparative rates of development (Fig. 10). The five females may have originated from one of several mature nests within 30 m of the site, one being less than 150 m distant. Two colonies were left in nature and one was examined 2 years later, just before it was poisoned. The first broods of the five colonies emerged to the soil surface on June 28 and July 2. They may have spent 1 or 2 weeks below the surface after maturing in June. The account which follows deals with the 10-year-old colony (colony No. 4425) with brief references to the sister colonies No. 4423 and No. 4424.

The colony activities above ground were followed before it was excavated July 15 and transferred to the United States July 17. The nest entrance was closed with approximately 10 ml of dried, curled leaflets each previous day by 1600–1830 hr. The following morning it was opened after receiving about 30

TABLE XII

**Annual Use of Substrate and Annual Maximum
Growth of Gardens by the Ten-Year Colony of**
Atta cephalotes

Year	Substrate used (gm)	Maximum garden size (ml)
1965[a]	1513	3200
1966	5718	4350
1967	6290	7800
1968	7008	10,000
1969	6915	11,600
1970	6142	11,700
1971	5800	11,300
1972	3949	7100
1973	5468	6500
1974	2189	7900
1975	—[b]	3600

[a] From July.
[b] Not measured.

min of sun, but as the days passed the time varied according to the morning cloudiness. In opening the entrance, a few small workers penetrated the loose debris from below, then cleared the leaflets. A full 30 min was necessary to clear the 10–20 mm entrance in dry weather. More than a score of minima to small media would be engaged. The earliest the entrance was opened was 0750 hr; the latest was 1115 hr.

The excavated garden and colony was placed in the standard 2.25-liter plastic box for transfer to the United States. After 2 days of rebuilding, the garden bulk July 19 was an estimated 220 ml.

All substrate furnished the ants in the laboratory was fresh and was weighed immediately. While using dried substrate would have facilitated some computations, it was felt that natural substrate, supplied fresh, duplicated the situation found in nature. This also furnished the normal water protoplasm essential to the metabolism of the fungus and perhaps added some volatile nutrients imbibed by the ants as they manipulated the cut sections. Dried leaves were supplied to the ants on occasion in the winter and were acceptable as they are in the natural dry season.

Rose flowers were given to the ants at first and July 24–25 they used 20 gm. By August 5 the garden was 500 ml, increasing to 1800 ml by September 8. Conditions in the laboratory were as described previously for the other attine colonies.

Rose flowers and leaves continued to be the principal substrate until November, when rhododendron leaves were supplied. In following years the annual sequence of major substrate was: December–February, rhododendron; March, forsythia flowers and green leaves; May–June, flowers of wisteria, azalea, flowering cherry and rhododendron (in limited amounts); July–September, ash (*Fraxinus americana*); and through summer to November, rose. In Florida, pecan, *Sambucus americana,* and *Stachys floridana* leaves were supplied. Leaves of a variety of other plants were accepted on occasion and stems of the succulents, *Kalanchoe tubiflora* and *Echeveria pulvinata.*

In July–December 1965, the ants used 1513 gm substrate; in January–June 1966, 1913 gm; in July–December 1966, 3805 gm; and in January–June 1967, 2773 gm for a total of 10,004 gm in 2 years. The first 6 months of colony usage probably compares well with that in nature, but the colonies probably expand much faster in nature later and use many more leaves.

The ratio of volume to weight for representative gardens was determined for five species (Table XVII). These were normal gardens in the laboratory, quickly isolated from the rest of the nest so that all ants at work in them were included. Most were full size, largely filling the standard plastic chamber. One was the same garden after it has grown 200 ml in 79 days. With one exception the ratio were 1:8 or 1:9 (gm:ml) the exception being 1:7.3 (gm:ml) and probably it was in the postmaturity phase. These correspond to the ratios found in *Atta vollenweideri* gardens in nature (Jonkman, 1977). The ants included in the gardens weighed less than 20 gm as shown by the date presented in Tables IV–VI.

4. Rates of Garden Growth

The rates of garden growth compared with the intake of substrate shows significant differences among the species. A direct comparison was repeatedly made over many days when the same quantity of the same leaves was given to various colonies and species.

Two colonies of each of four species (*Acromyrmex octospinosus, Atta cephalotes, A. sexdens,* and *A. colombica tonsipes*) are compared with one another in their third and fourth year for monthly garden size (ml) and weight of fresh substrate (gm) in Tables XIV and XV. Within a species the colony records differ and the tables permit an evaluation of the vigor of one species with another. As earlier described, *Atta cephalotes* is generally the most vigorous, followed by the *Acromyrmex,* then by *A. sexdens,* and last by *A. colombica tonsipes.* Table VII records growth of three *A. sexdens* colonies. Table XVI gives the rate of cutting 20 gm grapefruit pith by six attine species in the laboratory, and Table XVII gives the volume and weight of entire gardens in five species.

Growth of Panamanian colonies of *Atta colombica tonsipes* Santschi and other features of their biology have recently been published (Weber, 1977a)

TABLE XIII

Comparison of Monthly Garden Size (ml) and Weight of Substrate Used (gm), *Atta cephalotes* Ten-Year Colony of Weber (1976b)

	Jan	Feb	Mar	Apr	May	June	July	Aug	Sept	Oct	Nov	Dec	Total Year
1965													
garden							200	500	1800		2400	2900	
substrate							128	139	305	263	323	355	1513
1966													
garden	3200	3300	3180	3200	3500	3550	3200	3500	3800	3700	4100		
substrate	215	159	255	290	394	600	620	840	860	500	520	465	5718
1967													
garden	4150	3800	3800	3800	3800	4750	5000	5900	7200	7750	7800	7500	
substrate	340	270	308	350	755	750	655	640	580	605	500	537	6290
1968													
garden	5600	4600	4500	4750	4600	3900	4200	6400	8000	8800	9800	10,000	
substrate	365	370	417	581	600	540	930	700	560	850	615	480	7008
1969													
garden	9900	9700	9500	8700	8000		9000	9200	11,500	11,600	11,000	10,700	
substrate	550	365	305	395	435	560	880	850	650	750	595	490	6825

Year		1	2	3	4	5	6	7	8	9	10	11	12	13
1970	garden	11,300	11,500	11,300	11,000	10,000	9800	7900	8900	9400	9400	8800	8800	6442
	substrate	230	675	380	476	485	580	755	670	595	607	399	590	
1971	garden	8500	8600	9000	9100	9200		10,400	11,300[a]	5100	5000	5600	6100	5800
	substrate	500	387	595	490	485	610	668	410	430	485	450	290	
1972	garden	6800	7100	6400	6900	5600	6000	5700	4900	4400	4100	4400	4000	3940
	substrate	450	310	340	400	300	320	330	420	180	400	260	230	
1973	garden	4200	4300	4500	4700	5400	5400[b]	3800		5800	6500	6500	6500	5468
	substrate	110	130	150	260	663	530	540	965	734	525	320	541	
1974	garden	7200	7600	6000	6200	6500	6200	5700			7600	7900	6600	2189[c]
	substrate	325	241	395	426	562	240							
1975	garden	3600	1900	1700	1600	1500	1200	600	died					
	substrate[d]													

[a] Removed two chambers.
[b] Old garden cast out by ants.
[c] To July 1.
[d] None measured.

TABLE XIV

Comparison of Monthly Garden Size (ml) and Weight of Substrate (gm) Used in the Third Year by Laboratory Colonies of *Acromyrmex* and *Atta*

Species	Jan	Feb	Mar	Apr	May	June	July	Aug	Sept	Oct	Nov	Dec	Total Year
Acromyrmex octospinosus													
garden A	4300	4800	4100	3300	500	1250	1750	2500	3600	4200	4000	4000	
substrate A	325	257	330	434	410	310	295	435	320	433	380	310	4239
garden B	3600	3300	3700	3900	2600	1950	2550	4300	5500	6000	6100	5400	
substrate B	290	190	260	510	432	360	530	485	450	470	415	330	4722
Atta sexdens													
garden A	3500	1600	1800	1900	1900		4000	5100	6000	5300	5700	5300	
substrate A	120	173	100	233	183	210	410	485	450	500	300	265	3429
garden B	3300	3500	3800	3300	2600	1700	2400		4600	4600	3800	4800	
substrate B	205	190	160	190	300	260	480	800	582	390	320	285	4162
Atta colombica													
garden A	1500	1600	1600	1800	1600		2500	2600	3500	3700	3400	2900	
substrate A	138	96	94	230	210	182	200	318	580	210	215	210	2683
garden B	1600	1300	1500	1300	1000		1700	2800	3400	1900	700	500	
substrate B	86	99	80	212	170	180	240	310	265	62	22	10	1736
Atta cephalotes													
garden A	4900	6200	5700	5350	5600	6200	5100	5100	5700	6365	5600	7400	
substrate A	350	260	326	345	590	690	583	485	540	590	457	415	5631
garden B	3950	4500	3900	3900	2700	1900	1600	1900	2600	3700	3900	4100	
substrate B	290	220	269	202	430	359	230	250	218	319	250	193	3230

TABLE XV

Comparison of Monthly Garden Size (ml) and Weight of Substrate (gm) Used in the Fourth Year of Laboratory Colonies of *Acromyrmex* and *Atta*

Species	Jan	Feb	Mar	Apr	May	June	July	Aug	Sept	Oct	Nov	Dec	Total Year
Acromyrmex octospinosus													
garden A	4200	4600	5100	5200	4500		3900	4400	4800	5200	4700	4600	
substrate A	345	290	320	262	255	325	405	475	410	380	325	295	4087
garden B	5300	5000	5200	5200	4800		3900	4600	4800	5400	4700	4800	
substrate B	340	307	295	307	250	285	325	485	410	380	325	260	3969
Atta sexdens													
garden A	5500	5600	5200	5300	4900	5700	4300	2600	4300	5600	5700	5600	
substrate A	205	316	243	270	395	305	305	387	452	470	264	375	3987
garden B	4600	4500	4000	3500	2500	2700				1400	1200	150	
substrate B[a]	153	153	191	265	300								
Atta colombica													
garden A	2600	2300	2100	1400	1200	1700	1500	2000	4300	5100	5300	4300	
substrate A	127	132	92	84	230	205	240	280	380	435	230	180	2815
garden B	600	800	900	600	700	2300	1900	2000	2900	3100	2500	2000	
substrate B	79	56	75	45	162	182	175	265	250	152	75	137	1653
Atta cephalotes													
garden A	6900	5000	5100	5200	4700	3400	4100	7000	9100	10,000	10,600	11,000	
substrate A	422	380	320	478	540	520	590	860	540	850	670	457	6627
garden B	3600	3700	4200	3600	2300	1700	2500	4100	6200	6800	9400	10,000	
substrate B	327	200	230	288	495	260	285	605	400	810	585	330	4815

[a] See text.

TABLE XVI

Rate of Cutting 20-gm Grapefruit Pith by Attine Species in the Laboratory

Species	Hours	Rates (gm/hr)	Garden size (ml)	Colony
Sericomyrmex urichi	15	1.3	2200	A
Acromyrmex (A.)	10	2	800	A
lobicornis	10	2	800	A
Acromyrmex (A.)	10	2	2000	A
octospinosus	6	3.3	2000	A
	6.7	3.0	4500	B
	6.5	3.1	5400	C
	4.5	4.4	3700	D
	6.5	3.1	3700	D
Atta cephalotes	6	3.3	7800	A
	8	2.5	7800	A
	10	2	7800	A
	15	1.3	1000	B
	12	1.7	2700	C
	8	2.5	8400	D
	6.5	3.1	10,600	E
Atta sexdens	9.5	2.1	1200	A
	7	2.9	2200	B
	23	0.9	2200	B
	9.3	2.2	2000	C
Atta colombica	8	2.5	1800	A
tonsipes	9.3	2	1800	A
	10	2	1700	B
	12	1.7	1700	B

TABLE XVII

Weight and Volume of Entire Gardens Containing Ants

Species	Weight (gm)	Volume (ml)	Ratio
Atta cephalotes[a]	130	1200	1:9
Atta cephalotes[b]	203	1800	1:9
Atta cephalotes[b]	227	2000	1:9
Atta cephalotes	232	1850	1:8
Atta cephalotes	258	2000	1:8
Atta cephalotes	277	2100	1:8
Atta colombica tonsipes	177	1650	1:9.3
Atta sexdens	274	2000	1:7.3
Acromyrmex octospinosus	70	550	1:8
Acromyrmex octospinosus	226	1800	1:8
Acromyrmex octospinosus	241	2150	1:9
Sericomyrmex urichi	270	2150	1:8

[a] Weber (1966a, p. 598).
[b] Same garden on April 20, 1966, and July 8, 1966, respectively.

and permit a comparison with *A. cephalotes* and *A. sexdens. Atta columbica tonsipes* is adapted to tropical forests and is the common species of the Panama Canal Zone, where *A. sexdens* and *A. cephalotes isthmicola* also occur. A census of an area 15.24 m² was made in 1971, 1973, and 1976, at one site under a tree and the location of numerous incipient colonies was mapped. Many were taken close together. There were 118 independent nests in 1973, the gardens being at depths of 5–8 cm in the clay soil and of volume 30–65 ml. Living weights of the workers were comparable to those of *A. sexdens* and *A. cephalotes,* but the queens were lighter. The average weight of six was 172.1 mg. Three colonies had growth increases from 120 to 650 ml, from 20 to 800 ml, and from 20 to 2100 ml for their first full year, which demonstrates the disparity in growth of individual colonies as in other species. A 1971 colony showed the following growth of gardens correlated with monthly substrate intake (see tabulation below):

Year	Substrate use (gm)		Total for year	Garden size (estimated in ml)	
	Minimum	Maximum		Minimum	Maximum
1971	2	35	85	2.5	300
1972	23	155	841	220	1000
1973	32	640	2515	200	4100
1974	25	58	177	250	1600
1975	Not measured		—	500	1500

The garden size fluctuated during January–July, 1976, from 330 to 1020 ml.

The substrate furnished during the first month was first rose, then hibiscus petals or forsythia leaves; the sequence followed later was the same as for the *A. sexdens* and *A. cephalotes* colonies.

This is the species used by Martin and associates for biochemical studies (*loc. cit.*) and other features are described in Weber (1969b, 1972a,b). It nests somewhat lower than *A. cephalotes isthmicola* in the hilly part of the Panama Canal Zone and has alternated with both this and *A. sexdens* in Panama City at the same sites, although the latter usually occupies more open and drier places. An *A. c. tonsipes* colony was taken 43.4 m from a mature *A. sexdens* nest at the Pacific end of the Panama Canal and 35 m inland from the beach.

I. Guests, Parasites, and Predators

The large nests and populations of the leaf cutters offer many opportunities for guests, parasites, and predators to avail themselves of food or shelter. Guests are species whose status has not been clearly defined but appear to be primarily facultative, making more or less temporary use of the nests as shelter. They may be simply transient. Parasites are clearly related and derive more benefit from the ants than the ants do from them. If of small size, such as mites (Table XVIII), they affix themselves to the appendages or the body of the ant (Fig. 12) and both organisms survive. A social parasite, such as *Monomorium* (*Megalomyrmex*) *wheeleri* (Weber, 1940, 1941) with *Cyphomyrmex costatus* appears to live in the nest of the latter and feed on the garden or ant brood. The parasite category has not been thoroughly explored and it is anticipated that many tiny flies and wasps (such as the *Trichopria* of Weber, 1957a) will be found to be associated with the entire range of genera. Predators attack the ants in various ways. They range in size from the ant bear (*Myrmecophaga*) to Doryline ants to tiny mites. This category also includes predation on the ant fungus or the substrate.

An unusual relationship that is a tribute to the magnitude of the *Atta* nest may be termed an ecological guest. These are the mammals such as the armodillo, deer, and peccary that are attracted to the salty soil brought up by *Atta* or that tunnel into the nests for shelter or prey (Brandt and Mayer, 1977, *in litt.* Gallardo (1964) and Vaz-Ferreira *et al.* (1970) have described amphibians and reptiles using *Acromyrmex* nests for incubating their eggs.

A list of 157 species of animals (Weber, 1972a) indicates the range of known animals involved in some relationship with attines. This will undoubtedly be markedly extended with further study. Eidmann (1938) has listed guests of *Acromyrmex* nests in Brazil.

Nematodes are commonly found in soil and occur with annelids in wet cassava substrate of experimental nests. In one Panama nest of *Trachymyrmex zeteki* they were of the *Rhabditis* type and may have evolved adaptations to this life (Weber, 1964). Kermarrec (1975) found nematodes in nests of *Acromyrmex octospinosus,* mainly of the genera *Rhabditis* and *Diplogaster.* They occurred principally in abandoned nests, as in the colonies in this author's laboratory. However, five inhabited gardens were infested. Large

TABLE XVIII

Location of Mites on Live *Atta cephalotes* Workers

Ant size (mm)	No. on head	No. on thorax	No. on gaster	No. on appendages	Total mites	Total ants
2	2	2	1	1	6	2
3	9	2	4	3	18	5
4	49	11	17	7	84	13
5	44	15	16	6	81	8
6	2	0	1	0	3	2
Totals	106	30	39	17	192	30

worker ants had nematodes in the buccal cavity and the other ants were found to remove those ants infected. Pale annelid worms (*oligochaetes*) of small size may become abundant in observation nests and infest the abandoned substrate. Small gastropods occurred in one observation nest and appeared to be an important pest, possibly preying on the brood (Weber, 1957a). Pale white sowbugs were believed to be the chief cause for the disintegration of a colony of *Apterostigma dentigerum*. Small millipeds that were snow white when young and pale brown when older were associated with deteriorating *Cyphomyrmex costatus* laboratory colonies (Weber, 1957a). A larger species feeding on *Atta* exhausted substrate is shown in Weber (1972a). A small brown spider, *Triaeris patellaris* Bryant, was seen to prey on the collembola, *Cyphoderus inaequalis,* in *Cyphomyrmex costatus* nests and undoubtedly also killed the ants (Weber, 1957a). Roaches (Fig. 13) may adapt to nests.

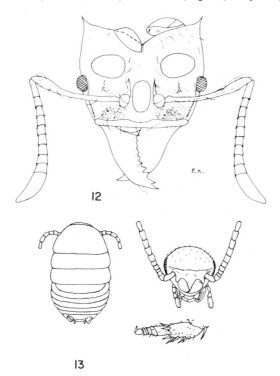

Figs. 12 and 13. Fig. 12. Head of *Atta cephalotes* worker with five mites on the front and rear of head. An incidence of mites such as this may be found in deteriorating colonies. Mites in some attine nests are known to feed on the fungus. The mites are 0.25 mm in length. Fig. 13. Specialized roach, *Attaphila bergi* Bolivar, from the nest of *Acromyrmex lundi* in Argentina (after Bruch). Their smooth bodies, compact antennae, and backward directed spiny legs permit them to dart through the nests and avoid the ants.

The fungus growers lack the distinctive relations with birds that the army ants show with the ant thrushes (Formicariidae). Both hoatzin birds and *Atta cephalotes* may harvest the same leaves, as shown by the stuffed crops of several birds that were shot above an ant nest.

Army ants (Dorylinae) prey on the brood of the smaller attines, but files of *Eciton* and of the large attines generally ignore one another. A file of *Acromyrmex octospinosus* carrying both fresh and withered sections of leaves and flowers took a direct path to their nest and this led them through an *Eciton vagans* file. The leaf cutters unhesitatingly marched through the *Eciton* files, each attine bearing a large section of yellow petals or green leaves. Many of the *Ecitons* threatened (by advancing or opening the mandibles) and even tried unsuccessfully to grasp the attines but nearly all of the latter kept on; few turned back.

The mites are of considerable variety. Under the microscope, several were observed eating the fungus as it was being ingested by the ants. The mites held on to the mouthparts of the ants as the latter were feeding (Weber, 1955c, 1972a). The bizarre shapes of mites and their positions on other ants have been described by Rettenmeyer (1963).

Small, shiny, and furtive roaches (Blattaria) are occasionally seen in the nests of the leaf cutters. They are so smooth and slippery (lacking protruding wings) that the ants are unable to secure a grip on them. They generally dart from cell to cell of the gardens and avoid the ants that might attack them. It is probable that the minima workers are not quick enough. They have been known since the days of Bruch (1916) who showed *Attaphila bergi* Bolivar (Fig. 13) in the nests of *Acromyrmex lundi* in Argentina. Moser (1963) has taken roaches with *Atta texana*.

Flies of the genera *Microdon* (or taxonomically close, e.g., Syrphidae) were taken repeatedly in the Argentine nests of *Acromyrmex lobicornis* as larval stages (Weber, unpublished). They resembled turtlelike discs firmly attached to surfaces. Probably they fed on the brood. Flies of the family Phoridae appear to be regular parasites of the leaf cutters and many species described by Borgmeier are in the list of 157 species noted previously (Weber 1972a).

One class of predators that evolved with the attines in tropical America has a dim future because of man's modification of his environment. These are the edentate mammals, *Myrmcecophaga* and *Tamandua,* the ant eaters that prey on termites and ants. The latter is arboreal and is meeting with a peculiar hazard that is decimating its numbers. As forests are cut down, and asphalt roads are laid, the ant eater cannot come down to the ground and cross over to an inviting wooded area without risk of being run over by vehicles. On a traverse in 1972 between Lake Maracaibo and the Eastern Andes in Venezuela a number of *Tamandua* were killed on new sections of road laid through hitherto largely virgin forest. Road kills of vertebrates are numerous throughout the world, but the ant eaters may be locally exterminated before they can adapt. They are vulnerable because of their extremely slow gait when they take to the ground. The ant eaters have never been a seious threat to the increase of leaf cutters.

In summary, while many parasites of leaf cutters are known, none is known to infest a nest in quantity or to be a handicap to the ants. None appear when nests are reared in the laboratory from the young colony with a queen, which is evidence that it is the growing nest that attracts them.

II. THE ANT FUNGUS

A. Historical Account and Definition of Terms

No one has recorded what the Indians of the Americas thought the leaf cutters were doing with their files of cut leaves and Europeans assumed that they were using the leaves directly for food or for some aspect of their nests. The exact use was not known until Thomas Belt published his book (1874) *The Naturalist in Nicaragua,* as noted earlier, and specifically stated that the "*Atta* are in reality, mushroom growers and eaters." The spongy mass of "speckled, brown, flocculent material" clearly referred to the fungus growing on the finally divided leaf sections.

This was an exciting theory to Möller, who received a grant from the Berlin Academy of Science to go to south Brazil to study these ants. His report (1893) as noted earlier, established clearly the nature of the ant fungi and introduced some complexities that remain today. These involve especially his interpretation of "strong versus weak" forms and conidial forms. He was unable to rear sporophores in artificial culture (something not accomplished until 1956) (Weber, 1957a).

Urich (1895) and others of this period were quick to adopt Möller's interpretations of ant fungi. Urich noted that the fungus formed like "shot" in the cells of *Sericomyrmex urichi* (as *opacus*). Spegazzini (1899, 1921) in Argentina described a number of fungi associated with attine nests. Wheeler (1907) diagrammed the yeast grown by *Cyphomyrmex rimosus* and named it *Tyridiomyces formicarum.* Weber (1938) reviewed the early ant fungus literature, publishing photographs of a large sporophore of *Lentinus* or similiar taxa associated with the nest of *Atta cephalotes.* Cultures of ant fungi were shown in photographs and described in 1945 and 1946. In 1957, the first sporophore to be reared in culture was described (Weber, 1957a). In 1966, forms of ant fungi were figured (Weber, 1966a) and in 1972 experiments with artificial cultures were described (Weber, 1972a). Other aspects were later published (Weber, 1979a,b).

Definition of Terms

The terms that are commonly applied to ant fungi are the following:

Ant fungus: a fungus growing in a mutualistic or symbiotic relationship with attine ants.

Bromatium: cheeselike cluster of packed yeast cells cultured by *Cyphomyrmex rimosus* and its taxa. Term formerly applied to masses of gongylidia.

Cell: The gross cavities of a fungus garden as in the cells of a sponge. New substrate is always added to the rims of the upper and upper-lateral cells.

Fungus garden: the mass of substrate and fungus formed by the ants and in which they keep their brood.

Hypha: individual filament of the vegetative ant fungus.

Gongylidium: the individual swelling at the apex or subapex of a hypha.

Mycelium: the collective term for the mass of hyphae and staphylae of the ant fungus.

Staphyla: an aggregate mass, often entertwined, of the gongylidia-bearing hyphae; these occur naturally in pure cultures as in the garden in nature. They are approximately 0.20–0.30 mm in diameter in *Atta* and *Acromyrmex* gardens.

Substrate: organic material, usually vegetal, on which the fungus grows.

B. The Fungus Garden, Life History, and Substrate

The development of the fungus garden in *Atta* has been described in previous Section I,3. The numbers of gardens and some dimensions are recorded in Tables III, VII, X–XV, and XVII. The garden is strictly the creation of the ants and does not continue to grow by itself in any species. A comparative and more detailed account follows. The garden in the primitive attines consists mostly of insect droppings and particles of partially decomposed vegetal material such as dead leaves or wood. In the intermediate attines it may include fallen fruit parts, such as the pith or albedo, and flower or leaf parts.

The garden of *Cyphomyrmex rimosus minutus* and its close relatives consists of cheeselike masses called bromatia, which are approximately 0.25–0.5 in diameter, that are placed by the ants on pellets of insect excrement brought into the cavity, in the soil or rotted wood. The fungal masses consist of tightly packed cells that are similar in appearance to ordinary yeast. They absorb nutrients from the moist pellets and grow by cell division to form larger masses. From time to time such a mass is picked up by the ants and partially eaten. The remainder is placed back on the pellets and allowed to grow by budding into other bromatia. If the fungus is neglected for long periods of time it starts to send out short filaments or hyphae from the exposed surfaces. In any case, these tiny masses never grow much larger than 0.025–0.05 mm.

The garden of other species of *Cyphomyrmex* and that of other genera of ants of small to medium size are very different. Instead of bromatia dotting insect excrement, they consist of fine hyphae growing from small pieces of substrate. This could be insect excrement, insect carcasses, small fragments of rotted wood, or pieces of green leaves and flowers. The nest and garden of *Cyphomyrmex longiscapus* in a densely wooded ravine contained an unusual form. The ants had started their garden above the soil level on exposed rootlets. Then as the garden grew, they covered it with pellets of muddy humus that were 2–7 mm thick. The garden consisted of a dense white mycelial growth on a garden 40 mm high and 30 mm wide. The cells were 3–8 mm in diameter and the substrate consisted of vegetal matter.

Such a garden is usually formed by this species and other small genera in a small cavity in damp earth. It may be 2–9 cm in diameter according to the species or age of nest.

As a rule, small ants have small gardens and cells, and the substrate of insect excrement has a different consistency than one of fresh leaf sections. The hypha in all species is colorless, appearing white *en masse,* but if the substrate is dark the whole garden has a darkish cast and may appear brown or gray. The entire garden is enclosed in an extremely thin and fragile veil of hyphae in certain *Apterostigma.*

Despite the general similarity it is often possible to recognize generic differences.

The gardens of species other than *C. rimosus* in *Cyphomyrmex* are very small (20–50 mm in diameter) and the size corresponds to the size of the ant. Those of *Mycocepurus* are also small but looser. The gardens of the small ants of *Myrmicocrypta* are generally much larger than the size of the ants would suggest, often 50–80 mm in diameter, which is the size built by ants two or three times larger. All three genera have gardens of fragments of vegetal matter and with cells a few millimeters in diameter. Sometimes carcasses of insects are incorporated as substrate. The garden may rest broadly on the floor of an earthen cavity, or on small stones or rootlets left in place.

The ants of *Apterostigma* use insect excrement; the particles of substrate are correspondingly often rounded. Such gardens are frequently suspended in irregular spaces in rotted wood and are difficult to dissect.

The gardens of *Trachymyrmex* tend to have a laminated structure; septa are hung from rootlets entering the ceiling of the chamber. In other cases the fragile garden may consist of masses suspended independently and loosely held together by mycelial threads. Some gardens rest completely on stones on the floor of the excavation.

The gardens of *Sericomyrmex* tend to be disproportionately large as are those of *Myrmicocrypta* and the substrate is often pale golden brown, fruity, and succulent. These two genera have gardens with the most solid appearance; relatively thick septa separate the cells.

The most fragile gardens are those of species such as *Mycetophylax conformis* and *Acromyrmex* (*Moellerius*) *landolti;* these ants use grass for substrate. The pieces are relatively long and narrow, creating an irregular mesh of cells, and the mycelium is scanty.

The largest gardens are those of the largest ants, *Acromyrmex* and *Atta*. The former may have a single very large garden or may have a number of gardens of 10–12 cm. Those of *Atta* appear to be more consistently globular and frequently are 10–12 cm. In several species gardens are much larger, but all are flattened at the base. Many species of various genera show versatility in suspending the garden from the flat underside of buried rocks.

1. Preparation by the Ants for Support for the Garden

A light weight fungus garden is always supported by hyphae. The walls, foundation, roots, or whatever supports the hyphae are thoroughly cleaned of all debris. The surface of the stones that are in contact with the hypha is cleaned of loose foreign matter. In all cases, the ants appear to lick or swab the surface thoroughly. In the laboratory, the ants treat their observation nests in a similar manner. They may spend hours at this before they bring in the first nucleus for a new garden or before an old garden is extended.

2. Life History of a Fungus Garden

The life history of a fungus garden of *Acromyrmex* or *Atta* is best studied in the laboratory. Since the temperature, humidity, and substrate in nature can be duplicated, the history should be comparable. The laboratory results can be checked with those in nature by excavating several of the latter over a period of time; however, excavation destroys the garden.

A mature garden of *Atta cephalotes* in nature is a honeycombed structure which is far from homogeneous (Fig. 14). Its sides and upper surface are formed of coarse cells with thin walls. These are dark in color, appearing superficially grayish brown because of the mycelium. The underside is yellow to pale brown and the inner cells average only a fraction of the size of the outer cells in diameter (Fig. 11D). This part of the garden, roughly the lower one-half or one-third, is much more solid than the outer part. The compactness is a result of the gradual packing in of the substrate from the weight above and from replacement of the loose mycelial mesh by compact fungal masses or staphylae. The septa, therefore, are thicker compared with their breadth and height.

New substrate is added only to the upper and outer parts of the garden.

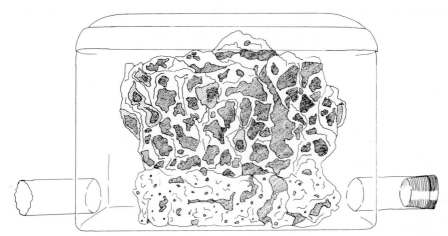

Fig. 14. Garden of *Atta cephalotes* in standard chamber of 2250 ml capacity (188 mm wide, 135 mm deep × 118 mm high, including domed roof). The lower one-third of the garden is past maturity, has small and collapsed cells with thick septa, and will be cast out piece by piece as exhausted substrate.

Whether this substrate is of particles of originally bright green leaves or of colorful flowers, they all become dark brown in the process of handling. These particles are added to the coarse outer cells on their rims, creating knifelike ridges.

A young colony of *Atta cephalotes* built a fungus garden in the standard plastic container of 2250-ml capacity that duplicated the size and form of a large earth chamber in nature. It became 70 mm in diameter in 2 weeks and by the end of 1 month it was approximately 90 mm in its three dimensions. By this time the lower part had become pale brown and with small cells, many in the 3–6 mm range. The upper part had large and irregular cells as in nature; the lateral diameter of one was 30 × 16 mm while the others were 10–15 mm. Staphylae grew in large numbers, were moved into the lower one-half of the garden and were fed to the larvae. By the end of the third month, it had become approximately 110 × 117 mm in maximum lateral dimensions and 97 mm high. It reached the plastic ceiling of the container and was clearly a mature garden. The basal part was removed in the first week of the fourth month, signifying the postmaturity phase. During this month the entire yellowish base was removed. This garden of natural size and shape, therefore, had a life of 4 months.

A second *A. cephalotes* colony at this time, slightly older when collected, started a new garden under excellent conditions of visibility, when there was no room for expansion in the container housing the original garden. The new garden was started on September 2, on the plastic floor in front of a pile of sand. The ants had first given the area a thorough examination, licking it minutely, as was described by Autuori (1941, 1942). It was only a fraction of a milliliter in bulk by September 13, but in the next 24 hours it grew to an estimated 10 ml because of the abundant fresh rose flower substrate that was furnished. The garden was $68 \times 42 \times 40$ mm in maximum dimensions by September 28, and, on October 1, it had the pale brown base with small cells and thick walls characteristic of mature gardens. The garden was $70 \times 50 \times 55$ mm in its maximum dimensions by October 2. The corresponding bulk of 192.5 ml from these measurements would be the capacity of the smallest container in which it would fit. This young garden, therefore, was fully developed except in size in its fourth week.

Three months after it was started, on December 2, the garden was approximately $110 \times 110 \times 100$ mm and the basal 20–40 mm was the yellow color, as is observed in nature. The cells were small and compact, with thick walls. Brood was confined to the upper and newer dark grayish-brown cells. It had all of the characteristics of a mature garden. The foregoing two records appear to be the normal sequence for colonies of *Atta* of different species in their early years: a garden having a life span of 3–4 months at 25°–27°C.

3. Fungus Gardens in Dry Seasons and Dry Climates

Fungus gardens require a constantly high humidity because the hyphae collapse quickly in dry air. Nevertheless the gardens have to survive dry seasons in many climates and the ants have developed several strategies.

In the tropical dry season the ants can excavate deeper chambers below the ones inhabited in the wet season and move their gardens downward. For example, a nest of *Trachymyrmex urichi* in Panama had seven small chambers in a vertical series that permitted them to move upward or downward according to the season, with no additional excavation. During an unusually long dry season in Panama (December, 1956–March, 1957) (Weber, 1957b, 1969b) the species of *Atta* abandoned their upper tier of chambers. Ants of smaller genera that ordinarily nested under shallow rocks moved deeper and others formed a turret entrance of hard clay that may have acted to conserve moisture.

In the *Atta* nests the fungus was of a sparser than usual filamentous growth and had a particularly dense form of staphylae. These were similar to miniature golf balls of compressed gongylidia and were carried about by the ants.

In temperate climates the northernmost attine, *Trachymyrmex septentrionalis,* in the New Jersey Pine Barrens, moved its nest in sand to a slightly lower elevation, closer to the water table and to a stream.

In arid California and Arizona the species nested deeply, or, in the case of *Trachymyrmex arizonensis,* nested shallowly in arroyos under rocks in what appeared to be dry soil, but the chambers must have been closed to the outside and drew on subsurface moisture.

4. Substrate

As noted previously, the substrate generally used by the smaller and more primitive attines is insect feces or excrement and partially decayed fragments of wood or other vegetal material. That of the leaf cutters consists at times of green leaves, flowers, and woody stems.

Commonly the ant brood itself is treated as substrate; the mycelium coats the integument. Brood of *Atta* is less likely to be covered than those of the other genera. Insect carcasses regularly are used by smaller genera.

The versatility of the ants and their fungus is shown by the acceptability of exotic plants as described partially in Section I,E.

Less than 500 years ago Europeans came to the tropical New World and the leaf cutters adapted remarkably to the plants that man brought. It is known that Columbus took seeds of many Old World plants on his second voyage. He stopped in the Canary Islands October 5–13, 1493, and took seeds of oranges, lemons, and other plants. These were taken first to Hispaniola and the citrus that he planted thrived mightily according to descriptions of some 30 years later. By 1559, Bartolome de las Casas recorded plagues of ants on Hispaniola, including one that stung badly, which may have been *Solenopsis geminata,* and one that may have been *Acromyrmex* with nests and eggs in the soil as white as snow (the fungus garden). Citrus was carried to Mexico and Central America in the early years of the sixteenth century and the Portuguese had established them in Brazil by 1540. Grapefruit apparently originated in the West Indies from a mutant of the pummelo (*Citrus grandis*) that came probably from the Malay area. The grapefruit was first described in 1750 from Barbados. The leaf cutters, therefore, developed the habit of cutting these exotic plants over a period of some 200–400 years. Other plants, not native to the New World, include the banana and the olive. The latter belongs to the family that includes forsythia, whose leaves are accepted in quantity by laboratory ants even though unrelated to plants they normally harvest. An excellent example of an exotic tree much favored by *Atta sexdens rubropilosa* and *A. laevigata* in Brazil is the Australian *Eucalyptus,* widely planted for its wood and paper pulp.

C. Artificial Cultures

When a fungus is cultured artificially on sterile media it is imperative that the inoculum be taken directly in nature from a normal, healthy garden before alien organisms start to grow. If a part of an ant colony with a garden is taken to another country or another area, there is always a possibility that the true ant fungus will be replaced by something else. This may be the explanation for some of the differences of opinion in the literature on what an ant fungus is.

For purposes of verification it is advisable to test an artificial culture from time to time by giving samples to the ant species that originally cultured it, to make sure that it is the same. The ants immediately accept and eat the sample if it is their own.

Standard mycological and bacteriological techniques were used in culturing by this author in Trinidad since 1934-1935. A transfer chamber where the air can be sterilized is ideal. However, cultures were made in Guyana in 1936 from samples taken in the interior of the country. In this case, a room was used, and fungus transfers were made to sterile agar tubes in the early dawn, when the air was still and saturated with moisture. In the New Jersey Pine Barrens some success was obtained by transferring pieces of garden directly to sterile nutrient agar petri dishes at the nest sites, then quickly making later transfers from these in the laboratory. Sketches of ant fungi parts appear in Fig. 15 and, in more detail, in such studies as Weber, 1966a, 1972a, and 1979b.

The hyphae in nature form a tangle of threads covering the ant brood and the substrate. They grow irregularly close to the surface because of the ant activities. In artificial culture or when the ants are removed the hyphae may grow straight into the air. These threads, which are white in appearance when viewed by the unaided eye, are nearly transparent when placed in fluid on a microscope slide (Fig. 15A-F). They are usually 2-6 μm in diameter among the smaller ant genera and 6-10 or more μm in *Atta*. Usually a hyphae will course irregularly, branching occasionally and, when producing staphylae, will suddenly or gradually develop inflations (gongylidia), all without producing septa. Cultures from three ant genera had hyphae growing at the rate of 9-28 μm/hr at 23°-26°C. Clamp connections (typical of basidiomycetes) were rare and appeared to be most common in an *Apterostigma* fungus. The individual gongylidia were easily visible to the unaided eye as bright pinpoints of light. In *Trachymyrmex septentrionalis* these may be five times the diameter of the hypha or about 27 × 46 μm as an average.

The staphylae are easily seen with the unaided eye as masses of compact fungus. The average size was 0.40 × 0.48 mm in ten examples of the fungus of *Atta cephalotes isthmicola*.

The culturing experiments in Trinidad were performed with the collaboration of the mycologists, C. W. Wardlaw and H. Briton-Jones, who made the identifications of non-ant fungi. These were the first cultures to be made using these ant species.

Clusters of bromatia from *Cyphomyrmex rimosus* were placed in sterile tubes and flasks of potato dextrose and of maltose-peptone agar (the latter pH 2.8). They were overwhelmed by alien fungi, but before they were, in several containers, the bromatia had multiplied by budding. A vermiculate growth of the ant fungus grew in one (photograph in Weber, 1945). Contaminants included *Trichoderma lignorum*. Later inoculations produced pure cultures, one of which was sent to England by Dr. Briton-Jones; however, no further information is available.

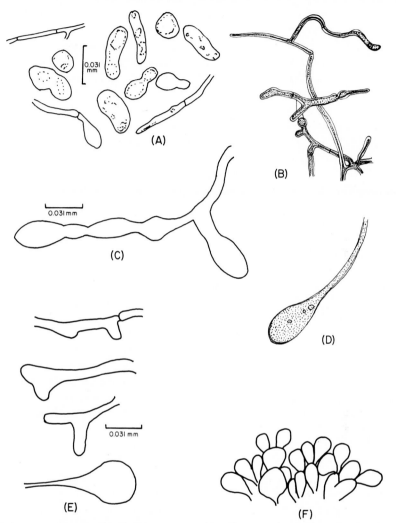

Fig. 15. (A) Yeast cells of *Cyphomyrmex rimosus* with sections of hyphae. (B) Normal mycelium of the fungus of *Trachymyrmex septentrionalis* from a fresh garden fragment in Florida. (C) A forked gongylidium from a culture of the fungus of *Acromyrmex octospinosus*. (D) A gongylidium from the normal garden of *Trachymyrmex septentrionalis* (Florida). (E) Gongylidia and septate hypha from a culture of the fungus of *Atta cephalotes*. (F) A compact staphyla consisting of gongylidia from the culture of the fungus of *Atta sexdens*.

Containers of the above-mentioned agars and of Dox's agar were in-oculated with the fungus from *Mycocepurus smithi*. Ten of 15 tubes became contaminated by *Rhizopus nigricans,* (synonym *Mucor racemosus*).

Similar agar cultures from *Mycetophylax conformis* fungus developed sporangia of several contaminants, including *Penicillium*. Later cultures contained *Rhizopus nigricans*.

Contaminants of cultures from the fungus of *Myrmicocrypta buenzlii* included *Aspergillus versicolor, Fusarium,* and *Cladosporium* (Wardlaw). They were ignored when the latter two were returned to the ants.

The fungi from *Myrmicocrypta urichi, Apterostigma urichi, Sericomyrmex urichi, Trachymyrmex urichi, T. ruthae, Acromyrmex octospinosus,* and *Atta cephalotes* were cultured and similar contaminations developed.

Brief references to these culturing results have been published (Weber, 1945, 1946, 1979a,b).

Culturing of ant fungi was resumed in the 1950s from many ant species from Panama and was conducted particularly at the University of Wisconsin Bacteriology Department in 1955-1956. This work is summarized in the following discussion.

A variety of culture media was used and a number were considered satisfactory for growing the vegetative form. Sabouraud's dextrose agar, for example, permitted a growth in which the staphylae occurred as in nature. These were as equally acceptable to the ants as those from their fungus garden in nature. Details of a 2-month experiment with various media appear in Weber (1972a).

Duplicate cultures on Sabouraud's and potato–dextrose agar were maintained during the aforementioned experimental period at three temperatures, 20°C, room temperature kept at approximately 25°C, and 30°C. The cultures came from ants of six genera (*Mycocepurus, Apterostigma, Sericomyrmex, Trachymyrmex, Acromyrmex,* and *Atta*) totaling 11 species. Both 20°C and 30°C proved to be too extreme, although 20°C was generally better than 30°C. Only the fungus of *Apterostigma mayri* grew well at all three temperatures and the best for all was room temperature (25°C).

In 1963, culturing of ant fungi was resumed at Swarthmore College, Swarthmore, Pennsylvania. Sample cultures were furnished to Dr. W. J. Robbins. The New York Botanical Gardens through Robbins set up a culture collection termed the Weber Ant Fungi (Robbins, 1969). These have all been tested repeatedly by returning samples to the ants that originally grew them. All have been repeatedly accepted as the ants natural fungus since the first culture was prepared in 1963. All continue to grow in culture as they do in the ant nests, with similar hyphal developments, such as stayphlae.

Reactions between several ant fungi growing in the same flask were examined, especially in 1955-1956 (q.v. Weber, 1966a, 1972a). Cultures from different ant genera or species would grow normally and there were no obvious reactions when they grew up to each other.

Cultures developed significant stages as experimentation continued during 1955-1956. Some developed incipient sporophores and finally mature sporophores bearing spores.

A December, 1957, Panamanian culture of the fungus of *Cyphomyrmex rimosus* produced a stromatic form resembling *Xylaria*.

A December, 1957, Panamanian culture of the fungus of *Apterostigma mayri* that had been transferred to a succession of media for a year suddenly developed an apparently incipient sporophore during January, 1959. This was examined by the mycologist, Dr. L. Batra, and was then sent to another mycologist who reported it to be an *Auricularia*. It did not mature further (Weber, 1966a, 1972a).

A December, 1957, Panamanian collection from *Myrmicocrypta ednaella* also produced an incipient sporophore. This was transferred to a succession of media and during April, 1960, produced an immature form that was sent to an agaric specialist after being air dried; however, it was not identifiable. Had this author not left the United States the next month for a foreign assignment for more than 2 years, other and more identifiable cultures might have been produced. Robins and Hervey later produced *Lepiota* sporophores from a culture from *Myrmicocrypta ednaella* and from the Trinidad *buenzlii* (Weber, 1966a). Recent publications (Hervey *et al.,* Weber, 1979a,b) describe fungus culturing in the laboratory and in nature.

1. Mature Sporophores from Cultures

The first mature sporophores to be produced in cultures resulted from a Panamanian collection of the fungus of *Cyphomyrmex costatus* (Weber, 1957a, 1972a).

The 2-mm-long ants culture the fungus in the form of fine hyphae that lack the gongylidia of higher attines. The gardens are 20–30 mm in diameter and in earth cells, commonly under shallow stones. Four colonies were collected on Barro Colorado Island, Canal Zone, June 25–27, 1955, and taken to the laboratory.

Pure cultures of the fungus were started from two of the ant colonies the morning after collecting them. The other two colonies served as useful tests for the artificial cultures from time to time and finally for verification of the sporophores (basidiocarps).

Sporophores were reared from cultures of both of the first two colonies, further reducing the chance that a sporophore could have come from some accidental fungus.

The cultures were grown on a variety of nutrient agars (potato–dextrose, Sabouraud's–dextrose, Czapek's, malt, etc.) and other media (wood blocks, sterile soil, ground rose petals, oats, and wheat). The cultures on oats and wheat produced the sporophores. A 125-ml Erlenmeyer flask of sterilized oats, inoculated November 15, 1955, and kept in a darkened cabinet, developed the first sporophores in January, 1956. Two sporophores which developed January 26 proved to be mature basidiocarps. Gills are white and the upper surface of the pileus was white except for scales which were grayish brown. These scales consisted of swollen hyphae comparable to those in natural ant gardens, but with reddish brown pigment. The basidiocarps were sent at the suggestion of Dr. K. B. Raper to the mycologist, Dr. Alexander Smith. The latter made careful measurements and descriptions, then referred them to the French authority on *Lepiota* in Paris, Dr. Marcel Locquin. Locquin pronounced them to be a new species of *Lepiota*.

However, Heim (1957) concluded from a study of this record that this fungus was the same as Möller's (1893) and proposed the name *Leucoprinus gongylophorus* (Möller) for them. *Leucoprinus* is considered to be a segregate of *Lepiota*.

Sporophores of the same fungus were produced in the New York Botanical Garden from cultures of the Trinidad ant, *Myrmicocrypta buenzlii* (as noted previously and later in Hervey *et al.,* 1977). These matched the description of Smith's of the Panamanian *Cyphomyrmex costatus* in every detail (W. J. Robbins and C. T. Rogerson, personal communication).

If these two ant genera and Möller's *Acromyrmex* of Brazil are indeed culturing the same fungus it suggests that the evolution of the ant genera took place after their ancestor developed the symbiotic relation with a fungus.

2. The First Ant Fungus Student

Alfred Möller is the pioneer in mycological investigations of ant fungi. He was trained in Germany under Professor Brefeld, an outstanding mycologist, and received a grant to travel to Brazil to study the ant–fungus association. He went to Blumenau (27°S lat.) in the southern state of Santa Catarina and established his laboratory. He then corresponded with the Swiss myrmecologist, Auguste Forel, and received identifications of the ants with which he was working.

Field work began at the end of 1890, and concluded the following year. He accomplished a great deal of painstaking and fascinating research during this period, not only of the mycology but of the general biology of the fungus growers. Some ten representative species of ants and their gardens from *Cyphomyrmex* to *Acromyrmex* were studied, including four species of the latter genus. He apparently did not examine *Atta* in the modern sense and therefore, did not report on the really large colonies attained by *Atta*.

The most abundant *Acromyrmex* (referred to then as *Atta*) near his laboratory was *A. discigera* and it is from the fungus of this ant that definitive stages were reared and to which the name of *Rozites gongylophora* clearly applies. His technique was to transfer gardens to clean glass plates surrounded by petroleum moats and pick off the ants, one by one, with forceps. Then parts of the garden were removed to watch glasses, using as sterile means as possible. Mycological details of culturing are summarized in Weber (1972a). A Portuguese translation of Möller's work (Viegas and Zink) (see Möller, 1941) appeared in Brazil.

Möller gives the first detailed description of the staphylae ("Kohlrabi bodies") growing in the garden and in artificial culture. The heads or gongylidia were never found to be separated by a dividing wall from the general course of the hypha. He had success using these staphylae as inoculae for his cultures as was this author's experience years later; they are regularly uncontaminated with other organisms. New gongylidia formed in artificial culture, a statement that was misunderstood by subsequent investigators who stated that these were found only in the ant garden.

He concluded that no conidial form would develop under the normal influence of the ants. Two different conidial forms appear when the ants are removed.

His original material was brought to him by a man who reported to him that he had found a large mushroom growth on an ant nest. It was firmly united with the ant garden. A similar growth was later found on a nest of *Acromyrmex discigera* (called here *Atta discigera)*. Two similar masses were seen in the same area and Möller observed that two of these were slowly developing into mushrooms or sporophores. One of them took 10–12 days to mature. The mushrooms were as much as 16 cm in diameter and were wine-red in color in mature condition. The basidia were about 30 μm long and clavate. The four short sterigmata were 8 μm long. Spores were oval and ochre colored.

Ripe mushroom spores were germinated in nutrient solutions. Those few that germinated produced hyphae on day 9 that were 10 μm thick and had side branches of 3 μm. Similar developments occurred in garden fungus cultures. In 5 weeks the mycelium, that was produced from spores, formed the staphyla type of aggregates.

Similar results were obtained by culturing small fragments of the mushroom cap (pileus). These, too, formed the same type of staphylae as in the garden. Conidial forms apparently also appeared. Möller found that the ants would accept the staphylae so produced as well as their own. They would also eat fragments of the mushrooms.

Möller was somewhat reluctant to give a name to this ant fungus because of his limited library facilities, but did name it *Rozites gongylophora*. He concluded "I do not know for sure if the mushroom has not been described somewhere before."

South American investigators have accepted these studies without qualification. Stahel and Geijskes reported (1939) that their own studies in Surinam of *Atta cephalotes* and *A. sexdens* nests and ant fungi tended to confirm Möller's studies. They found three typical fruiting bodies in culture gardens of *A. cephalotes* that were covered with rough scales and resembled Möller's *Rozites* but without inner differentiation in stipe and pileus.

Later they found a damaged sporophore growing from an *Atta cephalotes* nest and took parts of the pileus and stipe for cultures. Staphylae developed after the first 10 days and they felt that this sporophore was truly the nest fungus of *Atta cephalotes*. They did not doubt that the *A. cephalotes* fungus belonged to *Rozites gonglyophora* itself, or at least to a near relation. Such sporophores could not be found again in 15 *Atta* nests.

3. Conidial Forms in Ant Gardens

The fungus gardens may rarely develop conidia in a part of the garden that is relatively unattended.

Stahel and Geijskes (1939) brought portions of a colony and garden to the laboratory and sometimes found that the formerly busy workers, attempting to rebuild the garden, would suddenly stop work. They would stand inert on the walls of the container and abandon the garden. The whole top part of the garden would then develop a white mold. After 2 days a high, snowy-down of mycelium formed and 1 day later everything was covered by a mass of brown conidia. The garden in the meanwhile shrunk badly. This author's experience with *Trachymyrmex, Acromyrmex,* and *Atta* gardens has been similar in a few cases. However, from such cultures from gardens of *Acromyrmex* and *Atta* tentative identifications have been received twice of an unusual *Aspergillus* species, so that caution is indicated in interpreting these developments as being those from *Rozites*. This may also be the explanation of such recent reports as those of Lehmann (1975). Kreisel (1972) has described the conidial form of abnormal gardens of *Atta insularis* in Cuba.

D. Experimental Transposition of Fungi and Gardens between Ant Species

Experiments in which ants of one species have been given pieces of the fungus garden or of fungus cultures from another ant genus or species were initiated in Trinidad during 1934–1935. The experiments were generally successful when the fungus in artificial culture was clearly similar to natural fungus, i.e., produces staphylae or other forms that are clearly alike. This is most likely to be true in interchanges between the species of *Sericomyrmex, Trachymyrmex, Acromyrmex,* and *Atta*. It is not enough that the ants eat the fungus of an alien species but they must either adopt a bit of the other's fungus garden or tend some of a culture from the other ant. Early experiments showed that: *Myrmicocrypta buenzlii* workers adopted *M. urichi fungus,* and *Trachymyrmex urichi* workers adopted pieces of the fungus garden of *T. cornetzi, T. ruthae,* and *Acromyrmex octospinosus*.

Such experiments were continued in later years. In no case was the unique yeast fungus of *Cyphomyrmex rimosus* cultured by any other ant although this ant would at the least taste the fungus of another, such as *Atta sexdens*. Another fungus that was ignored by all other ant genera was the extremely fine hyphal form of *Apterostigma mayri* and *A. auriculatum*. However, these ants did taste the fungus of *Sericomyrmex urichi*.

Colonies of *Trachymyrmex,* such as of *T. urichi* or *T. septentrionalis,* repeatedly accepted a fungus garden fragment from an *Atta cephalotes* colony. The ants would then proceed to care for the fragment and construct a viable garden.

The following ants ate or cultured the fungus of one another: *Trachymyrmex septentrionalis, Trachymyrmex urichi, Trachymyrmex cornetzi, Acromyrmex octospinosus, Atta sexdens, Atta colombica tonsipes,* and *Atta cephalotes isthmicola.*

Some inconsistencies were shown: for example, *Trachymyrmex zeteki* ate the fungus of the ants in the preceding list but the other ants did not eat the *T. zeteki* fungus. Ants of *Sericomyrmex amabalis* ate the fungus of most of the listed ants, but its fungus in turn was eaten only by two of the *Atta*. A colony of *Acromyrmex lobicornis* of Argentina adopted the garden of *Atta cephalotes* of Trinidad when it lost its own and built it up in the form of its original type. Further study along this line will probably show greater flexibility among some of the ants but actual development of a viable fungus garden will usually only proceed with ants of the closely related higher genera.

There was no evidence that the ants would culture anything that was not an obvious ant fungus, contrary to the report of Goetsch on South American ants and gardens brought back to Germany. Goetsch and Stoppel (1940) reported the consistent isolation of *Hypomyces ipomoeae* and *Fusarium* species from the fungus gardens of the Brazilian *Atta sexdens,* while those of *Acromyrmex striatus* from Patagonia yielded closely related forms. It is not clear that these were normal gardens that produced gongylidia and staphylae known to be the normal fungus. This author's experience has always been that the ants are repelled by *Aspergillus, Penicillium, Trichoderrna, Mucor,* or any other fungus forming conidiophores.

Maintaining Isolates of Gardens with Ants

The use of sterile, nutrient-agar, petri dish plates for isolated garden fragments with a few ants was initiated in 1954 (Weber, 1955a, 1956a,b) and more recently demonstrated in photographs (Weber, 1972a,c). A small part of a normal garden of *Trachymyrmex septentrionalis* with its ants was quickly introduced, lifting the cover momentarily and picking up the fragment with sterile forceps. The ants maintained the garden intact despite eventual massive contamination of the peripheral agar surface from other fungi and bacteria. Despite caution, these were inevitably introduced when the cover was lifted to introduce substrate. The fungus of the garden grew normally under the care of the ants. The continual licking of the fungus, manipulation of the cassava granules used for substrate, and constant manuring were plainly visible under the microscope. Brood also matured normally. Workers lived as long as 19 months in such isolation, thus establishing precise longevity records of individuals. Regularly the ants of various species of attines excavate the pale, opalescent agar from the immediate vicinity and under the garden fragment, thus isolating it in the cell as in nature. It remained free of contaminants much longer than the general nutrient agar surface, indicating that the saliva transferred in the process of cutting had an inhibitory effect on alien organisms, as was long suggested.

When plain, nonnutrient agar was used, the same mining operation took place; the ants cut all agar blocks in the vicinity of the garden. In some cases the ants placed the agar blocks on the garden. These were manured and became light brown instead of the original opalescent color.

The most successful experiments were those with relatively monomorphic species, such as *Trachymyrmex* and *Sericomyrmex,* than with the highly polymorphic *Atta* and *Acromyrmex.* In the latter instances, the right proportion of the castes may have been unsuited to garden maintenance. These are also the most "nervous" ants, responding to outside stimuli so quickly as to interfere with care of the garden. In all cases the continual application of saliva and fecal droplets could readily be viewed under the microscope.

E. Biochemistry of Ant Excretions and Ant Fungi

The statements noted in Section I,E on the suggested roles of the ant saliva and fecal droplets induced studies by other investigators. The continual applications of these excretions was clear during the 1934–1935 year in Trinidad and noted by this author in Wheeler (1937, p. 14) as "nutritive fecal droplets and the mycelium which grows on them." In Weber 1947b, it was more plainly stated "that the chemical environment created by the malaxated substrate and ant fecal droplets favors the particular fungus grown by the ants."

Since then the studies by M. M. Martin and associates (1969–1975; Boyd and Martin 1975a,b) have expanded knowledge of the subject and shown that different interpretations arose at different stages of investigations.

First, the ant fecal droplets were investigated (M. M. Martin *et al.,* 1969a; M. M. Martin, 1970; J. and M. Martin, 1970; M. M. Martin *et al.,* 1973–1975). The two major nitrogenous components in the rectal fluid of *Atta colombica tonsipes* were allantoic acid and allantoin. Significant quantities of ammonia and free amino acids were present, including 21 of the common natural amino acids. Glutamic acid, histidine, arginine, proline, lysine, and leucine composed 82% (by weight) of the total amino acids.

These substances were clearly beneficial to the growth of the ant fungus, and they must be supplied repeatedly. The saliva and the fecal material were also examined for proteolytic enzymes. The investigators detected no protease activity in homogenates of the salivary glands, mandibular glands, maxillary glands, or postpharyngeal glands. Significant protease activity was detected only in the contents of the midgut and rectum. Seven attine species were examined by them in this author's laboratory (in 1969) and in every case the level of protease activity in the rectal fluid was several times higher than in the contents of the midgut.

Next, the chemical contributions of the fungus to the ants were examined. Martin and Weber (1969) concluded that the fungus cultured by *Atta colombica tonsipes* provided the ants with a rich and complete diet. More than 50% of the dry weight of the fungus was available as soluble nutrient. Carbohydrates composed 27% of the dry weight of the fungus; free amino acids, 4.7%; protein-bound amino acid, 13%; and lipid, 0.2.%. The carbohydrates consisted of trehalose, mannitol, arabinitol, and glucose. No polysaccharides were present. The lipid fraction contained ergosterol as the major sterol. This evidence thus substantiated the well-known reliance in nature of the ants on their fungus.

The fungus at first was cultured on nutrient agar as described previously; a synthetic diet (Robbins and Hervey, 1960) was found to promote faster growth. The growth and biochemical characteristics of the fungus were investigated by the Martins (1970–1971). It grew poorly in culture media in which the nitrogen was supplied as a protein (such as casein lactalbumin or zein) but well in media in which the nitrogen source was a hydrolyzate of the same proteins. They concluded that the ant fungus lacked the proteolytic enzymes necessary to digest the polypeptides of the normal substrate. However, the ant fungus contributes the cellulose-degrading apparatus to this mutualistic association (Martin and Weber, 1969). This had been determined earlier (Weber, 1957a) by growing the fungi from several ant species on several types of wood blocks as the sole nutrient source and finding that the fungus definitely rotted them.

In work summarized recently by Martin (1974), Schildknecht and Koob (1970, 1971) and Maschwitz *et al.* (1970) detected phenylacetic acid, D-3-hydroxydecanoic acid and indolylacetic acid in the thoracic metapleural glands of the attine ant *Atta sexdens*. These workers have suggested that the substances play an important role in fungus culturing: phenylacetic acid through its bactericidal properties, D-3-hydroxydecanoic acid through its capacity to inhibit the germination of the spores of alien fungi, and indolylacetic acid as a consequence of its properties as a plant hormone. This interesting suggestion awaits confirmation through a demonstration that the components of the metapleural glands exhibit their bacteriostatic and fungistatic properties at the still undetermined levels at which they are present in the fungus gardens.

The explanation for the role of insect carcasses in the nests of the small attine species was determined by Martin *et al.* (1973). Investigating seven species of ants from *Cyphomyrmex* to *Atta* they found that their fecal enzymes produced α-amylase and chitinase in significant quantity. These degrade the insect cuticle, rendering it available for the fungus growing on it. They suggest that selection should have favored those ants which could excrete larger quantities of chitinase.

The conclusions by Martin (1970) that "in biochemical terms the ants contribute their enzymatic apparatus to degrade protein and the fungus contributes its enzymatic apparatus to degrade cellulose" were significantly changed by Boyd and Martin (1975b) as follows:

> The finding that the faecal enzymes of the leaf-cutting ants are derived from the fungus, and that they are not digestive enzymes produced by the ants, significantly clarifies the nature of this intriguing symbiosis. Contrary to our earlier suggestion (Martin, 1969; Martin *et al.*, 1973), it is not their biochemical machinery which the ants are contributing to this mutualistic association, but rather their capacity to serve as vehicles of transport. The ants are simply moving fungal enzymes from a region of the garden in which the fungus is in a state of rapid growth and the enzymes are in ample supply, to the site of inoculation where the enzymes are in short supply.

Thus one comes to the general contention held by biologists for a century or more that life on Earth depends on green plants. Animals, fungi, and other organisms ultimately depend on them. The fungi of the ants depend on living or dead green plants and the attines depend on their fungi, something we have long known.

F. Epilogue

Many of the intricacies of this mutualistic arrangement between ants and fungi have been learned through the cooperation and independent studies of ecologists, entomologists, biochemists, mycologists, and field biologists. In all these fields, the possibilities of finding new facts and developing theories have not been exhausted. All continue to offer rich possibilities for adding new knowledge, especially of the evolution of both the ants and the fungi and their interrelationships. Their importance in fragile ecosystems that are being profoundly influenced by man introduces a note of urgency.

REFERENCES

Amante, E. (1967a). A Saúva *Atta capiguara,* Praga Das Pastagens. Secr. Agr. Depart. *Prod. Veg. Instr. Pract., Sao Paulo* **41**, 1-12.
Amante, E. (1967b). Prejuizos causados pela formiga saúva em plantactes de *Eucalyptus* e *Pinus* no estado de São Paulo. *Silvicult. Sao Paulo, Rev. Tec. Serv. Flor. Sao Paulo* **6**, 355-363.
Amante, E. (1967c). *Biologico* **33**, 113-120.
Amante, E. (1968). "Nova Metodo Para Criação Da Formiga Saúva *Atta* spp., Em Laboratorie

E. Contribuição A Biologio E Comportamento De *Atta laevigata* Em Condiçoes Artificiais (Hymenoptera, Formicidae)," pp. 1-9. São Paulo, Brazil.

Amante, E. (1972). Influencîa de algunas Fatores Microclimáticas Sobre a Formiga Saúva *Atta laevigata* (F. Smith, 1858), *Atta sexdens rubropilosa* Forel, 1908, *Atta bisphaerica* Forel 1908, E *Atta capiguara* Gonçalves, 1944 (Hymenoptera, Formicidae), em Formigueiros Localizadoa No Estado de São Paulo. Unpubl. thesis, University of São Paulo, Piracicaba, 175 pp. (maps pp. 12-13).

Autuori, M. (1941). Contribução para o conhecimento da Saúva (*Atta* spp.) (1)-Evoluçao do saúveiro (*Atta sexdens rubropilosa* Forel, 1908). *Arq. Inst. Biol., Sao Paulo* **12**, 197-228.

Autuori, M. (1942). "Contribução para o conhecimento da Saúva (*Atta* spp.). (II), O saúveiro inicial (*Atta sexdens rubropilosa* Forel, 1908)." Arq. Inst. Biol. São Paulo 13: pp. 67-86.

Autuori, M. (1950). Contribução para o conhecimento da Saúva (*Atta* spp.). (V). Numero de formas aladas e redução dos saúveiros iniciais. *Arq. Inst. Biol., Sao Paulo* **19**, 325-331.

Autuori, M. (1956). Contribução para o conhecimento da Saúva (*Atta* spp.). (VI). Infestação residual da saúva. *Arq. Inst. Biol., Sao Paulo* **23**, 109-116.

Barrer, P. M., and Cherrett, J. M. (1972). Some factors affecting the site and pattern of leaf-cutting activity in the ant *Atta cephalotes* L. *J. Entomol., Ser. A: Gen. Entomol.* **47** (1), 15-27.

Bates, H. W. (1863). "The Naturalist on the River Amazon," 1st ed. Clodd, London (rev. ed., 1891).

Bazire-Benazet, M. (1974). Relations interindividuelles dans la jeune colonie d'*Atta sexdens* L (Hym. For.). Note (*). *C. R. Hebd. Leances Acad. Sci., Ser. D* **279**, 1713-1716.

Belt, T. (1874). "The Naturalist in Nicaragua." Murray, London.

Blanche, D. (1960-1961). "La Fourmi-Manioc," pp. 12-21. Phytoma, Paris.

Blum, M. S., and Portocarrera, C. A. (1966). Chemical releasers of social behavior. X. An attine trail laying substance in the venom of a non-trail laying myrmicine *Daceton armigerum* (Latreille). *Psyche* **73**, 150-155.

Blum, M. S., Moser, J. C., and Cordero, A. D. (1964). Chemical releasers of social behavior. II. Source and specificity of the odor trail substances in four attine genera (Hymenoptera: Formicidae). *Psyche* **71**, 1-7.

Bonetto, A. A. (1959). "Las hormigas "cortadoras" de la Provincia de Santa Fe, Argentina," pp. 1-79.

Borgmeier, T. (1950). Estudos sobre *Atta* (Hym. Formicidae). *Mem. Inst. Oswaldo Cruz* **48**, 239-392.

Borgmeier, T. (1959). Revision der Gattung *Atta* Fabr. (Hym. Formicidae). *Stud. Entomol.* [N.S.] **2**, 321-390.

Boyd, N. D., and Martin, M. M. (1975a). Faecal proteinases of the fungus-growing ant *Atta texana:* Properties, significance and possible origin. *Insect Biochem.* **5**, 619-635.

Boyd, N. D., and Martin, M. M. (1975b). Faecal proteinases of the fungus-growing ant, *Atta texana:* Their fungal origin and ecological significance. *J. Insect Physiol.* **21**, 1815-1820.

Brandt, P., and Mayer, J. (1977). Letter on mammals attracted to *Atta* nests.

Bruch, C. (1916). Contribucion al estudio de las hormigas de San Luis. *Rev. Mus. La Plata* **23**, 291-354.

Bruch, C. (1917a). Costumbres y Nidos de Hormigas. *An. Soc. Cient. Argent.* **83**, 302-316.

Bruch, C. (1917b). Costumbres y Nidos de Hormigas. *An. Soc. Cient. Argent.* **84**, 154-168.

Bruch, C. (1928). Estudios Mirmecológicos. *An. Mus. Nat. Hist. Nat. Buenos Aires, Entomol.* **141**, 341-360.

Bucher, E. H. (1974). Observaciones Ecologicals sobre los Artropodos del Bosque Chaqueno de Tucumán. *Rev. Fac. Cienc. Exactas Fis. Nat. Cordoba* [N.S.]; *Biologia* **1**, 35-122.

Bucher, E. H., and Montenegro, R. (1974). Hábitos forrajeros de cuatro hormigas simpátridas del genero *Acromyrmex* (Hymenoptera, Formicidae). *Ecología, Assoc. Argentina Ecol.* **2**, (1), 47-53.

Bucher, E. H., and Zuccardi, R. B. (1967). Significacion de los hormigueros de *Atta vollenweideri* Forel como al teradores del suelo en la Provincia de Tucumán. *Acta Zool. Lillouna* **23**,83–95.

Buckley, S. B. (1860). The Cutting ant of Texas (*Oecodoma mexicana Sm.*) *Proc. Acad. Nat. Sci. Phila.*, 1860. 9–10, 233–236. *Ann. Mag. Nat. Hist.* (3) VI, 1860, 386–389.

Carbonell, Mas, C. S. (1943). "Las hormigas cortadoras del Uruguay". *Rev. Asoc. Ing. Agron. Montevideo* **3**, 3–12.

Cherrett, J. M. (1968). The foraging behavior of *Atta cephalotes* L. (Hymenoptera, Formicidae). Foraging patterns and plant species attacked in tropical rain forest. *J. Anim. Ecol.* **37**, 387–403.

Cherrett, J. M. (1972). Chemical aspects of plant attack by leaf-cutting ants. In J. B. Harbourne (Ed.). *Phytochem. Ecol.* 13–74. Academic Press, London and New York.

Cherrett, J. M., and Seaforth, C. (1968). Phytochemical arrestants for the leaf-cutting ants, *Atta cephalotes* L and *Acromyrmex octospinus* (Reich), with some notes on the ant's response. *Bull. Ent. Res.* **59**, 615–625.

Crewe, R. M., and Blum, M. S. (1972). Alarm pheromones of the Attini: Their phylogentic significance. *J. Insect Physiol.* **18** (1), 31–42.

Da Silva, U. P. (1975). Contribution to the study of populations of *Atta sexdens rubropilosa* and *Atta laevigata* from the State of São Paulo Brazil (Hymenoptera-Formicidae). *Stud. Entomol.* **19** (1–4), 201–250.

Duffy, S. S. (1977). Arthropod allomones: Chemical effronteries and antagonists. *Proc. Int. Congr. Entomol., 15th, 1976* pp. 323–394.

Eibl-Eibesfeldt, von J. (1967). On the guarding of leaf-cutters by minima workers. *Naturwissenschaften* **13**, 346.

Eibl-Eibesfeldt, J., and Eibl-Eibesfeldt, E. (1967). Das Parasitenabwehren der Minima-Arbeiterinnen der Blattschneider-Ameisen. *Z. Tierpsychol.* **24**, 278–281.

Eidmann, H. (1935). Zur Kenntnis der Blattschneiderameise *Atta sexdens*. *Z. Angew. Entomol.* **22**, 185–436.

Eidmann, H. (1938). Zur Kenntnis der Lebensweise der Blattschneiderameise *Acromyrmex subterraneus* For. var. *eidmanni* Santschi und ihrer Gäste. *Rev. Entomol.* **8**, 291–314.

Emery, C. (1905). Revisione delle specie del genere *Atta* appartenenti al sottogeneri Moellerius e *Acromyrmex*. *R. Acad. Sci. Ist. Bologna* **2**, (6), 1–18.

Evans, M. A., and Evans, H. E. (1970). "William Morton Wheeler, Biologist." Harvard Univ. Press, Cambridge, Massachusetts.

Gallardo, J. M. (1951). Sobre un Teiidae (Reptilia, Sauria) poco conocido para la fauna Argentina. *Commun. Inst. Nac. Invest. Cien. Nat.* **2**, 8.

Gallardo, J. M. (1964). Los anfibios de la Provincia de Entre Rios, Argentina, y algunas notas sobre su distribucion geograficos y ecologia. *Neotropica* **10**, 23–28.

Gamboa, G. J. (1975a). Ant carrying in the desert leaf-cutter ant *Acromyrmex versicolor versicolor* (Pergande) (Hymenoptera: Formicidae). *Insectes Soc.* **22**, 75–82.

Gamboa, G. J. (1975b). Foraging and leaf-cutting of the desert gardening ant *Acromyrmex versicolor versicolor*. *Oecologia* **20**, 103–110.

Goeldi, E. (1905a). Beobachtungen über die erste Anlageiner neuen Kononie von *Atta cephalotes*. *C. R. Congr. Int. Zool., 6th, 1905* pp. 457–458.

Goeldi, E. (1905b). Myrmecologische Mittheilung das Wachsen des Pilzgartens bei *Atta cephalotes* betreffend. *C. R. Congr. Int. Zool., 6th, 1905* pp. 508–509.

Goetsch, W., and Stoppel, R. (1940). Die Pilze der Blattscheiderameisen. *Biol. Zentralbl.* **60**, 393–398.

Gonçalves, C. R. (1944). Descrição de una nova saúva brasileira (Hym. Form). *Rev. Bras. Biol.* **4**, 233–238.

Gonçalves, C. R. (1960). "Distribuicão, Biologia e Ecologia Das Saúvas. *Divulgação Agronomica* (Shell)," Vol. 1, pp. 2–10. Rio de Janeiro.

Gonçalves, C. R. (1961). O genero *Acromyrmex* no Brasil (Hymenoptera: Formicidae). *Stud. Entomol.* **4** (1–4), 8–180.

Haines, B. (1973). Impact of leaf-cutting ants on vegetation development at Barro Colorado Island. *Trop. Ecol.* **2**, 398.

Heim, R. (1957). A propos du *Rozites gongylophora* A Moller. *Rev. Mycol.* **22**, 293–299.

Hermann, H. R., Moser, J. C., and Hunt, A. N. (1970). The nymenopterous poison apparatus. X. Morphological and behavioral changes in *Atta texana* (Hymenoptera: Formicidae). *Ann. Entomol. Soc. Am.* **63** (6), 1553–1558.

Huber, J. (1905). Über die Kononiegrundung bei *Atta sexdens* L. *Biol. Centralbl.* **25**, 606–619; 624–635.

Hervey, A., C. T. Rogerson, and I. Leong. (1977). Studies on fungi cultivated by ants. *Brittonia* **29**, 226–236.

Jonkman, J. C. M. (1976). Biology and ecology of the leaf cutting ant *Atta vollenweideri* Forel, 1893. *Z. Angew. Entomol.* **81**, 140–148.

Jonkman, J. C. M. (1977). Thesis, p. 132. University of Leiden (unpublished)

Kempf, W. W. (1972). Catalago abreviado das Formigas da Regiao Neotropical (Hym. Formicidae). *Stud. Entomol.* **15** (1–4), 3–344.

Kermarrec, A. (1975). Etude des relations synécologiques entre les nematodes et al Fourmimanioc, *Acromyrmex octospinosus* Reich. *Ann. Zool.—Ecol. Anim.* **7** (1), 27–44.

Kreisel, H. (1972). Pilze aus pilzgarten von *Atta insularis* in Kuba. *Allg. Mikrobiol.* **12**, 643–654.

Kusnezov, N. (1949). El genero "*Cyphomyrmex*" en la Argentina. *Acta Zool. Lilloana* **8**, 427–456.

Kusnezov, N. (1963). Zoogeographia de las hormigas en Sudamerica. *Acta Zool. Lilloana* **19**, 25–186.

Labrador, J. R., Martinez, J., and Rubio, E. (1972). University of Zulia, Maracaibo, Venezuela (unpublished work).

Lehmann, J. (1975). Ist der Nahrungopilz der pilzzuchtenden Blattschneiderameisen und Termiten ein *Aspergillus) Waldhygiene* **10**, 252–255.

Lewis, T., Pollard, G., and Dibley, G. (1974a). Rhythmic foraging in the leaf-cutting ant *Atta cephalotes* (L) (Formicidae: Attini). *J. Anim. Ecol.* **43**, 129–141.

Lewis, T., Pollard, G., and Dibley, G. (1974b). Micro-environmental factors affecting diel patterns of foraging in the leaf-cutting ant *Atta cephalotes* (L). (Formicidae: Attini). *J. Anim. Ecol.* **43**, 143–153.

Lewis, T. (1975). Colony size, density and distribution of the leaf-cutting ant, *Acromyrmex octospinosus* (Reich) in cultivated fields. *Trans. R. Ent. Soc. London* **127** (1), 51–64.

Littledyke, M., and Cherrett, J. M. (1976). Direct ingestion of plant sap from cut leaves by the leaf-cutting ants *Atta cephalotes* (L) and *Acromyrmex octospinosus* (Reich) (Formicidae, Attini). *Bull. Entomol. Res.* **66**, 205–217.

Lugo, A. E., Farnsworth, E. G., Pool, D. G., Jerez, P., and Kaufman, G. (1973). The impact of the leaf-cutter ant *Atta colombica* on the energy flow of a tropical wet forest. *Ecology* **54**, 1292–1301.

McCook, H. C. (1880). Note on a New Northern Cutting Ant, *Atta septentrionalis. Proc. Acad. Nat. Sci. Philadelphia,* 359–360.

Mariconi, F. A. M. (1974). A contribution to the knowledge of the incipient nest of a Saúva ant (Parasol ant) *Atta capiguara* (Hymenoptera: Formicidae). *An. Soc. Entomol. Bras.* **3** (1), 5–13.

Markl, H. (1965). Stridulation in leaf-cutting ants. *Science* **149**, 1392–1393.

Markl, H. (1970). Communication by Stridulatory Signals in leaf-cutting ants. Part 3. The sensitivity to substrate vibrations. *Z. Vergl. Physiol.* **69** (1), 6–37

Marshall, L. G., Pascual, R., Curtis, G. H., and Drake, R. E. (1977). South American geochronology: Radiometric time scale for middle to late Tertiary mammal-bearing horizons in Patagonia. *Science* **195**, 1324–1328.

Martin, J. S., and Martin, M. M. (1970). The presence of protease activity in the rectal fluid of attine ants. *J. Insect Physiol.* **16**, 227-232.

Martin, M. M. (1970). The biochemical basis of the fungus-attine ant symbiosis. *Science* **169**, 16-20.

Martin, M. M. (1974). Biochemical ecology of the Attine ants. *Acc. Chem. Res.* **7**, 1-5.

Martin, M. M., and Martin, J. S. (1970). The biochemical basis for the symbiosis between the ant, *Atta colombica tonsipes,* and its food fungus. *J. Insect Physiol.* **16**, 109-119.

Martin, M. M., and Martin, J. S. (1971). The presence of protease activity in the rectal fluid of primitive attine ants. *J. Insect. Physiol.* **17**, 1897-1906.

Martin, M. M., and Weber, N. A. (1969). The cellulose-utilizing capability of the fungus cultured by the attine ant, *Atta colombica tonsipes.* *Ann. Entomol. Soc. Am.* **62**, 1386-1387.

Martin, M. M., Carls, G. A., Hutchins, R. F. N., MacConnell, J. G., Martin, J. S., and Steiner, O. D. (1967). Observations on *Atta colombica tonsipes* (Hymenoptera: Formicidae). *Ann. Entomol. Soc. Am.* **60** (6), 1329-1330.

Martin, M. M., Carman, R. M., and MacConnell, J. G. (1969a). Nutrients derived from the fungus cultured by the fungus-growing ant *Atta colombica tonsipes.* *Ann. Entomol. Soc. Am.* **62**, 11-13.

Martin, M. M., MacConnell, J. C., and Gale, G. R. (1969b). The chemical basis for the attine ant fungus symbiosis. Absence of Antibiotics. *Ann. Entomol. Soc. Am.* **62**, 386-388.

Martin, M. M., Gieselmann, J., and Martin, J. S. (1973). Rectal enzymes of attine ants. Amylase and chitinase. *J. Insect Physiol.* **19**, 1409-1416.

Martin, M. M., Boyd, N. D., Gieselmann, M. J., and Silver, R. G. (1975). Activity of faecal fluid of a leaf-cutting ant toward plant cell wall polysaccharides. *J. Insect Physiol.* **21**, 1887-1892.

Maschwitz, U., Koob, K., and Schildknecht, J. (1970). Beitrag zur funktion der Metathoracaldruse der Ameisen. *J. Insect Physiol.* **16**, 386-404.

Mattson, W. J., and Addy, N. D. (1975). Phytophagous insects as regulators of forest primary production. *Science* **190**, 515-522.

Meggus, B. J. (1975). Application of the biological model of diversification to cultural distributions in tropical lowland in South America. *Biotropica* **7**, 141-161.

Möller, A. (1893). "Die Pilzgarten eniger sudamericanischer Ameisen," No. VI. Schimper's *Bot. Mitth. aus d. Tropen,* 127 p.

Möller, A. (1941). "As Hortas de fungo de algunas formigas sul-americanas" (transl. by A. P. Viegas and E. M. Zink), Rev. Entomol., Suppl. No. 1. Rio de Janeiro.

Montenegro, R. A. (1973). La Fundacion del Hormiguero en *Acromyrmex striatus* (Rog.). (Hymenoptera, Formicidae). *Stud. Entomol.* **16** (1-4).

Moser, J. C. (1963). Contents and structure of *Atta texana* nest in summer. *Ann. Entomol. Soc. Am.* **56**, 286-291.

Moser, J. C. (1967). Mating Activities of *Atta texana* (Hymenoptera, Formicidae). *Insectes Soc.* **14**, 295-312.

Moser, J. C., and Blum, M. S. (1963). Trail marking substance of the Texas leaf-cutting ant: Source and potency. *Science* **140**, 1228.

Moser, J. C., and Silverstein, R. M. (1967). Volatility of trail making substance of the town ant. *Nature (London)* **215**, 206-207.

Moser, J. C., Silverstein, R. M., and Brownlee, R. C. (1968). Alarm pheromones of the ant *Atta texana.* *J. Insect Physiol.* **14**, 529-535.

Pasteels, J. M. (1977). Evolutionary aspects in chemical ecology and chemical communication. *Proc. Int. Congr. Entomol., 15th, 1976* pp. 281-293.

Rettenmeyer, C. W. (1963). Behavioral studies of army ants. *Univ. Kans. Sci. Bull.* **44**, 281-465.

Riley, R. G., and Silverstein, R. M. (1973). Trail-marking and alarm pheromones of some ants of the genus *Atta.* U. S. NTIS, PB Rep. 235562.

Robbins, W. J. (1969). Current botanical research. *In* "Current Topics in Plant Science" (J. E. Gunckel, ed.), pp. 71–73. Academic Press, New York.

Robbins, W. J., and Hervey, A. (1960). Light development of *Poria ambigua*. *Mycologia* **52**, 231–247.

Robinson, S. W., and Cherrett, I. M. (1973). Studies on the use of leaf-cutting ant scent trail pheromones as attractants in baits. *Proc., Int. Congr.—Int. Union Study Soc. Insects, 7th, 1973* pp. 1332–1338.

Robinson, S. W., and Cherrett, J. M. (1974). Laboratory investigations to evaluate the possible use of brood pheromones of the leaf-cutting ant *Atta cephalotes* (L.) (Formicidae, Attini) as a component in an attractive bait. *Bull. Ent. Res.* **63**, 519–529.

Rockwood, L. L. (1971). Population ecology of leaf-cutter ants in Guanacaste. *Org. Trop. Stud. News* **71** (5), 4–5.

Rockwood, L. L. (1973). Distribution, density, and dispersion of two species of *Atta* (Hymenoptera: Formicidae) in Guanacaste Province, Costa Rica. *J. Anim. Ecol.* **42**, 803–817.

Rockwood, L. L. (1975). The effects of seasonality and host plant selection on foraging of two species of leaf-cutting ants (*Atta*) in Guanacaste Province, Costs Rica. Biotropica 7(3),176–193.

Rockwood, L. L. (1976). Plant Selection and Foraging Patterns in Two Species of Leaf-Cutting ants (Atta). *Ecology* **57** (1), 48–61.

Sampaio De Azevedo, A. G. (1894). "Sauva ou Manhuaara," Monogr. Sao Paulo.

Schildknecht, H., and Koob, K. (1970). Plant bio-regulators in the metathoracic glands of myrmicine ants. *Angew. Chem.* **9**, 173.

Schildknecht, H., and Koob, K. (1971). Myrmicine ants secrete herbicidal chemical. *Chem. Eng. News,* 39.

Schreiber, J. R. (1974). Isolation of attine brood from the social environment (Hymenoptera: Formicidae). *Entomol. News* **85**, 300–314.

Shorey, H. H. (1977). The adaptiveness of pheromone communication. *Proc. Int. Congr. Entomol., 15th, 1976* pp. 294–307.

Stahel, C., and Geijskes, D. C. (1939). Ueber den bau der nester von *Atta cephalotes* L und *Atta sexdens* L. (Hym. Formicidae). *Rev. Entomol.* **10**, 27–78.

Tumlinson, J. H., Silverstein, R. M., Moser, J. C., Brownlee, R. G., and Ruth, J. M. (1971). Trail pheromone isolated and identified. *Nature (London)* **234**, 348–349.

Urich, F. W. (1895). Notes on some fungus-growing ants in Trinidad. *Jour. Trinidad Field Nat. Club* **2**, 175–182.

Vaz-Ferreira, R., Covelo de Zolessi, L., and Achaval, F. (1970). Oviposicion y desarrollo de ofidios y lacertilios en hormiguereos de *Acromyrmex*. *Physis (Buenos Aires)* **29**, 431–459.

Vaz-Ferreira, R., Covelo de Zolessi, L., and Achaval, F. (1973). Oviposicion y desarrollo de ofidios y lacertillios en hormigueros de *Acromyrmex*. II. *Trab. Congr. Latinoam Zool., 5th, 19* Vol. 1, pp. 323–344.

von Ihering, H. (1882). Ueber Schichteubildung durch Ameisen (*Atta cephalotes*). Briefl. Mitth. aus Mundoro, Rio Grande do Sul, Brasilien, Oct. 1881. *Neues Jahrb. Mineral.* **1**, 156–147.

Weber, N. A. (1937). The biology of the fungus-grown ants. Part II. Nesting habits of the bachac (*Atta cephalotes* L.). *Trop. Agric. (Trinidad)* **14**, 223–226.

Weber, N. A. (1938a). The biology of the fungus-growing ants. Part III. The sporophore of the fungus grown by *Atta cephalotes* and a review of reported sporophores. *Revista Entomol.* **8**, 265–272.

Weber, N. A. (1938b). "The Biology of the Fungus-growing Ants. Additional New Forms. Part V. The Attini of Bolivia." Revista Entomol. 9. pp. 154–206.

Weber, N. A. (1940). The Biology of the Fungus-growing Ants. Part VI. Key to *Cyphomyrmex,* new Attini and a new Guest Ant. *Revista Entomol.* **II**, 406–427.

Weber, N. A. (1941). The biology of the fungus-growing ants. Part VII. The Barro Colorado Island, Canal Zone, species. *Revista Entomol.* **12**, 93–130.

Weber, N. A. (1945). The biology of the fungus-growing ants. Part VIII. The Trinidad. B.W.I. species. *Revista Entomol.* **16**, 1–88.

Weber, N. A. (1946). The biology of the fungus-growing ants. Part IX. The British Guiana species. *Revista Entomol.* **17**, 114–172.

Weber, N. A. (1947a). Binary Anterior Ocelli in Ants. *Bol. Bull.* **93**, pp. 112–113.

Weber, N. A. (1947b). Lower Orinoco River fungus-growing ants (Hymenoptera: Formicidae, Attini). *Bol. Entomol. Venez.* **6**, 143–161.

Weber, N. A. (1954). Fungus-growing ants and their fungi. *Anat. Rec.* **120**, 735.

Weber, N. A. (1955a). Pure cultures of fungi produced by ants. *Science* **121**, 109.

Weber, N. A. (1955b). Fungus-growing ants (*Atta* and *Trachymyrmex*) of Panama and their fungi. *Anat. Rec.* **122**, 425.

Weber, N. A. (1955c). Fungus-growing ants and their fungi. *Cyphomyrmex rimosus minutus* Mayr. *J. Wash. Acad. Sci.* **45**, 275–281.

Weber, N. A. (1956). Fungus-growing ants and their fungi: *Trachymyrmex septentrionalis* McCook. *Ecology* **37**, 150–161; 197–199.

Weber, N. A. (1957a). Fungus-growing ants and their fungi. *Cyphomyrmex costatus. Ecology* **38**, 480–494.

Weber, N. A. (1957b). Dry season adaptations of fungus-growing ants and their fungi. *Anat. Rec.* **128**, 638.

Weber, N. A. (1958). Evolution in fungus-growing ants. *Proc. Int. Congr. Entomol., 10th, 1956* **2**, 459–473.

Weber, N. A. (1959). Isothermal conditions in tropical soil. *Ecology* **40**, 153–154.

Weber, N. A. (1960). The behavior of the queen as compared with the worker in fungus-growing ants. *Entomol. News* **71**, 1–6.

Weber, N. A. (1963). Ten kilometers of swarms of an ant. *Proc. Entomol. Soc. Wash.* **65**, 109.

Weber, N. A. (1964). A five-year colony of a fungus-growing ant, *Trachymyrmex zeteki. Ann. Entomol. Soc. Am.* **57**, 85–89.

Weber, N. A. (1966a). The fungus-growing ants. *Science* **153**, 587–604.

Weber, N. A. (1966b). Fungus-growing ants and soil nutrition. *Actas Prim. Coloq. Latinoam. Biol. Suelo. Monogr. I, Cent. Coop. Cien. Am. Latina* pp. 221–256.

Weber, N. A. (1966c). Development of pigmentation in the pupa and callow of *Trachymyrmex septentrionalis* (Hymenoptera: Formicidae). *Entomol. News* **76**, 241–246.

Weber, N. A. (1967a). The Fungus-growing Ant, *Trachymyrmex jamaicensis,* on Bimini Island, Bahamas (Hymenoptera: Formicidae). *Entomol. News* **78**, 107–109.

Weber, N. A. (1967b). The Growth of Young *Acromyrmex* Colonies in their First Year (Hymenoptera: Formicidae). *Ann. Entomol. Soc. Am.* 60, pp. 506–508.

Weber, N. A. (1967c). Growth of a Colony of the Fungus-growing Ant, *Sericomyrmex urichi* (Hymenoptera: Formicidae). *Ann. Entomol. Soc. Am.* **60**, 1328–1329.

Weber, N. A. (1969a). Three years of growth of a *Sericomyrmex* colony. *Ann. Entomol. Soc. Am.* **62**, 244–245.

Weber, N. A. (1969b). Ecological relations of three *Atta* species in Panama. *Ecology* **50**, 141–147.

Weber, N. A. (1969c). Comparative Study of the Nests, Gardens and Fungi of the Fungus-growing Ants, Attini. *Proc. 6th Cong. Intern. Union Stud. Soc. Insects, Bern,* 299–307.

Weber, N. A. (1970). Northern extent of attine ants (Hymenoptera: Formicidae). *Proc. Entomol. Soc.* Washington **72**, 414–415.

Weber, N. A. (1972a). Gardening ants, the Attines. *Mem. Am. Philos. Soc.* **92**, xvii + 146.

Weber, N. A. (1972b). The attines: The fungus-culturing ants. *Am. Sci.* **60**, 448–456.

Weber, N. A. (1972c). The fungus-culturing behavior of ants. *Am. Zool.* **12**, 577–587.

Weber, N. A. (1976a). A ten-year colony of *Sericomyrmex urichi* (Hymenoptera: Formicidae). *Ann. Entomol. Soc. Am.* **69** (5), 815–819.

Weber, N. A. (1976b). A ten-year laboratory colony of *Atta cephalotes*. *Ann. Entomol. Soc. Am.* **69** (5), 825–829.

Weber, N. A. (1977a). Recurrence of *Atta* colonies at a Canal Zone site (Hymenoptera: Formicidae). *Entomol. News* **88**, 85–86.

Weber, N. A. (1977b). A ten-year colony of *Acromyrmex octospinosus* (Hymenoptera: Formicidae). *Proc. Entomol. Soc. Wash.* **79**, 284–292.

Weber, N. A. (1979a). Historical Note on Culturing Attine-Ant Fungi. *Mycologia* **61** (2), 633–634.

Weber, N. A. (1979b). Fungus Culturing by Ants. *In* "Insect Fungus Symbiosis, Mutualism and Commensalism" (L. R. Batra, ed.), pp. 77–116. Allanheld, Osmun and Co., Halsted Press.

Weyrauch, W. (1942). Las hormigas cortadoras de hojas del valle Chanchamayo. *Bol. Dir. Agric. Ganad. (Peru)* **15**, 204–259.

Wheeler, G. C. (1948). The larvae of the fungus growing ants. *Am. Midl. Nat.* **40**, 664–689.

Wheeler, G. C., and Wheeler, J. (1974). Ant larvae of the Myrmicine tribe Attini: Second supplement (Hymenoptera: Formicidae). *Proc. Entomol. Soc. Wash.* **76**, 76–81.

Wheeler, W. M. (1907). The fungus-growing ants of North America. *Bull. Am. Mus. Nat. Hist.* **23**, 669–807.

Wheeler, W. M. (1910). "Ants: Their Structure, Development and Behavior." Columbia Univ. Press, New York.

Wheeler, W. M. (1925). A new guest-ant and other new Formicidae from Barro Colorado Island, Panama. *Biol. Bull. (Woods Hole, Mass.)* **49**, 150–181.

Wheeler, W. M. (1937). "Mosaics and Other Anomalies among Ants." Harvard Univ. Press, Cambridge, Massachusetts.

Wilson, E. O. (1953). "The Origin and Evolution of Polymorphism in Ants". *Q. Rev. Biol.* **28**, 136–156.

Wilson, E. O. (1971). "The Insect Societies." Belknap Press, Cambridge, Massachusetts.

Wilson, E. O. (1975). "Sociobiology, The New Synthesis." Belknap Press, Cambridge, Massachusetts.

Wilson, E. O. (1980a). Caste and Division of Labor in Leaf-Cutter Ants (Hymenoptera: Formicidae: *Atta*). I. The Overall Pattern in *A. sexdens*. *Behav. Ecol. Sociobiol.* **7**, 143–156.

Wilson, E. O. (1980b). Caste and Division of Labor in Leaf-Cutter Ants (Hymenoptera: Formicidae: *Atta*). I. The Overall Pattern in *A. sexdens*. *Behav. Ecol. Sociobiol.* **7**, 157–165.

Zolessi, L. C., and de Abenante, Y. P. (1973). Nidificacion y mesoetologia de *Acromyrmex* en el Uruguay III. *Acromyrmex* (A) *hispidus* Santschi, 1925 (Hymenoptera: Formicidae). *Rev. Biol. Uruguay.* **1**, 151–165.

Species Index

Subject Index

A

Acanthostichini, 237
Aculeata, 109
Adaptive syndrome, of army ants, 240
Aenictini, 163, 166, 167, 174, 178, 179, 238
  geographical distribution of, 235
Aenictogitini, 164
Age, polyethism and, in ants, 142–144
Aggressive behavior, in polistines, 39
Alarm pheromones
  of army ants, 195–196
  of fungus ants, 302
  of wasps, 66, 68, 69
Aleyrodidae, 15
Alitrunk, in army ants
  males, 182
  queens, 178–179
  workers, 174
Allometry, in army ants, 172–173
Ant(s), 107–109 *see also* Army ants
  as army ant predators, 233
  brood care in, 137
    recognition of brood and, 137–138
    transport of brood and, 138–139
  communal defense in, 81–82
  diet of, 108
  food collection in
    honeydew and nectar, 121–124
    prey recognition and, 111–113
    recruitment and, 117–121
    searching for food and, 113–117
  genera of, 304, 306
    distribution and appearance of, 306–309
  nests of, 108
    compounding of materials and, 136–137
    material and, 135–136
  polyethism in
    age and, 142–144

    behavioral repertoires and, 139–142
    caste and, 144–145
    specialists and, 145–148
  taxonomy of, 108
Ant eaters, 338–339
Ant guards, of stenogastrines, 8
Aphids, *see* Honeydew
Apoidea, 108
Army ants, 158–160
  behavioral characteristics of, 160–161
  brood of
    biology of, 188–191
    morphology of, 186–189
  classification of, 161–164
  colony founding in
    doryline ants, 227–229
    nondoryline ants, 227
  emigration in
    colony movement and, 221–224
    functional cycle and, 224–227
    nests and, 215–219
    origin of, 219–221
    population dynamics and, 221
  foraging in
    diet and, 208–214
    group predation and, 197–203
    group retrieval of prey and, 205–208
    hypogaeic and epigaeic, 203–205
  keys to tribes, genera, and subgenera of,
    166
    Dorylinae and Ecitoninae, 166–167
    Dorylus, 168–172
    Ecitonini, 167–168
  longevity of
    caste-related determinants of, 229–230
    predators and, 230–234
  males
    characteristics of, 165
    Dorylinae and Ecitoninae, 167